The Struggle for Ecological Democracy

DEMOCRACY AND ECOLOGY
A Guilford Series

Published in conjunction
with the Center for Political Ecology

JAMES O'CONNOR
Series Editor

THE STRUGGLE FOR ECOLOGICAL DEMOCRACY
ENVIRONMENTAL JUSTICE MOVEMENTS IN THE UNITED STATES
Daniel Faber, Editor

NATURAL CAUSES
ESSAYS IN ECOLOGICAL MARXISM
James O'Connor

WORK, HEALTH, AND ENVIRONMENT
OLD PROBLEMS, NEW SOLUTIONS
Charles Levenstein and John Wooding, Editors

THE GREENING OF MARXISM
Ted Benton, Editor

MINDING NATURE
THE PHILOSOPHERS OF ECOLOGY
David Macauley, Editor

GREEN PRODUCTION
TOWARD AN ENVIRONMENTAL RATIONALITY
Enrique Leff

IS CAPITALISM SUSTAINABLE?
POLITICAL ECONOMY AND THE POLITICS OF ECOLOGY
Martin O'Connor, Editor

The Struggle for Ecological Democracy

ENVIRONMENTAL JUSTICE MOVEMENTS IN THE UNITED STATES

A Project of the Boston *Capitalism, Nature, Socialism* Editorial Group

EDITED BY
Daniel Faber

Foreword by Carl Anthony

THE GUILFORD PRESS
New York London

A Division of Guilford Publications, Inc.
72 Spring Street, New York, NY 10012
http://www.guilford.com

Printed in the United States of America

This book is printed on acid-free paper.

Last digit is print number: 9 8 7 6 5 4 3 2

Library of Congress Cataloging-in-Publication Data

The struggle for ecological democracy: environmental justice movements in the United States / edited by Daniel Faber.
p. cm.—(Democracy and ecology)
Includes bibliographical references and index.
ISBN 1-57230-341-7 (hc.). —ISBN 1-57230-342-5 (pbk.)
1. Environmental justice—United States. 2. Environmental policy—United States. 3. Environmentalism—Political aspects—United States. I. Faber, Daniel J. II. Capitalism, nature, socialism. III. Series.
GE180.S84 1998
363.7′056′0973—dc21 98-25564
CIP

This book is dedicated
to the memory of a great
environmental and social justice warrior,
mother, labor organizer,
singer-songwriter, Earth First!er activist,
and much, much more

Judi Bari (1949–1977)

(Judi often quoted Joe Hill:
"Don't mourn. Organize!")

and, lovingly,
to my dear mother

Patricia Jane Faber

whose sweetness, humor, and love
I cherish every day of my life

Introduction to the Democracy and Ecology Series

This book series titled "Democracy and Ecology" is a contribution to the debates on the future of the global environment and "free market economy" and the prospects of radical green and democratic movements in the world today. While some call the post-Cold War period the "end of history," others sense that we may be living at its beginning. These scholars and activists believe that the seemingly all-powerful and reified world of global capital is creating more economic, social, political, and ecological problems than the world's ruling and political classes are able to resolve. There is a feeling that we are living through a general crisis, a turning point or divide that will create great dangers, and also opportunities for a nonexploitative, socially just, democratic ecological society. Many think that our species is learning how to regulate the relationship that we have with ourselves and the rest of nature in ways that defend ecological values and sensibilities, as well as right the exploitation and injustice that disfigure the present world order. All are asking hard questions about what went wrong with the worlds that global capitalism and state socialism made, and about the kind of life that might be rebuilt from the wreckage of ecologically and socially bankrupt ways of working and living. The "Democracy and Ecology" series rehearses these and related questions, poses new ones, and tries to respond to them, if only tentatively and provisionally, because the stakes are so high, and since "time-honored slogans and time-worn formulae" have become part of the problem.

JAMES O'CONNOR
Series Editor

Foreword

Easily the most striking thing to emerge in the U.S. environmental movement in the past decade has been the infusion of energy from poor people, working people, and people of color.

From Buffalo, New York, to Brownsville, Texas; from Watsonville, California, to Warren County, North Carolina; from the Hopi reservation on the Black Mesa to Humboldt County in the Pacific Northwest African Americans, Asian immigrants, Chicanos, Latinos, Pacific Islanders, and Native Americans are fighting for health and safety, fishing rights, and protection of ecosystems in barrios, ghettos, forests, fields, and factories. Atomic and chemical workers, janitors, farm workers, public service employees, uranium miners, and transit workers are all demanding the right to know about dioxin, endocrine disrupters, pesticides, and lead poisoning in the places where they live, work, and play.

Companies such as Browning Ferris, Chevron, Dow Chemical, Georgia Pacific, Intel, Monsanto, Union Carbide, and Waste Management, Inc., are being challenged by workers and communities to clean up their act, to internalize the full costs of their production, distribution, and waste rather than passing these costs of pollution, health risks, brownfield sites, and damaged ecosystems to poor and vulnerable communities.

In 1987, the United Church of Christ, in its civil rights outreach, published the landmark study *Toxic Wastes and Race in the United States*, which documented racial discrimination in the siting of hazardous waste facilities across the nation. The report produced a flurry of denials by companies that any racial bias was intended and by activists who argued that class rather than race was the primary determinant of corporate and public agency decisions about sites for locally undesirable land uses. Whatever the merits of these arguments, after a 1992 study by *The National Law Journal*, there could be no denying that there were glaring inequities in the way the federal EPA enforced environmental protection laws. Penalties for infraction of environmental laws are 500 percent higher, the government responds more rapidly, and chooses preferred cleanup technologies more frequently in white communities than in communities of color.

In October 1991, the United Church of Christ's Commission on Racial Justice convened the First People of Color National Environmental Leadership Summit. Since that meeting regional environmental and economic justice networks have sprung up in the southeast and southwest, northeast, midwest, and northwest, including networks focusing on the special needs of Asian Americans and Pacific Islanders, farm workers, and indigenous communities. On Earth Day 1993, President Clinton announced an Executive Order on Environmental Justice requiring federal agencies to assess the impact of their environmental activities on low-income communities and communities of color and establishing a National Environmental Justice Advisory Council to comment on federal action to achieve environmental justice for all Americans.

Despite these impressive successes, the environmental justice movement faces enormous challenges. To date, most environmental justice struggles have been very local—fighting an incinerator here, a prison, highway expansion, or a waste dump elsewhere. Rural and land-based struggles over mining, timber, and other natural resources, for example, are often quite different from urban struggles against corporate polluters. Focusing on single issues such as breast cancer or lead poisoning in a local neighborhood or work place makes it difficult to build coalitions with other groups concerned about economic opportunity. Communities face the task of linking protest strategies to the building of positive alternatives. Conflicts between racial and class politics remain.

In the meantime, global decision making by large multinational corporations affecting all aspects of community and work life is becoming increasingly centralized. Chemical, pharmaceutical, electronic, computer, mining, and petroleum industries are radically restructuring every aspect of everyday life—from the food we eat, where we live, and the clothes we wear, to how we raise our children and spend our time. They are dividing the world into consumer markets, sources of cheap labor and raw materials, and ecological sacrifice zones. Most vulnerable, more often than not, are disenfranchised communities of poor and working class people.

The city, state, and national governments that should be protecting the interests of such communities now compete with one another in offering tax breaks, financing schemes, infrastructure improvements, and "streamlined" environmental permits. Instead of addressing these issues, local politicians scapegoat immigrants, communities of color, and unemployed people as the sources of our greatest problems.

How can environmental justice struggles succeed against such powerful opposition? Never before have the relationships among public and private institutions, communities, and proponents for the natural world called on us to face our collective predicament with so much honesty and awareness. Maintaining vital struggles and connections with others challenges us to free ourselves from old habits and blind spots and to develop the full range of our powers, talents, and sensitivities as activists for change. In former times, activists had ready access to alternative ideologies of social change that provided a framework, if not illusions, about how best to go forward.

With the collapse of the Soviet Union, and the Berlin Wall, the agonies of Chernobyl, acknowledgment of the pollution and other excesses, the illusion of bureaucratic state socialism as an automatic alternative to the free market has disappeared.

On the other hand, it has become clear that social and environmental problems cannot be cured by giving free markets even freer play. The rapid deterioration of the quality of life for the poor, the working class, and even middle-class families under a regime of shrinking public intervention has become all too apparent. It should be clear by now that achieving environmental justice will require new forms of social cooperation and collaboration. For many of us today, destruction of the environment and erosion of community have become a new wilderness, bringing us face to face with all our gods and demons.

Since neither the political economy nor the natural world can be counted on as a predictable source of comfort or security, we have come to a new crossroads where we face pivotal and unprecedented choices. We can struggle to hold on to fanciful fantasies and obsolete formulas that neither correspond to reality nor provide any useful direction. Or, instead, we can learn to use the environmental and social difficulties our communities face as opportunities to awaken and bring forth our finest human qualities: awareness, compassion, humor, wisdom, and fearless dedication to truth. If we choose this path, we can deepen connections within and among communities, and between communities and the natural world, in our common struggle for ecological democracy.

In this collection, Daniel Faber has assembled essays by leading scholars that look to the emerging principles of ecological democracy to provide a vision of both the means and the end goals of environmental justice. By demonstrating the political–economic links between the exploitation of both nature and working people as they exist in all walks of life under American capitalism, and exploring the various political strategies that embrace ecological democracy in response to such exploitation, this new volume may well assist in the formation of a radically new environmental politics capable of finding lasting remedies to America's social and ecological ills.

CARL ANTHONY
Urban Habitat Program
San Francisco

Acknowledgments

The contributors to this collection represent a diverse group of activists, scholars, researchers, and professionals, each of whom offers a perspective that situates social and environmental justice issues in the context of struggles for ecological democracy. The majority of writers and essays making up this collection are drawn from the international journal *Capitalism, Nature, Socialism: A Journal of Socialist Ecology*, edited by James O'Connor. It is the seventh publication to appear in the "Democracy and Ecology" book series jointly sponsored by *CNS* and The Guilford Press.

There are many people who provided invaluable assistance in the preparation of this publication. First, I would like to offer my sincerest appreciation to Barbara Laurence for assisting me with the organization of the book project itself, including computer layout and other logistical problems inherent in such an undertaking. Without her assistance the book project would not have been possible. Secondly, I would like to extend my gratitude to James O'Connor, Alan Klein, Victor Wallis, and Joshua Karliner for the extensive time and energy each of them spent reviewing drafts of the introduction. Their comments proved extremely helpful in revising and clarifying my thoughts. A word of thanks also to Peter Wissoker at The Guilford Press for his encouragement and patience in the preparation of the book, as well as to William Meyer. And of course, special thanks to Carl Anthony and all of the contributors for sharing their words with us in the "Democracy and Ecology" book series.

I am deeply indebted to Frank Ackerman, Eric Bougeois, Cathy Crumbley, Pat Dalendina, Debbie McCarthy, Tom Estabrook, Kevin Gallagher, Roger Gottlieb, Colin Hay, Rebecca Herzig, George Katsiaficas, Eric (Luke) Krieg, Chuck Levenstein, Greg DeLaurier, Jay Moore, Joan Roelofs, Alan Rudy, Victor Wallis, John Wooding, and other members of the Boston Editorial Group of *Capitalism, Nature, Socialism* for the many hours spent reviewing, editing, and discussing manuscripts contained in this collection. I am proud to say that this project belongs to the editorial group as much as it does to me.

To my parents, Chuck and Pat, as well as Debbie, Mindy, Zach, Kyffin, Kendra, Zane, Jay, Paul, and Sue, and all my family members and friends whom I cherish so much, I reaffirm my eternal love. Thanks also to Izzy and Clara, Judy and David, Marsha and Brad, David and my late sister-in-law Shelly, and Max, Noah, Eli, Jessica, and Hannah. I am proud to be part of your family as well.

Finally, I would like to offer my affection to my wife, Laura, for her emotional and spiritual guidance as we travel down the road of life together. Let our efforts combine to help make this world a better place for Emma Sophie and Jonah Andrew.

Contents

The Struggle for Ecological Democracy

Introduction

The Struggle for Ecological Democracy and Environmental Justice

Daniel Faber

ABOUT THIS COLLECTION

A new wave of grassroots environmentalism is building in the United States. In poor African American and Latino neighborhoods in the inner cities, white working-class suburbs and small towns, depressed Native American reservations, and Chicano farming communities all across the country, people who have traditionally been relegated to the periphery of environmental policy are beginning to challenge the wholesale depredation of their land, water, air, and community health by corporate polluters and indifferent governmental agencies. Combining elements of civil rights, social justice, and respect for the environment, these new community-based movements for environmental justice are committed to reversing past practices that have had the effect of placing disproportionately large ecological and economic burdens on working-class families and communities of color.[1]

To help remedy ecological racism and class-based environmental inequities, movements for environmental justice are increasingly embracing the principles of ecological democracy. Of these principles, the most fundamental is the claim that those communities of people suffering ecological injustices must be afforded greater participation in the decision-making processes of capitalist industry and the state (at all levels), as well as the environmental movement itself, if the social and ecological problems plaguing *all* Americans are ever to be resolved. This demand by marginalized communities for greater participation in the management of their own affairs is providing a growing number of activists with a vision of both the means and the end result by which environmental justice will be achieved. For not only does such a conception of ecological democracy begin to provide models of the forms of organizational practice and political strategies needed to build a transformative popular movement, it also provides an outline of the types of ultimate political–economic goals and solutions for which the movement must strive. Ecological democracy is emerging as both the *praxis* and *telos* of movements for environmental justice. But what is meant by the term "democratic participation" itself is often ambiguous, even for those who embrace it, and has sparked a critically important debate among scholars and activists for environmental justice. Exactly what is ecological democracy, and how can it be achieved?

The various approaches to this question expressed by the authors in this collection, as well as in the pages of the journal *Capitalism, Nature, Socialism,* from which most of the following chapters were drawn and revised, emanate from an ecosocialist perspective. In addition to examining ecological antagonisms systemically grounded in the workings of American capitalism, socialist ecology emphasizes the concrete material and class interests of those who benefit and suffer from the social and environmental inequities blighting this nation. By demonstrating the political–economic links between the exploitation of nature and working people and people of color, as they exist in all walks of life under U.S. capitalism, and exploring various political strategies that embrace ecological democracy in response to such exploitation, it is hoped that this collection will assist efforts to overcome divisions currently separating America's popular classes and dividing major social movements from one another. Only when this is accomplished can we begin to find lasting solutions to America's social and ecological crises.

THE ECOLOGICAL INJUSTICES OF AMERICAN CAPITALISM

The roots of America's ecological problems and injustices are grounded in the expansionary dynamics of the global capitalist system. Motivated by the growth of market competition and the drive to obtain ever greater returns for shareholders, U.S. capital must continuously increase investment and productive capacity, introduce more efficient technologies that produce commodities more cheaply, introduce new consumer products and enlarge markets, and undertake many other measures that promise high rates of economic growth and ever expanding profits, that is, capital accumulation. And in the current post-Cold War era, a period characterized by the regionalization and globalization of capitalist production, heightened international competition, and relatively stagnant markets in Europe and Japan, maintaining healthy growth rates is becoming increasingly problematic. Thousands of American companies are finding themselves embroiled in a life-and-death economic struggle. For the less "efficient" producers unable to keep pace, global market forces, especially financial markets, are driving them out of business. Adam Smith's "invisible hand" is bearing its knuckles. Now, more than ever, "accumulate or die" has become the sine qua non of American capitalism.[2]

To sustain the process of capital accumulation and higher profits in the new global economy, American capital is increasingly relying on ecologically and socially unsustainable forms of production. This theme is developed in Chapter 1, which I authored, "The Political Ecology of American Capitalism: New Challenges for the Environmental Movement." Motivated by the growing costs of regulation and processes of global economic restructuring, which are making U.S. capitalism's opposition to ecological protection obvious, corporate America is leading an unparalleled assault on the environmental movement. Corporate "greenwashing" is part of this offensive, as big business is promoting the view that the country's ecological problems are not all that serious (or do not exist at all). Business interests also argue that, if government would just get off the backs of industry and offer the proper incentives, the marketplace would solve most ecological problems.[3] The goal of this corporate campaign is "regulatory reform," the rollback of environmental laws, worker health and safety, consumer protection, and other state regulatory burdens that impinge upon the free market and profits of capital. Faithfully endorsed in various forms by the neoliberalist power blocs that now dominate both the Republican and Democratic parties (including the present Clinton administration), the result has been a worsening of major environmental problems and ecological inequities for working people and communities of color. I conclude that unless the environmental movement can integrate principles of social justice and ecological democracy into a more holistic strategy aligned with the interests of labor and other social movements, America's social and ecological crises will deepen.

With the ascendancy of neoliberalism and persistence of slow economic growth (relative to 1950–1970s standards), American capital is responding to the government's invitation to more ruthlessly exploit natural resources and cut spending for environmental and consumer protection. Business is also spearheading an assault on organized labor and the civil rights and women's movements in the attempt to reduce wages and benefits, increase labor productivity and discipline, and cut expenses related to worker and public health and safety. In so doing, America's corporate ruling class, the 1 percent of the population that owns 60 percent of all corporate stock and business assets,[4] is serving its own narrow material interests at the expense of the environment, communities, and the health of working people. The reason is that corporate expenses related to human health and environmental quality do not typically increase labor productivity (hence potential profits) sufficiently to outweigh such expenditures. Without prohibitions and the threat of punitive actions by state regulatory agencies or the courts, it is simply more profitable for corporations to pollute. So, rather than paying money (or what economists call "internalizing" costs) for pollution abatement technology, such as $10 million for the installation of a "scrubber" to clean the air of chemical particulates as it leaves the smokestacks of the factory, businesses seek to avoid or lower this cost (such as by not installing the scrubber) and instead to displace (or "externalize") this expense onto the larger public in the form of air pollution and other environmental problems. In addition to the over 60,000 Americans killed each year by air pollution,[5] these social losses also take the form of long-term damage to human health; the destruction or deterioration of property values and the premature depletion of natural wealth (such as with acid rain); and the impairment of less "tangible" values associated with environmental quality,[6] and loss of community. In essence, through this systematic strategy of cost displacement, American capitalism is killing hundreds of thousands of U.S. citizens and causing hundreds of billions (if not trillions) of dollars in damage to property, human health, the community, and the environment every year—all in the pursuit of higher profits.[7]

But not all Americans bear these social and ecological "costs" of capitalist production equally. In order to bolster profits and competitiveness, businesses typically adopt strategies for the exploitation of nature that offer the path of least political resistance (and therefore the greatest opportunity for continued economic success). The less political power a community of people possesses; the fewer resources a community has to defend itself with; the lower the level of community awareness and mobilization against potential ecological threats; the more likely they are to experience arduous environmental and human health problems at the hands of capital and the state. That the "disempowered" of America are to serve as the dumping ground for American business is often blatantly advertised. A 1984 report by Cerrell Associates for the California Waste Management Board, for instance, openly recommended that polluting industries and the state locate hazardous waste facilities in "lower socioeconomic neighborhoods" because those communities had a much lower likelihood of offering political opposition.[8]

The weight of the ecological burden upon a community is dependent upon the balance of power and level of struggle between capital, the state, and social movements responding to the needs and political demands of the populace. And in the United States, it is (as it always has been) the least politically powerful segments of peoples of color, industrial laborers, the underemployed and the working poor (especially women), rural farmers and farm workers, and undocumented immigrants that are being damaged to the greatest extent by the ecological crisis. This is not to say that the white middle class is not also being significantly harmed by industrial pollution and other abusive corporate practices, because it too is impacted.[9] But in contrast to the salaried and professional classes, who can often buy themselves access to ecological amenities and a cleaner environment, there is significantly less investment of social capital (spending for education, job training, housing, wages and benefits, health care, and so forth) for the maintenance of the working poor. Therefore, it costs capital and the state much less to displace environmental health problems onto people who lack health care insurance, possess lower incomes and property values, and as unskilled or semiskilled laborers are more easily replaced if they become sick or die. In this sense, environmental inequalities in all forms, whether they be class, race, gender, or geographically based, are socially constructed features grounded in the systemic logic of capitalist accumulation.[10]

Evidence of environmental racism and other ecological injustice produced by the laws of capital abound.[11] According to the EPA, 57 percent of all whites nationwide live in areas with poor air quality, compared to 80 percent of all Latinos.[12] In Los Angeles, where organizations such as the Labor/Community Strategy Center, Mothers of East Los Angeles, and Concerned Citizens of South Central Los Angeles have come to the forefront to challenge ecological racism, 71 percent of the city's African Americans and 50 percent of the Latinos live in what are categorized as the most polluted areas, compared to only 34 percent of whites.[13] Due to inadequate housing, health care, public parks and safe spaces for children, and other social services, people of color and the poor also suffer most due to extremes of weather—heat waves and winter cold—as well as from floods, earthquakes, and other "natural disasters."

The ecological injustices experienced by millions of Americans in their communities are further compounded by the dangerous and unhealthy working conditions that exist on the job. In Chapter 2, "Dying for a Living: Workers, Production, and the Environment," Charles Levenstein and John Wooding uncover the political–economic dynamics of American capitalism that result in hundreds of workers being killed and tens of thousands more injured on the job *each week*. Because capital exploits both workers and the environment simultaneously in the pursuit of profit, there is an intimate relationship between threats to worker health and safety at the point of production and threats to the environment and nearby communities posed by these same production activities. Nevertheless, profound differences exist between workers and environmentalists, owing in part to business and government divide-and-conquer strategies. By providing lessons to be learned from the activities of labor, community, and environmental coalitions, Levenstein and Wooding hope to bridge the gap between worker health and safety struggles and the movements for environmental justice and ecological health. They conclude that a healthy workplace and environment cannot be achieved by one movement alone. Rather, coalescence between the environmental movement and labor for democratic control over the production process is, according to Levenstein and Wooding, a critical part of the campaign for ecological democracy.

The types of occupational hazards discussed by Levenstein and Wooding are even more profound for laborers lacking the minimal protections afforded by unions or formal rights of citizenship. Over 313,000 of the 2 million farm workers in the United States—90 percent of whom are people of color and mostly undocumented immigrants—suffer from pesticide poisoning each year. Of these, between 800 and 1,000 die.[14] The plight of such undocumented workers is spurring new coalitions between farm worker associations such as the United Farm Workers (UFW) and the Farmworker Network for Economic and Environmental Justice, immigrant groups, consumer and environmental organizations, and the labor movement.[15] Nevertheless, much of the environmental movement has been slow to draw the connections between an unhealthy environment and conditions on the job (where the work process for labor is almost completely undemocratic and hidden from public view).

As is the case with immigrant farm workers and pesticides in the United States today, the working class in general (and especially people of color) faces a greater "quadruple exposure effect" to toxic pollution and other environmental hazards. This takes the form of: (1) higher rates of "on the job" exposure to toxics used in the production process; (2) greater neighborhood and community exposure to toxic pollutants emitted from nearby factories and agricultural fields or buried or stored in nearby facilities; (3) faulty cleanup efforts implemented by the government or the waste treatment industry, such as through the increased use of permanent or mobile incinerators that burn these wastes in the community; and (4) higher exposure rates to toxic chemicals in commercial foods and consumer products.

A 1983 report by the U.S. General Accounting Office, for instance, found that three-quarters of the nation's hazardous waste landfills are sited in poor (mainly African American and Latino) communities.[16] According to a 1987 report by the United Church of Christ's Commission on Racial Justice, three out of five African Americans and Latinos nationwide live in communities that have illegal or abandoned toxic dumps. Communities with one hazardous waste facility have twice the percentage of people of color as those with none, while the percentage triples in communities with two or more waste sites.[17] A subsequent follow-up study conducted in 1994 has now found the risks for people of color to be even greater than in 1980, as they are 47 percent more likely than whites to live near these health-threatening facilities.[18] Additionally, Superfund toxic waste sites in communities of color are likely to be cleaned 12 to 42 percent *later* than sites in white communities. Also, in communities made up predominantly of people of color government penalties for violations of hazardous waste laws average only one-sixth ($55,318) what they do in predominantly white communities ($335,566).[19] Due in part to growing community resistance to such dumps in these communities, Chemical Waste Management Inc. and other toxic disposers have been at the center of a number of controversial plans to use poverty-stricken Native American reservations as alternative sites for toxic and radioactive waste disposal.[20]

In Chapter 3, "Risk and Justice: Capitalist Production and the Environment," Rodger C. Field examines the political–economic forces behind ecological racism and toxic dumping on people of color and poor working-class communities. In that they fail to address the systemic nature of capitalist production per se, Field sees liberalist environmental policy initiatives as actually making many environmental inequities worse. By requiring capital to *contain pollution* sources for more proper treatment and disposal, pollution becomes commodified for disposal by the waste treatment industry and the state on behalf of capital as a whole, hence a new source of profit. With the expansion of the waste circuit of capital, pollution becomes increasingly geographically mobile, as corporations search for ever more "efficient" (low cost and politically feasible) disposal sites. The locations in which these environmental hazards become manifest are increasingly distinct from the industrial facilities in which they are produced, and typically wind up in communities politically unable to resist. Hence, the explosion of environmental hazards in poor working-class communities and the corresponding growth of the environmental justice movement in the 1980 and 1990s. Field also concludes that only with the establishment of ecological democracy, which includes a democratization of the administrative apparatus of the state and also of the production process itself, can the contradictory nature of liberalist environmental policy be overcome and the goals of the environmental justice movement be realized.

ECOLOGICAL DEMOCRACY AS *PRAXIS*: MOVEMENTS FOR ENVIRONMENTAL JUSTICE ON THE RISE

In reaction to the failures of existing environmental policy and the mounting ecological inequities of American capitalism, a dynamic new grassroots environmentalism has been given birth.[21] Beginning in 1978, with Lois Gibbs's (who later founded the Citizen's Clearinghouse for Hazardous Waste, now the Center for Health, Environment, and Justice) battle for her family and community against the toxic terror of the Love Canal in New York[22]—and propelled by the massive protests (and more than 500 arrests) in 1982 against a proposed PCB landfill and other practices of ecological racism in the mostly African American Warren County, North Carolina[23]—movements for environmental justice and ecological democracy are now rising in every corner of the country.[24] Linking struggles for social/economic justice and environmental quality and invoking the direct-action tactics typical of local initiatives, these advocates of environmental justice are addressing issues of ecological racism, the economic and environmental exploitation of working people, and unequal capitalist political–economic power and control over community planning and natural resource development. Culminating with the landmark People of Color Environmental Leadership Summit in 1991, which formally adopted the "Principles of Environmental Justice," and represented by the Washington Office on Environmental Justice, the Environmental and Economic Justice Project, and numerous other regional and constituency-based strategic networks, these movements for environmental justice are now building a coordinated national presence.

As is not the case in much of mainstream environmentalism, many leaders and activists in the environmental justice movement are poor women and people of color. In Chapter 4, "Environmental Justice from the Grassroots: Reflections on History, Gender, and Expertise," Giovanna Di Chiro examines the manner in which a new breed of women activists is challenging the paternalistic approach of government agencies and mainstream environmentalism to the plight of working people in general and poor communities of color in particular. Emphasizing the multiple, local, and historically and culturally specific contexts in which people are struggling to improve their conditions, Di Chiro sees these new environmentalists as providing invaluable lessons and innovative political strategies for transforming the nature of environmentalism in the United States. Informed by the principles of ecological democracy and, to some degree, ecofeminism,[25] these women activists are challenging the dominant discourses and practices of unjust environmental decision making, gender-, race-, and class-based stereotyping, and oppressive notions of scientific expertise currently employed by much of the environmental movement, the state, capital, and the larger culture.[26]

One of the most important lessons to be learned centers around the pervasiveness of instrumental rationality, or the separation between political means and ends, that permeates mainstream environmentalism and environmental policy. Many in the environmental justice movement are now elevating questions of political *praxis* (means) to the same level as *telos* (ends). In the words of Greg DeLaurier, "How we get to where we want to go is as important as the destination itself, no matter how noble or liberating. So when we ask what is ecological socialism or ecological democracy, we ask not, or not only, what some future society might look like, but what we are [also] doing now in our practice and theory."[27]

One manner in which ecological democracy is being put into practice by environmental justice activists involves the development of new organizational structures that directly represent and promote the substantive democratic participation of community members victimized by environmental abuses. In Chapter 5, "Popular Epidemiology and the Struggle for Community Health in the Environmental Justice Movement," Patrick Novotny explores the formation of what he terms these "empowering" organizations. An example is the Environmental Health Network (EHN), which takes advantage of the knowledge that community residents, labor unionists, and others have of the health and environmental problems in their communities and workplaces. Emphasizing the view that broad-based political organizing is fundamental to solving the health problems of the workplace, community, and environment, these alternative networks promote a popular epidemiological approach. Such an approach not only promotes greater democratic control over health information and services but also integrates health care work into larger political strategies for community empowerment. In contrast, classic epidemiology is deeply flawed methodologically and leads detached state agencies to employ mechanisms that impede popular participation in the public health system. Many alternative public health networks utilize political strategies that reject a dependency upon the liberal democratic state, relying instead on developing counterauthorities under the direct control of local cooperatives, communities, or popular-based movements.[28] In Novotny's view, practicing ecological democracy in the form of community self-management of health services is critical to building coalitions that can underscore the commonality of interests between workers and community residents. Such a unity is necessary if the "incapacity" of the capitalist state to protect public health and the environment is ever to be fundamentally challenged.

As revealed in the case of popular epidemiology, environmental justice activists argue that the only manner in which the material interests of the working class and oppressed people of color can be safeguarded is through their incorporation into the decision-making processes of the major social, political, and economic institutions that impact their lives and the environment in which they live. Hence, a key component of the term "ecological democracy" is an emphasis on greater *participatory* democracy.[29] In terms of movement building, there are two important considerations that need to be addressed.

First, the goal of social and environmental justice cannot be achieved unless the environmental movement itself becomes more democratic. In Chapter 6, "The Network for Environmental and Economic Justice in the Southwest: An Interview with Richard Moore," Paul Almeida discusses with Richard Moore, a leading environmental justice activist, the fact that for too long the movement has weakened itself by failing to fight against ecological inequities and inadequate or regressive environmental policies.[30] In so doing, Moore sees mainstream environmental organizations as having neglected to incorporate oppressed people of color and broad segments of working people into a more democratic, mass-based environmental movement capable of addressing the root causes of environmental injustice. Far too many mainstream environmental organizations neglect the central social and environmental issues of poor people of color and working-class Americans,[31] and are often insufficiently accountable to their own membership as well.[32] The political agendas of many of these organizations instead express an allegiance to the dictates of more affluent financial backers, foundations, and corporate sponsors, thereby deflecting attention away from potentially more controversial issues and solutions that challenge the power of capital to exploit both workers and nature.[33] Only by creating community–labor alliances that practice ecological democracy organizationally can a transformative environmental politics be developed in America.

Second, by practicing ecological democracy organizationally, becoming more open and inclusive, movements for environmental justice could serve to unite the environmental movement firmly with other movements for social justice that normally do not identify with ecological issues. In Chapter 7, "The Limits of Environmentalism without Class: Lessons from the Ancient Forest Struggle in the Pacific Northwest," John Bellamy Foster examines the northern spotted owl controversy, where the battle to save the last stands of some of America's most pristine old-growth forests has left timber workers and single-issue environmentalists at each other's throats. For Foster, this case demonstrates that any ecology movement that ignores class exploitation and other social inequalities will likely displace environmental problems on to others and contribute little to the overall green goal of forming a sustainable relationship between human beings and nature. It may even have the adverse effect of splitting popular forces by creating more opposition to the environmental cause. He concludes that the only defense against the assault of the "wise use" movement and its corporate allies is through a stronger labor–environmentalist alliance and the building of an environmental justice movement in the widest sense. Here the concrete problems of building such a movement are evaluated.

The globalization of capital makes the emergence of such a "labor–environmental" alliance an absolute imperative. With the help of the North American Free Trade Agreement (NAFTA), the General Agreement on Tariffs and Trade (GATT), and other free trade agreements, the growing ability of multinational corporations and financial institutions to ignore unions, evade environmental regulations, and undermine worker/community health and safety struggles in the First World is being assisted by the flight of capital to politically repressive havens in the Third World. The result? Peoples, movements, and governments of the world are now being successfully pitted against one another by capital as never before in the bid to attract investment, leading to wholesale assaults on past protections won by U.S. environmental and labor movements. Unless popular movements in the United States *and* the Third World can unify into a larger international movement for social and environmental justice, living standards and environmental quality throughout the world will continue to deteriorate.[34]

In Chapter 8, "Remapping North American Environmentalism: Contending Visions and Divergent Practices in the Fight over NAFTA," Michael Dreiling analyzes the actual responses of the U.S. environmental movement and popular movements in Mexico to NAFTA. Pushing for "fair trade" alternatives and a "continental social pact," hundreds of cross-border links are being forged among grassroots groups, constituting a critical base for several national-level networks in opposition to NAFTA. In contrast to many mainstream organizations, the environmental justice movement has proved crucial in these organizing efforts, providing a strategy emphasizing grassroots mobilization, international solidarity, and cross-movement alliance building. For Dreiling, it is apparent that the NAFTA debate is helping to stimulate the construction of a new internationalist environmental–labor politics in both countries which emphasizes the subordination of unregulated trade and capital accumulation to the long-term needs of the greater society. In many ways, the "free trade" zone along the United States–Mexico border has become a testing ground for the opposed frameworks of neoliberalism and environmental justice.[35] And while the battle against "free trade" has been lost *for now,* left populism in Mexico and movements for environmental justice in the United States are creating a new popular strategy that broadens the call for ecological democracy and may one day underpin a reversal of the NAFTA accords.

Given the severe repression faced by popular movements in Mexico and the Third World, perhaps the most fundamental prerequisite in the quest for environmental justice and ecological democracy is the struggle for human rights.[36] But repression of environmental activists in the United States is also pervasive.[37] In Chapter 9, "Earth First! in Northern California: An Interview with Judi Bari," Douglas Bevington discussed with the late Judi Bari the economics of logging and the corporate/government strategies utilized in Northern California against community and worker organizing efforts. These included the attempted murder of herself and a coworker, Darryl Cherney, with a car bomb planted by an unknown assailant, and then the arrest of both activists by local police and the Federal Bureau of Investigation (FBI) under suspicion that *they* had planted the bomb. An internationally renowned environmentalist before her tragic death from cancer, Bari concluded that worker and environmental struggles must be united under the banner of Revolutionary Ecology.[38] Only then could such repressive tactics be overcome and the exploitative practices of American capitalism truly transformed.

In addition to political repression of activists, corporate capital and the state employ a number of other strategies designed to weaken movements for ecological democracy and environmental justice. Among the most common are public relations campaigns and ideological manipulations that unfairly blame working-class communities, labor unions, environmental organizations, and oppressed peoples of color as being the cause of their own social and environmental problems.[39]

In Chapter 10, "Racism and Resource Colonization," Al Gedicks explores the implementation of this strategy in the case of the Chippewa Indian controversy in Wisconsin. In their attempt to gain control over low-cost resources in the state's ceded Indian territory, multinational mining corporations have organized a campaign headed by antitreaty groups and the state that unfairly scapegoats the Chippewa tribe as being responsible for overfishing and a host of other environmental problems in the northern forests. By so doing, attention has been diverted from the significant environmental threats to both the Indian and non-Indian economy and culture posed by large mining interests. The Chippewa, along with the other Indian nations in northern Wisconsin, already suffer a disproportionate environmental risk of illness and other health problems from industrial pollutants. As part of a nationwide corporate attack on the rights of Native

Americans, which include renewed calls for Congress to terminate treaty agreements, landed capital interests hope to discredit the Chippewa and remove any potential legal obstacle that Indian treaty rights may pose to mining and commercial development. In place of the resource colony being imposed by multinational mining corporations and the state, however, the Chippewa are calling for ecological democracy in the form of an "environmental zone." Such a zone would be jointly managed by the state and the tribes, and promote environmental restoration, tourism, and rural community reinvestment. Because of the potential it holds for democratic self-management of a sustainable development plan, Gedicks concludes that both grassroots and mainstream environmental organizations have been attracted to the cause, leading to a growing alliance between Native Americans, environmentalists, and left/progressive movements.

Overcoming corporate strategies that blame the victims of ecological racism and injustice is one key to building a stronger and more broadly based environmental movement. In Chapter 11, "Ecological Legitimacy and Cultural Essentialism: Hispano Grazing in the Southwest," Laura Pulido explores the case of Ganados del Valle in northern New Mexico. Here, the business community has perpetuated a myth that attributes poor resource management and a lack of environmental concern by the Hispano community as the cause of growing poverty. For Pulido, not only do such narratives deny the "ecological legitimacy" of Hispanos,[40] they also exonerate the Anglo power structure and exploitative capitalist social relations truly responsible for the region's social and environmental problems. Clearly, one of the greatest barriers in the struggle for environmental justice and democratic planning is the denial of ecological legitimacy to precisely those peoples most impacted by environmental problems (and for whom no solution can be reached without their full participation) by elitist "resource managerialists" operating within the state and other environmental organizations.[41] In this case, the construction of what Pulido terms a "culturally essentialist" counterhegemonic ideology has proven helpful to Hispano efforts to reestablish their own ecological legitimacy and to build an ecological democracy in the form of a community land trust that links communal land ownership to sound environmental management. However, as Pulido points out, there are potential problems. Any essentialist oppositional ideology runs the danger of reifying cultural differences and therefore inhibits the formation of the political coalitions necessary to usher in a lasting ecological–democratic alternative.[42]

In Chapter 12, "The 'Brown' and the 'Green' Revisited: Chicanos and Environmental Politics in the Upper Rio Grande," Devon Peña and María Mondragon-Valdéz further explore the political and economic basis for the fierce land use battles currently raging in south central Colorado and northern New Mexico, including Ganados del Valle. Here the Amerindian and Chicano cultures have a long tradition of land stewardship and are organizing a challenge to the increasing environmental degradation posed by capitalist extractive industries. But, unlike more widely championed campaigns, such as northeastern Arizona's Navajo and Hopi Indians at Big Mountain,[43] which unify the struggles to conserve both the biological *and* cultural diversity of a region, the community-based struggles of rural Chicano communities are often overlooked by mainstream organizations. One of the main barriers to a union of many environmentalists and oppressed peoples is the inability of wilderness advocates to countenance the view of "humans in nature" even in cases where their agricultural practices are sustainable and maintain a balance between ecosystem integrity and biodiversity. But, while Chicano struggles in the southwest for cultural survival and social justice historically have been deprived of ecological legitimacy, many environmentalists are being increasingly attracted to their proposals for ecological democracy. One main goal is to establish a self-managed community land trust and viable local cooperatives (such as with livestock), where local resources are managed in partnership with the state in an environmentally and culturally sound manner. This plan includes community-based economic development to be founded on the return of historic land use rights and the revitalization of artisan skills and worker empowerment.

THE STRUGGLE FOR ECOLOGICAL DEMOCRACY: THE *TELOS* OF MOVEMENTS FOR ENVIRONMENTAL JUSTICE

For environmental justice activists, the most immediate mission is to dismantle the mechanisms by which capital and the state disproportionately displace the social and ecological costs of production onto working-class families and oppressed peoples of color. Despite the movement's infancy, there have been a number of successes in recent years. Engaging in public protests, lobbying, media relations, electoral work, and other direct-action tactics, including mass-based civil disobedience, environmental justice activists have helped to kill 80 percent of all planned municipal incinerators; have protected the natural resources and unique wilderness areas of local communities; have stopped ocean dumping of radioactive wastes and sewage sludge; have prevented food irradiation; and have won many other important victories at the local level.[44]

Although the tactics for attacking environmental inequities are quite varied, one common political demand of these movements is for greater democratic participation in the governmental decision-making processes affecting their communities. By gaining greater access to policy makers and agencies, environmental justice activists hope to initiate better governmental regulation of the discriminatory manner in which the market distributes environmental risks.[45] At the national level, this has led important segments of the environmental justice movement to draw upon liberal–democratic strategies aimed at reforming "EPA's institutional focus," particularly the manner by which the agency drafts and enforces environmental policy.[46] Already, President Clinton has signed Executive Order 12898 on Environmental Justice, ordering all federal agencies to begin initiatives aimed at reducing environmental inequities.

Despite these achievements, there are limitations to this reformist approach to environmental injustice. Aside from failing to transform the manner in which corporate money and power now dominate the electoral and policy-making processes,[47] including the mechanisms by which the Environmental Protection Agency, Department of the Interior, and related state agencies become captured and subsumed by capital,[48] liberalist-oriented environmentalism fails to address the "essential cause" of ecological problems in America—the workings of a capitalist economy and marketplace.

This is evident in a movement discourse that defines environmental justice in terms of eliminating the *discriminatory* or *unequal distribution* of ecological hazards rather than eliminating the root causes of the hazards for *all* Americans. President Clinton's Executive Order on Environmental Justice, issued on February 11, 1994, for instance, specifically associates environmental justice with federal agencies' "identifying and addressing, as appropriate, disproportionately high and adverse human health or environmental effects of their programs, policies and activities on minority populations and low income populations in the United States." Similarly, the Department of Energy, in announcing its "Environmental Justice" policy in 1994, defined its approach as promoting "non-discrimination among minority, American Indian, and low-income communities."[49]

But the struggle for environmental justice is not just about distributing environmental risks equally but about preventing them from being produced in the first place so that on one is harmed at all. A movement for environmental justice is of limited efficacy if the end result is to have all Americans poisoned to the same perilous degree, regardless of race, color, or class (as if this were even possible)? The struggle for environmental justice must be about the politics of capitalist production per se and the elimination of the ecological threat, not just the "fair" distribution of ecological hazards via better government regulation of inequities in the marketplace.

The benefits of greater participatory democracy are extremely limited if the choices for a marginalized community are to reject construction of a toxic waste disposal site that may pose significant health hazards, on the one hand, versus community acceptance of such a site because of the greater job opportunities and tax revenues it affords, on the other. Unless movements for environmental justice can address the political–economic dynamics of capitalism that force communities to make such tradeoffs, their conception of ecological democracy as "participatory democracy" will remain extremely limited. And while increased participatory democracy by popular forces in governmental decision making and community planning is desirable (if not essential), and should be supported, it is, in and of itself, insufficient for achieving true environmental justice. What is needed is a richer conception of ecological democracy.

From an ecosocialist perspective, organizing efforts against procedures that result in an unequal distribution of environmental problems (*distributional inequity*) cannot ultimately succeed unless environmental justice activists address the procedures by which the environmental problems are *produced* in the first place (*procedural inequity*). In the words of Robert Lake:

> Probing the nature of procedural environmental equity as self-determination suggests that solving the distributional problem may be necessary but will not be sufficient for producing environmental justice. Stated differently, environmental justice will not have been achieved in the event that marginalized communities are no longer subjected to a disproportionate share of environmental problems. We will not have eliminated environmental inequity when a benevolent Environmental Protection Agency succeeds in redistributing environmental problems such that no community is disproportionately burdened. Removing the environmental burden on a community, say through facility licensing and monitoring, site remediation, and environmental cleanups, may well be a significant accomplishment but it will not have empowered that community to control its environment. Redistributing outcomes will not achieve environmental justice unless it is accompanied and, indeed, preceded by a procedural redistribution of power in decision-making.[50]

In Lake's view, any attempt to rectify distributional inequities without attacking the fundamental processes that produced the problems in the first place focuses on symptoms rather than causes and is therefore only a partial, temporary, and necessarily incomplete and insufficient solution. What is needed is an ecosocialist politics for procedural equity that emphasizes democratic participation in the capital investment decisions through which environmental burdens are *produced* then distributed.[51] This is a very broad concept of ecological democracy and environmental justice. As Michael Heiman has observed, "If we settle for liberal procedural and distributional equity, relying upon negotiation, mitigation, and fair-share allocation to address some sort of disproportional impact, we merely perpetuate the current production system that by its very structure is discriminatory and non-sustainable."[52] It is precisely this distinction between *distributional justice* versus *productive justice* that many in the environmental justice are now beginning to address.

Rather than existing as a collection of isolated organizations fighting defensive "not-in-my-backyard" battles (as important as they may be), the environmental justice movement must evolve into a political force capable of challenging the systemic causes of social and ecological injustices as they exist "in everyone's backyard." Only by bringing about what Barry Commoner calls "the social governance of the means of production"[53]—a radical democratization of all major political, social, and *economic* institutions in society—can humanity begin to gain control over the course of social and environmental history. Such a program for social governance would require sublate into the institutions of workplace and local direct democracy, liberal democratic procedures and constitutional guarantees, state planning, and the initiatives of popular-based social and environmental movements into a genuine ecological democracy.[54]

Only by bringing about an ecological democracy in this form can the anarchy of the free market and the primacy of economic growth (which is oriented first and foremost to the short-term interests of corporate capital at the expense of nature) be subordinated to long-term democratic planning aimed at meeting the human and environmental needs of both present and future generations. Only then can the environmental movement succeed in moving beyond single-issue or band-aid policy approaches to embrace more comprehensive solutions to the ecological crisis. This would include adopting *pollution prevention* measures that eliminate the use of dangerous chemicals, production processes, and consumer goods altogether (source reduction) rather than relying on costly and ineffective *pollution control* measures aimed at "containing" and "fairly" distributing environmental hazards once they are produced. In this respect, ecological democracy is not only a form of *praxis* but must also become the *telos* of movements for environmental justice.

Perhaps the first and most fundamental step in the quest for ecological democracy involves the state. Because the state administers the social division of labor on behalf of capital as a whole, popular democratic control of the state apparatus (e.g., the Occupational Safety and Health Administration, EPA, Department of the Interior, etc.), is essential if social and environmental planning are to become truly viable. In a genuine ecological democracy, this would require that political parties, elected officials, and state bureaucrats at all levels share power with those social and environmental movements that represent the interests of the popular classes—as workers, consumers, community residents, and citizens.[55] Only under such a system of democratic state planning and administration of society does it becomes possible to replace the car culture with more equitable and accessible mass transportation systems;[56] to dismiss energy-intensive and pollutive technologies favored by multinational oil companies in favor of clean, renewable alternatives; or to rationally plan community development and natural resource extraction on a sustainable basis.[57]

THE FUTURE OF DEMOCRATIC ECOSOCIALISM

The quest for ecological democracy is facilitating a new type of environmental politics in the United States. In the words of Florence Gardner and

Simon Greer, "demanding corporate accountability and more stringent governmental regulation, gaining increased worker and community control of production processes, dismantling institutional racism and classism, and dismissing individual lifestyle and blame the victim solutions to environmental problems are the principles at the heart of environmental justice organizing."[58] Each contribution in this collection highlights these struggles for ecological democracy (in one form or another) as being key to developing political strategies and goals that can overcome the root causes of America's social and ecological crises. And while ecological democracy is not socialism, it may one day prove to be a precondition for the emergence of a viable ecosocialist politics. As witnessed in the legacy of the Soviet Union and other Eastern European countries, one of the great lessons of history is that social *ownership* of the means of production *without* genuine democratic social *control* by the people is doomed to be a political and ecological failure.[59] In light of the abuses of power previously carried out under systems of party/bureaucratic political rule in state socialist countries, it is clear that the ideals of ecological democracy are central to a revitalization of the socialist project in the United States.[60]

Likewise, given the ecological contradictions and exploitative class relations prevalent under a capitalist mode of production, it is also true that a genuine ecological democracy can only be fully constructed in a socialist society. True "social governance of the means of production" and long-term democratic environmental planning based on human need require social ownership over key sectors of the economy (health care, food distribution, banking, housing, transportation, natural resource development and distribution, energy and alternative technologies). Otherwise, the short-term irrationalities and profit-making mandates of the capitalist marketplace would undermine any attempt to build a just and sustainable society. In this sense, the ideals of socialism are essential to the construction of a viable ecological democracy. Movements for environmental justice and ecological democracy in the United States thus hold the potential for becoming the basis for a popular red-green alliance committed to building a democratic ecosocialism in America. Making this vision a reality is a task facing the Left today, and one which the editors and authors associated with *Capitalism, Nature, Socialism: A Journal of Socialist Ecology* and the "Democracy and Ecology" book series are dedicated to facilitating. (For a selected bibliography on environmental justice, see the concluding pages of this chapter.)

ACKNOWLEDGMENTS

For their thoughtful criticisms and suggestions with regard to this introduction, I thank James O'Connor and Alan Klein. I would also like to thank Victor Wallis, John Wooding, Pat Dalendina, Roger Gottlieb, Cathy Crumbley, Joan Roelofs, Alan

Rudy, Greg DeLaurier, Jay Moore, Eric Bourgeois, Eric Krieg, Kevin Gallagher, Frank Ackerman, George Katsiaficas, Tom Estabrook, Chuck Levenstein, and other members of the Boston Editorial Group of *Capitalism, Nature, Socialism* for their helpful comments as well.

NOTES

1. For a concise history of the U.S. ecology movement and the emergence of what he terms the fourth wave of activist politics, that is, the environmental justice movement, see Mark Dowie, *Losing Ground: American Environmentalism at the Close of the Twentieth Century* (Cambridge, MA: MIT Press, 1995).

2. For a theoretical discussion of the contradictions between capitalist accumulation and environmental protection, see James O'Connor, *Conference Papers on Capitalism, Nature, Socialism* (Santa Cruz: *CNS*/CPE Pamphlet 1, 1991), available from *CNS*, P.O. Box 8467, Santa Cruz, CA 95061. See also Martin O'Connor, ed., *Is Capitalism Sustainable?: Political Economy and the Politics of Ecology* (New York: Guilford Press, 1994); and Ted Benton, ed., *The Greening of Marxism* (New York: Guilford Press, 1996).

3. This view is given the mantle of respectability through a number of pseudoscientific studies carried out by corporate-supported organizations, policy institutes, and writers. See Ronald Bailey, *Ecoscam: The False Prophets of Ecological Apocalypse* (New York: St. Martin's Press, 1993), and *The True State of the Planet* (New York: The Free Press, 1995); Dixie Lee Ray, *Environmental Overkill: Whatever Happened to Common Sense?* (New York: Harper Perennial Library, 1994); Ben W. Bolch and Robert D. McCallum, *Apocalypse Not: Science, Economics and Environmentalism* (Washington, DC: Cato Institute, 1993); Gregg Easterbrook, *A Moment on the Earth* (New York: Viking Press, 1995); Martin W. Lewis, *Green Delusions: An Environmentalist Critique of Radical Environmentalism* (Durham: Duke University Press, 1992); Charles T. Rubin, *The Green Crusade: Rethinking the Roots of Environmentalism* (New York: Free Press, 1994); and Wallace Kaufman, *No Turning Back: Dismantling the Fantasies of Environmental Thinking* (New York: Basic Books, 1994). For critiques that expose the many falsehoods of such works, see Raymond A. Rogers, "Doing the Dirty Work of Globalization," *Capitalism, Nature, Socialism,* 6(3), September 1995; Tom Athanasiou, "The Age of Greenwashing," *Capitalism, Nature, Socialism,* 7(1), March 1996, pp. 1–36, and *Divided Planet: The Ecology of Rich and Poor* (New York: Little, Brown, 1996); and "How They Lie" (Parts 1–2), in *Rachel's Environment & Health Weekly,* Issues 503–504 (July 18–25, 1996), as well as 437 (April 13, 1995).

4. 800,000 people who comprise the country's capitalist class and their top managers have more money and wealth than the 184 million other Americans (over age 16) who make up the working class and salariat. See Michael Parenti, *Democracy for the Few,* 6th ed. (New York: St. Martin's Press, 1995), pp. 7–14.

5. According to a Natural Resources Defense Council study released on May 8, 1996, some 64,000 Americans die each year from air pollution.

6. See K. William Kapp, *The Social Cost of Private Enterprise* (Cambridge,

MA: Harvard University Press, 1950), p. 13, reissued by Schocken Books, New York, 1971.

7. For a discussion of the social and ecological costs of capitalist production, see Chapter 1, "The Political Ecology of American Capitalism," in this collection.

8. See Julie Roque, "Review of EPA Report: 'Environmental Equity: Reducing Risk for All Communities,' " *Environment, 35*(5), June 1993, pp. 25–28.

9. See Theo Colburn, Dianne Dumanoski, and John Peterson Myers, *Our Stolen Future* (New York: Dutton, 1996), and the resource newsletter *Rachel's Environment & Health Weekly* (Annapolis, MD: Environmental Research Foundation, 1990–1997).

10. As argued by Laura Pulido, "while it may be possible statistically to separate and analyze 'racial' and income groups, such a procedure does not necessarily help us understand the processes of class formation, the division of labor, and functionality of poverty under capitalism which result in the racialized nature of the U.S. economy." See Laura Pulido, "A Critical Review of the Methodology of Environmental Racism Research," *Antipode, 28*(2), 1996, pp. 149–162.

11. For instance, see Michael Gelobter, "Toward a Model of 'Environmental Discrimination,' " in Paul Mohai and Bunyan Bryant, eds., *Race and the Incidence of Environmental Hazards: A Time for Discourse* (Boulder, CO: Westview Press, 1992), pp. 64–81; and Eric J. Krieg, "A Socio-Historical Interpretation of Toxic Waste Sites: The Case of Greater Boston," *American Journal of Economics and Sociology, 54*(1), January 1995 pp. 1–14.

12. See D. R. Wernette and L. A. Nieves, "Breathing Polluted Air: Minorities Are Disproportionately Exposed," *EPA Journal,* March/April 1992, p. 16.

13. See "The EPA's Dirty Secret," *Boston Globe,* February 13, 1994, p. 1; and Eric Mann with the Watchdog Organizing Committee, *L.A.'s Lethal Air: New Strategies for Policy, Organizing, and Action* (Los Angeles: Labor/Community Strategy Center, 1991).

14. See Ivette Perfecto, "Farm Workers, Pesticides, and the International Connection," in Paul Mohai and Bunyan Bryant, eds., *Race and the Incidence of Environmental Hazards: A Time for Discourse* (Boulder, CO: Westview Press, 1992), pp. 177–203.

15. The Immigration and Environment Campaign of the Political Ecology Group (PEG) in San Francisco (415-777-3488), which helped to organize the participation of numerous environmental justice and immigrant rights groups in the Immigration and Environment National Strategy Meeting in March 1996, is an example of one such coalition in the making.

16. See U.S. General Accounting Office, *Siting of Hazardous Waste Landfills and Their Correlation with Racial and Economic Status of Surrounding Communities* (Washington, DC: U.S. Government Printing Office, 1983).

17. See United Church of Christ Commission for Racial Justice, *Toxic Wastes and Race in the United States: A National Report on the Racial and Socioeconomic Characteristics of Communities Surrounding Hazardous Waste Sites* (New York, 1987); Susan Zakin, "The Ominous Color of Toxic Dumping," *Sierra,* July/August 1987, pp. 14–17; and Robert D. Bullard, *Dumping in Dixie: Race, Class, and Environmental Quality* (Boulder, CO: Westview Press, 1990).

18. See Benjamin Goldman and L. Fitton, *Toxic Waste and Race Revisited:*

An Update of the 1987 Report on the Racial and Socioeconomic Characteristics of Communities with Hazardous Waste Sites (Washington, DC: Center for Policy Alternatives, National Association for the Advancement of Colored People, and the United Church of Christ Commission for Racial Justice, 1994).

19. It also takes an average of 20 percent longer for the government to place toxic waste dumps in minority communities on the National Priorities List (NPL), or Superfund list, for cleanup than sites in white areas. See Marianne Lavelle and Marcia Coyle, "Unequal Protection: The Racial Divide in Environmental Law," *National Law Journal*, September 21, 1992, pp. 2–12.

20. See "Toxic Empire: The WMX Corporation, Hazardous Waste and Global Strategies for Environmental Justice" (1995), available from the Political Ecology Group, 965 Mission St., Suite 700, San Francisco, CA 94103.

21. See Richard Hofrichter, ed., *Toxic Struggles: The Theory and Practice of Environmental Justice* (Philadelphia: New Society Publishers, 1993); Robert D. Bullard, ed., *Unequal Protection: Environmental Justice and Communities of Color* (San Francisco: Sierra Club Books, 1994), and *Confronting Environmental Racism: Voices from the Grassroots* (Boston: South End Press, 1993); Bunyan Bryant and Paul Mohai, eds., *Race and the Incidence of Environmental Hazards: A Time for Discourse* (Boulder, CO: Westview Press, 1992); and Bunyan Bryant, ed., *Environmental Justice: Issues, Policies, and Solutions* (Washington, DC: Island Press, 1995).

22. In 1994, a study by the New York State Health Department found a 60 percent higher incidence of breast cancer in postmenopausal women who live near chemical plants compared with women who live elsewhere. Cited in "Pollution Linked to Breast Cancer," *Boston Globe,* April 13, 1994, p. 4.

23. See Ken Geiser and Gerry Waneck, "PCBs and Warren County," in Bullard, ed., *Unequal Protection,* pp. 43–52.

24. Organizations such as the SouthWest Organizing Project (SWOP) in Albuquerque, the West Dallas Coalition for Environmental Justice, People United for a Better Oakland (PUEBLO), Tucsonians for a Clean Environment, the Political Ecology Group (PEG) and Urban Habitat Program of the San Francisco Bay Area, West Harlem Environmental Action, the Gulf Coast Tenant Organization, and the Berkeley Coalition for Environmental Justice are among the numerous groups to have emerged over the last few years.

25. The ecofeminist project is to link the manner in which patriarchal systems serve to oppress both women and nature, and to develop strategies for replacing or transforming hierarchical or patriarchal structures that negate the active participation of "nonexperts" (particularly women) in decision-making processes. Part of this process entails legitimizing and promoting the different manner by which women (and "the *other*" oppressed members of society) *experience* and *conceptualize* social and environmental problems. Such endeavors, for instance, have resulted in the creation of alternative women's health collectives or research programs devoted to exploring the impact on women of toxic substances found in the workplace and environment. See Irene Diamond and Gloria Feman Orenstein, eds., *Reweaving the World: The Emergence of Ecofeminism* (San Francisco: Sierra Club Books, 1990).

26. It should be noted, as by Barbara Epstein, that "in spite of the large numbers of women in the environmental movement, and their increasingly prominent roles in leadership, women in the movement have not made issues of gender central to their political practice in the way that people of color have made issues

of race." See Barbara Epstein, "Ecofeminism and Grass-Roots Environmentalism in the United States," in Richard Hofrichter, ed., *Toxic Struggles*, pp. 144–152.

27. Greg Delaurier, personal correspondence, February 1, 1997.

28. Much of the inspiration for such a strategy is provided by ecoanarchism and left-green populism. See Roy Morrison, *Ecological Democracy* (Boston: South End Press, 1995); John O'Connor, "The Promise of Environmental Democracy," in Richard Hofrichter, *Toxic Struggles,* p. 47; and Murray Bookchin, *Remaking Society* (Boston: South End Press, 1990), as well as many of his other works on social ecology. For a critique of social ecology and ecoanarchism, see Andrew Light, ed., *Social Ecology after Bookchin* (New York: Guilford Press, 1998).

29. In the words of one prominent scholar–activist in the environmental justice movement today, "What do grass-roots leaders want? These leaders are demanding a shared role in the decision-making processes that affect their communities. They want participatory democracy to work for them"; see Robert D. Bullard, "Introduction," in Bullard, ed., *Unequal Protection*, p. xvii.

30. See also Allan Schnaiberg, ed., "Social Equity and Environmental Activism: Utopias, Dystopias and Incrementalism," *Qualitative Sociology, 16*(3), Fall 1993, pp. 203–262; and Allan Schnaiberg and Kenneth Alan Gould, *Environment and Society: The Enduring Conflict* (New York: St. Martin's Press, 1994).

31. See also Richard Moore, "Confronting Environmental Racism," in *Crossroads, 11*(2), April 1992, pp. 7–9. In March of 1990, the Southwest Organizing Project (SWOP) sent a letter to ten of the largest national environmental organizations (at that time referred to as "the Group of Ten"), which stated:

> There is a clear lack of accountability by the Group of Ten environmental organizations towards Third World communities in the Southwest, in the United States as a whole and internationally. Your organizations continue to support and promote policies which emphasize the clean-up and preservation of the environment on the backs of working people in general and people of color in particular. In the name of eliminating environmental hazards at any cost, across the country industrial and other economic activities which employ us are being shut down, curtailed or prevented while our survival needs and cultures are ignored. We suffer from the results of these actions, but are never full participants in the decision-making which leads to them.

32. See Mark Dowie, *Losing Ground: American Environmentalism at the Close of the Twentieth Century* (Cambridge, MA: MIT Press, 1995).

33. See Timothy Luke, "Worldwatching at the Limits to Growth," *Capitalism, Nature, Socialism, 5*(2), June 1994, pp. 43–64.

34. For a discussion, see Daniel Faber, *Environment Under Fire: Imperialism and the Ecological Crisis in Central America* (New York: Monthly Review Press, 1993).

35. See also Devon Peña, *The Terror of the Machine: Technology, Work, Gender and Ecology on the U.S.–Mexico Border* (Austin: CMAS Books, University of Texas Press, 1997).

36. For an analysis of political repression of environmental activists in the Third World, see Andrew Rowell, *Green Backlash: Global Subversion of the Environmental Movement* (New York: Routledge, 1996); and Barbara Rose Johnston, ed., *Who Pays the Price? The Sociocultural Context of Environmental Crisis* (Washington, DC: Island Press, 1994).

37. See David Helvarg, *The War against the Greens: The "Wise Use" Movement, the New Right, and Anti-Environmental Violence* (San Francisco: Sierra Club Books, 1994).

38. See Judi Bari, "Revolutionary Ecology," *Capitalism, Nature, Socialism, 8*(2), June 1997.

39. For instance, see Richard Kazis and Richard Grossman, *Fear at Work: Job Blackmail, Labor and the Environment* (New York: Pilgrim Press, 1982).

40. For Pulido, ecological legitimacy is attached to a people when they are seen as valid actors practicing environmental stewardship or when their commitment to caring for the land in a sustainable manner is not regarded as suspect.

41. See also Timothy Luke, "Worldwatching at the Limits of Growth."

42. See also Laura Pulido, *Environmentalism and Economic Justice* (Tucson: University of Arizona Press, 1996).

43. See Kathy Hall, "Changing Woman, Tukanavi and Coal: Impacts of the Energy Industry on the Navajo and Hopi Reservations," *Capitalism, Nature, Socialism, 3*(1), March 1992, pp. 49–78.

44. For a short summary, see "Where We Are Now" in *Rachel's Environment & Health Weekly,* 500, June 27, 1996.

45. See Robert C. Paehlke, *Environmentalism and the Future of Progressive Politics* (New Haven, CT: Yale University Press, 1989).

46. See Deeohn Ferris, "A Call for Justice and Equal Environmental Protection," in Bullard, ed., *Unequal Protection,* pp. 298–319.

47. The major sources of campaign money are corporations and wealthy individuals (the corporate ruling class). Of the total $659 million spent on federal elections in 1992, about 80 percent ($527 million) came from these two sources. See James B. Raskin and John Bonifaz, *The Wealth Primary* (Washington, DC: Center for Responsive Politics, 1994); and Larry Makinson, *Follow the Money Handbook* (Washington, DC: Center for Responsive Politics, 1994).

48. For an analysis of the manner in which the corporate ruling class dominates the state, see William G. Domhoff, *Who Rules America Now?* (New York: Simon & Schuster, 1983).

49. Cited in Robert Gottlieb and Andrew Fisher, " 'First Feed the Face': Environmental Justice and Community Food Security," *Antipode, 28*(2), 1996, pp. 193–203.

50. See Robert W. Lake, "Volunteers, NIMBYs, and Environmental Justice: Dilemmas of Democratic Practice," *Antipode, 28*(2), 1996, p. 169.

51. In the words of Robert W. Lake (ibid., p. 165), self-determination is not realized simply through participation in decisions regarding the distribution of environmental burdens if it does not also extend to participation in decisions controlling their production.

52. See Michael K. Heiman, "Race, Waste, and Class: New Perspectives on Environmental Justice," *Antipode, 28*(2), 1996, p. 120.

53. See Barry Commoner, *Making Peace with the Planet* (New York: Pantheon, 1990).

54. For an original presentation of these ideas, see James O'Connor, "A Political Strategy for Ecology Movements," *Capitalism, Nature, Socialism, 3*(1), March 1992, pp. 1–5, and "A Red–Green Politics in the U.S.?" *Capitalism, Nature, Socialism, 5*(1), March 1994, pp. 1–20.

55. See Michael Parenti, *Democracy for the Few,* 6th ed. (New York: St. Martin's Press, 1995), ch. 17.

56. See Peter Freund and George Martin, "The Commodity That Is Eating the World: The Automobile, the Environment, and Capitalism," *Capitalism, Nature, Socialism,* 7(4), December 1996, pp. 3–30.

57. See Stefan Kipfer, Franz Hartman, and Sara Marino for the Toronto *CNS* Editorial Group, "Cities, Nature, and Socialism: Towards an Urban Agenda for Action and Research," in the special edition on "Urban Ecology" in *Capitalism, Nature, Socialism,* 7(2), June 1996, pp. 5–20.

58. See Florence Gardner and Simon Greer, "Crossing the River: How Local Struggles Build a Broader Movement," *Antipode, 28*(2), 1996, pp. 175–192.

59. See James O'Connor, "Political Economy of Ecology of Socialism and Capitalism," *Capitalism, Nature, Socialism, 1*(3), November 1989, pp. 93–106; Victor Wallis, "Socialism, Ecology, and Democracy: Toward a Strategy of Conversion," *Monthly Review, 44*(2), June 1992, pp. 1–22; and David Pepper, *Eco-Socialism: From Deep Ecology to Social Justice* (New York: Routledge, 1993).

60. As stated by Hilary Wainwright, the "focus on democracy flows from a belief that the crisis of socialism is not the result of any exhaustion of the desire for social justice. It is rather that product of the failure of state institutions—the institutions historically associated with socialism—to bring social justice about and bring it about under the rule of the people. Capitalism's reckless and exploitative drive to accumulate has been assailed from all sides: from the standpoint of women, blacks, dominated regions and suppressed nations, the victims of a collapsing ecosystem, as well as the traditional assailants at the point of production—all these, but rarely in the name of socialism. Socialism has historically associated the goal of social justice with the use of state power by the working class and/or its representatives. The viability of these means are at a turning point, not the values of equality and social justice. It is this burning urgency of the problem of political means and mechanisms that leads me to consider issues of democracy to be central to the renewal of socialism." See Hilary Wainwright, "New Forms of Democracy for Socialist Renewal," *Socialist Review, 90*(2), Special 20-Year Anniversary Issue, 1970–1990, p. 32.

SELECTED BIBLIOGRAPHY ON ENVIRONMENTAL JUSTICE

Alston, Dana, ed., *We Speak for Ourselves: Social Justice, Race, and Environment* (Washington, DC: The Panos Institute, 1991).

Athanasiou, Tom, *Divided Planet: The Ecology of Rich and Poor* (New York: Little, Brown, 1996).

Benton, Ted, ed., *The Greening of Marxism* (New York: Guilford Press, 1996).

Bookchin, Murray, *Remaking Society* (Boston: South End Press, 1990).

Bryant, Bunyan, ed., *Environmental Justice: Issues, Policies, and Solutions* (Washington, DC: Island Press, 1995).

Bryant, Bunyan, and Paul Mohai, eds., *Race and the Incidence of Environmental Hazards* (Boulder, CO: Westview Press, 1992).

Bullard, Robert D., ed., *Unequal Protection: Environmental Justice and Communities of Color* (San Francisco: Sierra Club Books, 1994).

Bullard, Robert D., ed., *Confronting Environmental Racism: Voices from the Grassroots* (Boston: South End Press, 1993).

Bullard, Robert D., *Dumping in Dixie: Race, Class, and Environmental Quality* (Boulder, CO: Westview Press, 1990).

Cable, Sherry, and Charles Cable. *Environmental Problems—Grassroots Solutions: The Politics of Grassroots Environmental Conflict* (New York: St. Martin's Press, 1995).

Colborn, Theo, Dianne Dumanoski, and John Peterson Myers, *Our Stolen Future* (New York: Dutton, 1996).

Commoner, Barry, *Making Peace with the Planet* (New York: Pantheon Books, 1990).

Diamond, Irene, and Gloria Feman Orenstein, eds., *Reweaving the World: The Emergence of Ecofeminism* (San Francisco: Sierra Club Books, 1990).

Domhoff, G. William, *Who Rules America Now?* (New York: Simon & Schuster, 1983).

Dowie, Mark, *Losing Ground: American Environmentalism at the Close of the Twentieth Century* (Cambridge, MA: MIT Press, 1995).

Dunlap, Riley E., and Angela G. Mertig, eds., *American Environmentalism* (Washington, DC: Taylor & Francis, 1992).

Faber, Daniel R., *Environment Under Fire: Imperialism and the Ecological Crisis in Central America* (New York: Monthly Review Press, 1993).

Gardner, Florence, and Simon Greer, "Crossing the River: How Local Struggles Build a Broader Movement," *Antipode, 28*(2), 1996, pp. 175–192.

Gedicks, Al, *The New Resources Wars: Native and Environmental Struggles against Multinational Corporations* (Boston: South End Press, 1993).

Gottlieb, Robert, *Forcing the Spring: The Transformation of the American Environmental Movement* (Washington, DC: Island Press, 1993).

Gottlieb, Roger, ed., *The Ecological Community: Environmental Challenges for Philosophy, Politics and Morality* (New York: Routledge, 1996).

Hall, Kathy, "Changing Woman, Tukanavi and Coal: Impacts of the Energy Industry on the Navajo and Hopi Reservations," *Capitalism, Nature, Socialism, 3*(1), March 1992, pp. 49–78.

Helvarg, David, *The War against the Greens: The "Wise-Use" Movement, the New Right, and Anti-Environmental Violence* (San Francisco: Sierra Club Books, 1994).

Heiman, Michael K., "Race, Waste, and Class: New Perspectives on Environmental Justice," *Antipode, 28*(2), 1996, pp. 111–121.

Hofrichter, Richard, ed., *Toxic Struggles: The Theory and Practice of Environmental Justice* (Philadelphia: New Society Publishers, 1993).

Johnston, Barbara Rose, ed., *Who Pays the Price?: The Sociocultural Context of Environmental Crisis* (Washington, DC: Island Press, 1994).

Kapp, J. William, *The Social Cost of Private Enterprise* (Cambridge, MA: Harvard University Press, 1950).

Krieg, Eric J., "A Socio-Historical Interpretation of Toxic Waste Sites: The Case of Greater Boston," *The American Journal of Economics and Sociology, 54*(1), January 1995, pp. 1–14.

LaBalme, J. A., *A Road to Walk: A Struggle for Environmental Justice* (Durham, NC: The Regulator Press, 1987).

Lake, Robert W., "Volunteers, NIMBYs, and Environmental Justice: Dilemmas of Democratic Practice," *Antipode, 28*(2), 1996, pp. 160–174.

Light, Andrew, ed., *Social Ecology after Bookchin* (New York: Guilford Press, in press).
Luke, Timothy, "Worldwatching at the Limits of Growth," *Capitalism, Nature, Socialism, 5*(2), June 1994, pp. 43–64.
Manes, Christopher, *Green Rage: Radical Environmentalism and the Unmaking of Civilization* (New York: Little, Brown, 1990).
Martell, Luke, *Ecology and Society: An Introduction* (Amherst: University of Massachusetts Press, 1994).
McMichael, A. J., *Planetary Overload: Global Environmental Change and the Health of the Human Species* (Cambridge, UK: Cambridge University Press, 1993).
Merchant, Carolyn, *Radical Ecology* (New York: Routledge, 1992).
Merchant, Carolyn, *Ecological Revolutions* (Chapel Hill: University of North Carolina Press, 1989).
Morrison, Roy, *Ecological Democracy* (Boston: South End Press, 1995).
National Research Council, *Environmental Epidemiology: Public Health and Hazardous Wastes* (Washington, DC: National Academy Press, 1991).
Norton, Bryan G., *Toward Unity among Environmentalists* (New York: Oxford University Press, 1991).
O'Connor, James, *Conference Papers on Capitalism, Nature, Socialism* (Santa Cruz: CNS/CPE Pamphlet 1, 1991).
O'Connor, John, "The Promise of Environmental Democracy," in R. Hofrichter, ed., *Toxic Struggles: The Theory and Practice of Environmental Justice* (Philadelphia: New Society Publishers, 1993), pp. 47–57.
O'Connor, Martin, ed., *Is Capitalism Sustainable?: Political Economy and the Politics of Ecology* (New York: Guilford Press, 1994).
Paehlke, Robert C., *Environmentalism and the Future of Progressive Politics* (New Haven, CT: Yale University Press, 1989).
Parenti, Michael, *Democracy for the Few*, 4th ed. (New York: St. Martin's Press, 1995).
Peña, Devon, *The Terror of the Machine: Technology, Work, Gender and Ecology on the U.S.–Mexico Border* (Austin: CMAS Books, University of Texas Press, 1997).
Pepper, David, *Eco-Socialism: From Deep Ecology to Social Justice* (New York: Routledge, 1993).
Press, Daniel, *Democratic Dilemmas in the Age of Ecology* (Durham, NC: Duke University Press, 1994).
Pulido, Laura, "A Critical Review of the Methodology of Environmental Racism Research," *Antipode, 28*(2), 1996, pp. 142–159.
Pulido, Laura, *Environmentalism and Economic Justice: Two Chicano struggles in the Southwest* (Tucson: University of Arizona Press, 1996).
Rowell, Andrew, *Green Backlash: Global Subversion of the Environment Movement* (New York: Routledge, 1996).
Schnaiberg, Allan, and Kenneth Alan Gould, *Environment and Society: The Enduring Conflict* (New York: St. Martin's Press, 1994).
Schwab, Jim, *Deeper Shades of Green: The Rise of "Blue-Collar" and Minority Environmentalism in America* (San Francisco: Sierra Club Books, 1994).
Szasz, Andrew, *Ecopopulism: Toxic Waste and the Movement for Environmental Justice* (Minneapolis: University of Minnesota Press, 1994).

United Church of Christ Commission for Racial Justice, *Toxic Wastes and Race in the United States: A National Report on the Racial and Socioeconomic Characteristics of Communities with Hazardous Waste Sites* (New York: Author, 1987).

U.S. General Accounting Office, *Siting of Hazardous Waste Landfills and Their Correlation with Racial and Economic Status of Surrounding Communities* (Washington, DC: U.S. Government Printing Office, 1983).

Chapter 1

The Political Ecology of American Capitalism

New Challenges for the Environmental Justice Movement

Daniel Faber

THE ECOLOGICAL CRISIS AND THE CRISIS OF ENVIRONMENTALISM

With the approaching dawn of the new millennium, the environmental movement in the United States is confronting what appears to be an immense paradox. On the one hand, over the course of the past three decades environmentalists have built one of most broadly based and politically powerful social movements in this country's history.[1] Many important victories resulting in the protection of endangered species, wildlife habitats, parks, and wilderness; reductions in some types of air, noise, land, and water pollution; some key improvements of public/consumer/worker health and safety, including reductions in human exposure to such highly dangerous substances as lead, asbestos, DDT, and other toxic chemicals; and other measures have been won during this time. Furthermore, dozens of major environmental statutes costing business hundreds of billions of dollars have been passed into law. As a result, U.S. governmental policies for protecting the environment and worker/human health are now among the most stringent in the world.

Yet, on the other hand, despite having won many important battles, it is becoming increasingly apparent that the modern environmental movement is losing the war for a healthy ecology. Despite a U.S. governmental pledge at the 1992 Earth Summit in Rio to reduce emissions of greenhouse gases, air pollution in the form of carbon dioxide, dust particles, and carbon monoxide (which causes respiratory disease) is increasing. Over 60,000 Americans die each year from air pollution alone. Half a million people living in the most polluted areas in 151 cities across the country face a risk of death some 15 to 17 percent higher than in the least polluted areas.[2] Some 164 million Americans are now at risk for respiratory and other health problems from exposure to excessive air pollution.[3] While more than $100 billion has been spent as a result of Clean Water Act regulations, the water quality of most major rivers is poor. More water tables are being poisoned, while more agricultural soil is being eroded or rendered unproductive because of salinization and other problems. The occurrence of nitrate, arsenic, and cadmium, all serious pollutants, is increasing considerably.[4] Ocean and beach pollution are a national scandal, as is the threat to coastlines, wetlands, and wildlife sanctuaries, including the Florida Everglades.

Most environmental regulations also largely fail to solve problems related to urban/suburban sprawl, nuclear waste disposal, the car culture, global warming, and the destruction of unique wilderness areas by logging and mining activities. Perhaps most alarming is the explosion in the production and dumping of dangerous pesticides and toxic chemicals into the environment. Since the publication of Rachel Carson's *Silent Spring* in 1962, which awoke the nation to the deadly health threats and ecological damage posed by organochlorine-based pesticides, the United States has increased its production of these chemicals fivefold.[5] Some 462 million pounds of pesticide are applied to the nation's cropland each year—a quantity equal to more than 1.5 pounds for every man, woman, and child in the country—where they contaminate our land, food, waterways, and groundwater, as well as farm workers and nearby communities, with poisonous chemicals. The impact of these and other pesticides on America's children is particularly severe.[6]

Meanwhile, there are over 41 million people who live within four miles of at least one of the nation's roughly 1,500 highly dangerous National Priority List (NPL, or Superfund) toxic waste sites.[7] Although these dumps are the worst of the worst, the Office of Technology Assessment estimates there are as many as 439,000 other illegal hazardous waste sites in the country.[8] For nearby residents to toxic waste sites, the National Research Council has found a disturbing pattern of elevated health problems, including heart disease, spontaneous abortions and genital malformations, and death rates, while infants and children are found to suffer a higher incidence of cardiac abnormalities, leukemia, kidney–urinary tract infections, seizures, learning disabilities, hyperactivity, skin disorders, reduced weight, central nervous system damage, and Hodgkin's disease.[9] Exposure to industrial chemicals is also contributing to the dramatic increases since the 1950s in cancer of the testis, prostate gland, kidney, breast, skin, and lung, as well as malignant myeloma, non-Hodgkin's lymphoma, and numerous childhood cancers—a cancer epidemic that kills half-a-million Americans each year. And, by most reliable accounts, these and other environmentally related health problems are growing worse.[10]

Given the magnitude of these and other ecological problems, it is clear that current environmental policy is not working. Most environmental laws are poorly enforced and overly limited in their application, emphasizing, for instance, *pollution control* measures that aim to limit public exposure to specific industrial pollutants at "tolerable" levels (a very difficult task) over *pollution prevention* measures that prohibit whole families of dangerous pollutants from being produced in the first place. The U.S. system of environmental regulation may be among the best in the world, but it is grossly inadequate for safeguarding human health and the integrity of nature. And while there is no doubt that ecological problems would be much worse absent the environmental movement and current system of regulation, it is also clear that the traditional political strategies and liberalist policy solutions embraced by the mainstream of the U.S. environmental establishment are proving to be increasingly ineffective, and even contradictory to their intended purposes.

These contradictions are becoming manifest politically in at least two major ways. On the one hand, even though ecoliberalist policy approaches fail to establish popular democratic control over the process of capital accumulation (investment, production, and distribution), such approaches do impose significant regulatory costs and inflexibilities that impinge on the profits of capital or cause inflation. In the current period of "slow" economic growth, increased competition, and the globalization of capital, this has prompted a profound corporate backlash against the liberal environmental establishment. The alternative proposal on behalf of business is for the adoption of neoliberal/corporatist forms of "free-market" environmentalism in state policymaking and enforcement.[11]

On the other hand, liberalist policy approaches have initiated some improvements in environmental quality as sought by broad sectors of the population, particularly the white middle class (or salariat). But because liberalist policy approaches are not embedded in popular democratic control over capitalist accumulation and state administration of society, they often neglect (or even worsen) problems for other sectors of the population that possess less political–economic power to defend themselves. In fact, one of the hallmarks of ecoliberalism is the manner in which it has corrected some single-issue environmental and human health problems for some people by transforming the ecological hazard into another form, which is then displaced onto other members of society, typically poorer segments of the white working class and communities of color. Thus, ecoliberalism has served to divide (as much as unite) the environmental movement from within and against other popular social movements. In the current period of growing social and ecological inequities, this has prompted a profound backlash on behalf of communities of color and elements of the labor movement against the regressive and fragmenting tendencies of mainstream ecoliberalism. The alternative proposal on behalf of the environmental justice movement is for the adoption of ecological democracy and greater social governance over state planning (at all levels) and the capitalist production process itself.

In short, the crisis of ecoliberalism is igniting a battle for the heart and soul of the environmental movement. What is required to overcome this paradox is a long step forward to create a radically new environmental politics in America that fundamentally challenges the political–economic power of capital to exploit working people and nature. Whether or not a unity of grassroots community organizations, labor unions, greens, and other movements fighting for social and economic justice and genuine environmental protection can unite in a campaign for ecological democracy remains to be seen. This is the challenge confronting the environmental justice movement today.

THE CORPORATE BACKLASH AGAINST THE ENVIRONMENTAL MOVEMENT

The current corporate backlash against the environmental movement cannot be understood without reference to the restructuring that is taking place in the U.S. economy. Despite a booming stock market,[12] low interest rates and unemployment, reduced trade and budget deficits, higher corporate earnings relative to Japan and Western Europe, and other apparent signs of financial health, the U.S. economy is in trouble. Plagued by slower GDP growth rates, since the mid-1970s, including one of the weakest decades (the 1990s) in this century, American capitalism is experiencing major economic problems. Over the past six years economic growth has averaged just 2.6 percent, which is about two-thirds the rate of the typical expansion and the slowest boom since World War II. High-tech industry is one of the few areas of the economy that is rapidly expanding, accounting for fully a third of today's economic growth. The rest of the economy continues to grow at less than 2 percent a year.[13]

Perhaps the most significant forces behind the slow-growth crisis of American capitalism reside in the profound changes taking place in the global economy.[14] Not since the pre-World War II period has American business faced such intense international competition from foreign capital, especially in East Asia. Fueled by innovations in communications, transportation, and production technologies, huge investments in infrastructure, as well as major improvements in the educational, skill, and productivity levels of labor power, overseas industry and agriculture (particularly in the newly industrializing countries of the Third World) is rapidly expanding to challenge the economic hegemony of the United States. Although this process, which is being facilitated in great part by a host of "free-trade" agreements brokered by the Bush and Clinton administrations, has launched a major trade boom for many sectors of the U.S. economy, particularly industries exporting high-tech and other capital goods and services of all kinds (hence the relatively low unemployment rates in the United States in comparison to Europe), other areas of the economy are suffering. Industries that have traditionally served as the backbone of the U.S. economy, as well as the trade union movement, continue to lose market after market for mass-produced consumer goods. Furthermore, the U.S. position as a competitive producer of raw materials and energy supplies is also being eroded by more profitable operations overseas.[15]

Plagued by growing trade deficits, weak reinvestment rates, the burden of government regulations and forced inflexibilities, low gains in labor productivity, and the growing costs of labor power, health care, housing, litigation, and other needs, U.S. capital is becoming increasingly unwilling to abide by the traditional accords brokered by the Democratic party on its behalf with the labor, civil rights, women's, and other progressive social movements.[16] As a result, the defining characteristics of liberal democratic capitalism that traditionally enlists the mass loyalty of the American working class—high wages, good benefits, job security and advancement, affirmative action, universal entitlements, and welfare protections for the majority of workers—are being eroded. But in this regard the United States is not alone. For not only has the triumph of the Asian model of export-led authoritarian capitalism dealt a blow to "New Deal" liberalism and welfare state capitalism in North America, but also it has undermined the economic viability of bureaucratic state socialism in the East, nationalist-based models of dependent development in the South, and Keynesian social democratic regimes in the West. In short, the U.S. economic crisis is part of a larger structural crisis in the world economy.[17]

With the globalization of capital and increased international competition, U.S. corporations (even monopoly capital) are less able to boost profits by passing along their increased costs to consumers in the form of higher prices that, along with a restrictive monetary policy implemented by the Federal Reserve at the behest of Wall Street, has maintained relatively low inflation rates in the 1980 and 1990s. As a result, the first imperative of American capital in the current economic climate is *not* to increase prices or even production (since growth is slow) but rather to lower production costs (since domestic and export markets are becoming more cutthroat). This has enormous implications for ecological protection. Along with labor costs (which include health insurance and other benefits), environmental regulations are considered by many U.S. industries to be some of the most expensive and burdensome. Companies are therefore seeking to protect profits not only by "downsizing" the labor force but also by cutting investments in pollution control, environmental conservation, and worker health and safety.

In the eyes of corporate America, therefore, the effects of antitoxic struggles on the part of the environmental movement in the current era of slow growth are proving increasingly toxic to short-term profitability (and perhaps even longer-term capital restructuring and reinvestment). Since the beginning of the "environmental decade" in 1970, spending for environmental protection has grown three times faster than the gross domestic product (GDP)—from $27.7 billion to about $170 billion annually—and now constitutes 2.8 percent of GDP.[18] These environmental expenditures typically add to the costs of capital but not to revenues. As pointed out by self-described environmentalist and vice-presidential candidate Albert Gore during the 1992 national election, most pollution control devices in the United States are simply added on to existing plant and equipment and fail to make industry more cost-efficient. Unlike new machinery that increases labor productivity and indirectly lowers the unit costs of wage goods, pollution abatement devices and cleanup technologies usually increase costs, hence, everything else being the same, reduce profits or increase prices. In short, "end of the pipe" pollution containment and environmental conservation measures are a luxury that American business is increasingly unwilling to absorb, especially when one considers the advantages enjoyed by foreign competitors with lower labor costs and less stringent regulations.

It is also important to note that when environmental regulation raises the costs of capital in raw materials (such as copper or oil) and capital goods industries (i.e., industries producing inputs for other industries), higher costs are generalized throughout the economy for all industries. The same regulations applied to consumer goods industries affect the cost of only one or a handful of commodities. These input-producing sectors, such as the petrochemical industry, are hardest hit by environmental regulations, and therefore the costs of capital are increased throughout the economy as a whole.[19] Furthermore, on top of increased costs of capital as perceived by business, liberalist environmental regulations also create enormous *inflexibilities* in the deployment of capital funds and labor, increase the circulation and turnover time of capital, and lower productivity, reinvestment, and accumulation. Legislation such as the recently renewed Endangered Species Act also disrupts free and easy access to both renewable and nonrenewable resources. Consumer Product Safety Commission and Food and Drug Administration efforts often create bottlenecks to faster market expansion. These and other effects on capital are creating a widespread perception in the business community that liberal environmental regulation is a major obstacle to corporate profitability.[20]

In contrast, many labor and environmental activists disclaim any responsibility for declining profitability, stagnating productivity, or slow growth. Labor's standard view is that the United States lacks sufficient markets—hence, the perceived need for a more progressive redistribution of wealth, more public spending on social programs, and increased wages and consumption levels to boost production and realize profits. The environmentalists' line is that ecological protection and cleanup create new industries, profits, and jobs, as well as reducing future costs by preventing further damage to natural resources. Both movements see American capitalism as "demand-constrained." In this view, measures designed to increase the demand for wage goods and capital goods through environmental regulation should help workers and the whole economy or at least harm economic growth only minimally.[21]

Both labor and environmentalists, however, are reluctant to discuss the "cost side" of the economic equation, that is, problems of labor productivity, investment, and profitability. And it is the "cost side" that has turned out to be capital's obsession in the 1980 and 1990s, a period of intensified international competition and slow economic growth. Under this kind of pressure, the old "jobs and community development" versus "the environment" debate is resurfacing. As a result, the labor and environmental movements have been unable to maintain much more than a symbolic unity since the big-business assault on the environmental and human welfare states began full tilt under the Reagan and Bush administrations. A major challenge facing the environmental and economic justice movement is to rebuild the bridges between these movements around a larger campaign for ecological democracy.

Acting on the perception that liberal environmentalism is a major factor contributing to the gathering economic crisis of U.S. capitalism, the Business Roundtable (an organization consisting of the chief executives of nearly 200 of the largest and most powerful companies in the land), National Association of Manufacturers, U.S. Chamber of Commerce, and other political coalitions organized by large corporations are launching an unrelenting assault. Heavily regulated companies, especially landed capital interests involved in real estate, chemicals, oil refining, timber, and other natural resources, are pouring money into antienvironmental organizations (such as the so-called Wise Use Movement) and election campaigns of neoliberal candidates in both major political parties.[22] The aim of these efforts has been to deregulate and reprivatize the economy ("liberalize" or free the market of "excessive" state intervention in behalf of the larger public interest) in the hope of restoring economic growth and corporate profits. By subordinating the regulatory state to the economic imperatives of capital, which in political terms means the reestablishment of corporate power vis-à-vis the environmental, labor, civil rights, women's, peace, and other social movements, neoliberals hope to transform the current economic crisis into a victory for U.S. business. As a result, traditional liberals in the Democratic Party are giving way to the more moderate Democratic Leadership Council types (such as Albert Gore and DLC cofounder Bill Clinton), while moderate politicians in the Republican party are giving way to more conservative elements who perhaps identify with Newt Gingrich.

At the forefront of this neoliberal assault at the behest of American business has been the executive and congressional branches of the federal government. Two strategies have dominated, although both are highly complementary.

The first strategy, as implemented by the Reagan and Bush administrations, as well as the 104th Congress, reflects a commitment to *direct confrontation* with the environmental movement and a *wholesale rollback* of existing environmental laws and regulations considered to be detrimental to business. Bolstered by a host of reports produced by right-wing corporate think-tanks such as the Heritage Foundation, this strategy attempts to discredit and delegitimate the mainstream ecology movement by convincing the American public that most ecological problems are not serious (or do not exist at all) and that the costs of environmental regulation to American businesses, taxpayers, and workers is too excessive. Instead, if the "free market" could be left to its own devices without governmental regulatory interference, business would thus have the flexibility to solve those environmental problems that did exist in more profitable and innovate manners. By harnessing the genius of free-market environmentalism, economic growth and prosperity could then be restored.

The second neoliberal strategy, as utilized by the Clinton–Gore administration, is characterized much more by attempts of *cooptation* of the mainstream environmental movement and preemptive "containment" of costly environmental laws. Rather than embracing a politics of confrontation, which has engendered an enormous political backlash on behalf of the American public against the Republican Party,[23] the Clinton–Gore strategy *accommodates* general environmental policy aims in exchange for the granting of major concessions to American industry. This strategy includes isolating more business-friendly mainstream environmental organizations from the rest of the movement in corporatist-type negotiating arrangements. The purpose of such "environmental mediation" and "dispute resolution" strategies is to enlist the support of this wing of the ecology movement in a number of highly symbolic policy initiatives that give the Clinton–Gore administration the appearance of being proenvironment. In exchange for such support, business is rewarded with a loosening of other regulations and granted forms of economic compensation (often in the form of "free-market" alternatives) that come at the expense of other environmental organizations and the issues over which they are battling.

In this respect, the Clinton–Gore administration is strategically instituting what Ryan Delcambre, Dow Chemical's Director of Environmental and Health Regulatory Affairs, calls "surgical fixes" to each environmental statute. The order is to help industry meet more "reasonable" environmental standards in more "cost-effective and more compliance-assisted" manners (i.e., incentives) rather than through "gotcha" governmental enforcement.[24] Even when the administration has passed laws supported by the environmental community, there are usually significant trade-offs.[25] As a result, many older regulations requiring across-the-board compliance with environmental laws are being replaced with "cost-effective" reforms—EPA's new Project XL, pollution taxes and credits, effluent charges, markets for pollution rights, and bubble schemes—all designed to increase capital's flexibility to meet regulation requirements but continue polluting in a profitable manner.

The Clean Air Act of 1990 is one such example. Supported by

Vice-President Albert Gore as a then senator from Tennessee, a key aspect of that legislation involves the commodification of pollution (which can be bought and sold on the stock market), which has allowed enterprises such as the Tennessee Valley Authority (TVA) to buy an estimated $2.5 million worth of "pollution credits" from Wisconsin Power and Light. These pollution credits allow the TVA to exceed federal limitations on sulfur dioxide and other toxic emissions in older facilities, mostly in poor working class communities of color in the South. Thus, the Act is a powerful reminder of the manner in which neoliberal, free-market environmentalism is likely to exacerbate, rather the resolve, the profound social and environmental injustices fostered by traditional regulatory approaches over the past thirty years.[26]

THE ECONOMIC RESTRUCTURING OF AMERICAN CAPITALISM AND THE CRISIS OF ENVIRONMENTALISM

Profits are key to the life and death battle of business in the global marketplace. Without an adequate rate of profit, and hence rate of capital accumulation, corporate America will lapse into economic stagnation. With the globalization of capital and slow growth of the U.S. economy, ruthless measures aimed at *cost minimization* now lay at the heart of business strategies for *profit maximization*. Greater efficiency (greater output per unit of input) becomes more important precisely because it leads to more profits. Increases in sales matched by increases in the cost of production are of less interest since they make corporations bigger but not necessarily more profitable.

In the 1970 and 1980s, Japanese and West German capitalism battered the American economy in large part by focusing more on market share maximization and less on profit maximization. These countries were willing to make investments in expanding capacity and market share than more profit-conscious American firms, which instead simply withdrew from overly competitive markets where profits became too low. In the 1990s, however, the American version of capitalism now looks much stronger than that in Japan, Germany, and Europe as a whole. The reason is that in the current period of slower worldwide growth American companies are proving to be more effective cost cutters. As a result, U.S. capital is now raising profits and expanding into stagnant markets more effectively than many European and Asian competitors.

Greater cost minimization of American capital is thus being achieved through a process of capital restructuring. The aim of this restructuring is to reestablish the necessary economic, social, political, and cultural conditions for renewed profitability, including new institutional arrangements congruent with the development of new technologies, production processes, work relations, and changing patterns of commodity demand. So, for example, by closing higher-cost facilities and moving to lower-cost production facilities offshore more rapidly than competing nations—particularly West German and Japanese market-share maximizers (who were left with so few profits in the 1990s that they found it difficult to finance expansions even when more profitable opportunities presented themselves), American business has been able to recapture some of the markets in the 1990s they had lost in the 1970 and 1980s.

The most important goal of capital restructuring for American business in the current period is to reestablish corporate "discipline" over trade unions and other social movements that are cutting into profits. Simply put, the key to cost minimization lies in processes of capital restructuring that enable American businesses to extract more value from labor power and nature in less time and at less cost. Thus, the primary means by which American capitalism is attempting to rescue itself is by simultaneously increasing the rate at which both working people (laborpower) and nature are economically exploited. And in the 1990s capital restructuring and deep cuts in labor and environmentally related costs are *boosting the earnings of American business at a much faster rate than revenue growth or increased sales.*

Generally speaking, increased rates of labor exploitation are being achieved by extracting more work (surplus-value) out of fewer workers in shorter periods of time and at less cost.[27] American business is achieving this result through a general assault on the past gains of the labor movement, which is taking numerous forms: workforce reductions or so-called corporate downsizing; the business offensive against unions; increased layoffs of permanent workers and the increased use of temporary or "contingent" workers at less pay; greater job insecurity, falling wages, benefits, and living standards; longer hours and a "speedup" of the production process; attacks on the minimal protections offered by the welfare state; deteriorating worker health and safety conditions; and a general assault on those private and public programs and policies that serve the interests of working Americans.[28] Combined with the monetarist actions of Federal Reserve Chairman Alan Greenspan to restrict the growth in labor's earnings—which might otherwise cut into corporate profit margins, spark inflation, and ruin the party atmosphere on Wall Street—the effect of the corporate offensive has been to add jobs and increase labor productivity without significantly raising the real wages of workers (hourly wages and benefits have gone up during the 1990s only half as much as in the typical expansion).[29]

On the other hand, increased rates of environmental exploitation are being achieved by such measures as: extracting greater quantities of natural resources of greater quality more quickly and at less cost; cutting production costs by spending less on pollution prevention and control, as well as environmental restoration; adopting new production processes that increase productivity but are also more pollutive or destructive of the environment; and so forth. American business is producing these results through a general assault on the past gains of the environmental movement and a general offensive upon the policies and programs that make up the environmental protection state. The result is increased dumping of ever more toxic pollution into the environment, particularly in poor working-class neighborhoods and communities of color; more destructive extraction of raw materials from this country's most unique and treasured landscapes; a deterioration in consumer product safety; the disappearance of more and more natural species and habitats; and a general assault on those programs and policies designed to protect the environment.

So, if increased profits are the economic engine pulling the train of American business in the world economy of the 1990s, then the increased exploitation of nature (both human and nonhuman) is providing the energy powering the locomotive. Neoliberal politicians stand at the controls, having engineered a loss of political power by organized labor, environmentalists, and other progressive social movements. The process of capital restructuring, which neoliberalism has helped facilitate, is thus responsible for the deterioration in ecological and working/living conditions. The hardships of both the American worker and their environment are thus two sides of the same political–economic coin and are now so dialectically related (if not essential) to each other as to become part of the same historical process—the restructuring and globalization of American capitalism. As a result, the issues of social–economic justice and ecological justice have surfaced together as in no other period in American history.

POLLUTION PAYS: THE SOCIAL AND ECOLOGICAL COSTS OF NEOLIBERALISM

Let us now examine the effects of capital restructuring and cost minimization on American workers, communities of color, and the environment in greater detail. Coupled with the neoliberal assault on the regulatory capacities of the state, business is now externalizing more costs and spending less on prevention of health and safety problems inside and outside of the factory, as well as on reducing pollution and the depletion of natural resources.[30] According to an EPA report released on May 20, 1997, American industry continues to generate more toxic waste each year. In 1995, industrial companies covered by the report generated 35 billion pounds of toxic waste, 3 percent more than in the previous year and 7 percent more than in 1991. Some 2.2 billion pounds of these toxic chemicals were directly released as pollution into the air, land, and water in 1995. Moreover, manufacturers reliance on deep-injection wells to dispose of toxic chemicals—a cheap way of dumping wastes on site that poses potential hazards to underground drinking water supplies—jumped nearly 20 percent (an additional 24.5 million pounds) during the four-year period.[31] Furthermore, according to a 1991 report by the National Academy of Sciences, it is estimated that American business annually produces some 4.5 billion *tons* of hazardous wastes, an amount equal to 36,000 pounds (or 100 pounds daily) of hazardous waste for every man, woman, and child in the United States.[32]

As is evident from the growing toxic waste problems, pollution, and other social/environmental costs of capitalist production, many liberalist policy initiatives in the current crisis are actually intensifying problems they were designed to cure. Most environmental laws require capital to *contain* pollution sources for more proper treatment and disposal (in contrast to the previous practice of dumping onsite or into nearby commons). Once the pollution is "trapped," the manufacturing industry pays the state or a company such as Chemical Waste Management for its treatment and disposal. The waste, now commodified, becomes mobile, crossing local, state, and even national borders in search of "efficient" (i.e., low-cost and politically feasible) areas for treatment, incineration, and/or disposal. More often than not, the waste sites and facilities are themselves hazardous and located in poor working class neighborhoods and communities of color.[33] As stated by one government report, billions of dollars are spent "to remove pollutants from the air and water only to dispose of such pollutants on the land, and in an environmentally unsound manner."[34]

For instance, in Sierra Blanca, Texas, the local economy has collapsed. Underemployment is so pervasive that 40 percent of the population lives below the poverty line. Since 1992, New York City and a "biosolids" company called Merco have shipped roughly 200 tons of processed sewage *a day* to the small town. Once dumped into the Atlantic Ocean before Congress banned the practice in 1988, the sludge is now transported by rail to be applied (in what industry terms "beneficial land application") as fertilizer to nearly 200 square miles of land around this mostly Latino community. Due to concerns that the sludge is poisoned with heavy metals, petroleum, and pathogens, community residents see this practice as posing a significant health threat and therefore a form of environmental racism. There are more than 200 other such sewage sludge sites in Texas alone.[35]

Current regulations are also largely failing to halt the expansion of already widespread practices of illegal hazardous waste dumping, which is cheaper than legal dumping and has more adverse health and environmental consequences. The higher costs arising from the EPA's regulation of the disposal of toxic wastes (based on the agency's authority under the 1976 Resource Conservation and Recovery Act) is leading some companies, exasperated by the shortage of low-cost legitimate hazardous waste hauling and disposal services, to employ organized crime to handle their wastes. The Mafia, traditionally active in garbage hauling and landfilling, is now also one of the largest toxic waste disposers in the country.[36] Again, it is communities of lower "socioeconomic" status that suffer the most from illegal dumping. Thus, liberalist environmental policy has fueled the rapid expansion of the waste circuit of capital (in both legal and illegal forms) that, perhaps more than any other phenomenon, has magnified problems of ecological racism and other inequities that the environmental justice movement is now challenging.[37] Since 1980, the risks for people of color have escalated dramatically, as they are now 47 percent more likely than whites to live near health-threatening toxic waste facilities.[38]

In addition to spending less on the prevention of environmental and community health problems outside of the factory, capital is also spending less on the prevention of health and safety problems that impact the working class inside the factory. In order to increase the rate of exploitation of labor, business is now reducing and eliminating safety equipment and procedures that lower worker productivity and cut into profits.[39] As a result, American workers are being exposed to greater hazards at the point of production. Some 16,000 workers are injured on the job *every day,* of which about 17 will die. Another 135 workers die every day from diseases caused by exposure to toxins in the workplace.[40] In short, over 55,000 deaths and almost 6 million injuries occur each year as a result of dangerous working conditions.[41]

Given the neoliberal assault on the Occupational Safety and Health Administration (OSHA) and the Labor Department, such high "accident" rates are inevitable. There are now only 800 inspectors nationwide to cover the 110 million workers in 6.5 million workplaces. As a result, the growth of "home work" and the comeback of new industrial "sweatshops" where health and safety conditions are more or less unregulated is accelerating on a scale that, in the words of former Secretary of Labor Robert Reich, we have not "seen since the turn of the century." The General Accounting Office estimates that at least 2,000 of New York City's 6,000 garment factories (employing 70,000 workers, mostly Latino and Chinese immigrants) are sweatshops—defined as apparel makers that systematically violate labor laws.[42]

Another example is the chicken processing industry, which has been one the fastest growing businesses in a number of southern states, including Arkansas and North Carolina. On September 3, 1991, a fire ripped through the Imperial Food Products poultry processing plant in Hamlet, North Carolina, killing 25 workers and injuring another 55. According to the Government Accountability Project, because the plant had never been inspected during eleven years of operation, safety violations were rife (emergency doors were locked from the outside to prevent theft, thus trapping workers inside during the fire).[43] Forming alliances with such workers, and the unions and labor organizations that represent their interests, is yet another challenge confronting the environmental justice movement.

The petrochemical industry, consisting of more than 12,000 manufacturing plants in the United States producing a total of more than 66,000 commonly used chemicals, is emblematic of the manner in which American business is simultaneously increasing the exploitation of labor and nature in a dialectical fashion. Profits for 23 of the largest oil companies in the United States are being boosted by what the trade paper *Oil Daily* calls "massive cost-cutting," including: large layoffs; cuts in pollution prevention, worker health and safety measures, and oil recovery programs; and less maintenance and repair of decaying equipment and infrastructure. Since 1982, some 450,000 jobs have been slashed, thus reducing the ability of the remaining workforce to safely monitor, operate, and maintain equipment, including pollution abatement technology.[44] These cuts are typical of capital's counterattack on organized labor, as petrochemical companies have increasingly replaced more highly trained and well-paid union workers with less costly contingent laborers—most of whom lack the knowledge and experience of their union counterparts. Some 30 percent of the hours worked in the petrochemical industry are now logged by contract employees, which, coupled with other cost cutting measures, are resulting in more worker/environmental "accidents."[45]

Since 1980, the beginning of the "antienvironmental decade," the frequency of *major* chemical accidents in the United States and Europe has increased tenfold.[46] It is estimated that there are now as many as 240,000 chemical industry accidents per year in the United States—some 658 per day.[47] More than 80 petrochemical workers have been killed and more than 900 injured in workplace accidents since 1984. At least 175,000 aboveground petro-storage tanks are leaking their poisons into soil, groundwater, streams, and sewage systems, and causing health and environmental problems for hundreds of communities. Some 46 million gallons of oil alone are discharged annually into the nation's waterways, as well as another 15–20 billion barrels of oil waste that are dumped, buried, or injected into the earth. And thanks to neoliberalist antiregulatory initiatives, there are no federal rules in the form of regular inspection mandates, required leak-detection systems, construction codes requiring double-containment bottoms on tankers (such construction would have likely prevented the Exxon *Valdez* oil spill disaster in Alaska), and many other procedures necessary to deter pollution.[48]

THE RAVAGED LANDSCAPE: CAPITAL'S QUEST FOR CHEAPER NATURAL RESOURCES

A key component of the neoliberal offensive against the environmental movement involves efforts to contain and roll back policies establishing national parks, as well as protections for wilderness, forests, wild rivers, wetlands, and endangered species. The reason is that capital restructuring is facilitating a much more aggressive and destructive scramble by American business for cheaper sources of renewable and nonrenewable natural resources. These include efforts to exploit the majestic old-growth forests in Alaska's Tongass National Forest and ancient redwoods in the Pacific Northwest habitat of the endangered spotted owl; or the rich deposits of low-sulfur coal that lie underneath the Black Mesa homelands of the Hopi and Navajo Indians in the Four Corners region of the American Southwest; or the vast oil and natural gas reserves that lay in the Arctic National Wildlife Refuge in Alaska, as well as along the southern, western, and New England coastlines of the continental United States; to hold down grazing fees for ranchers on environmentally sensitive federal lands; and to open up more wetlands and fragile ecosystems to agricultural, commercial, and residential developers.[49] Such schemes to exploit new resource reserves are motivated less by oil, coal, or timber shortages than by the need of U.S. landed capital to bring in lower cost oil, coal, timber, and other fuels and raw materials to more effectively compete in the world market, as well as to lower the cost of inputs utilized by American capital as a whole in the production process. The result has been the growth of offshore drilling, stripmining, and destructive timber harvests with all attendant adverse social and environmental consequences.

New resource wars are consequently developing in every corner of the country. Among the more infamous of these are the timber wars taking place in ancient forests of the Pacific Northwest (in the largest civil disobedience action ever in the forest protection movement over 1,100 activists were arrested in a 1996 mass action protesting Maxxam Corporation efforts to salvage log the Headwaters Forest in northern California).[50] In the summer of 1995, Congress and President Clinton passed the so-called salvage logging rider, which suspends all environmental protections for public forests and opens up hundreds of thousands of acres, including old-growth stands in the Pacific Northwest, to devastating clearcuts (the rider revokes all citizen appeal procedures and the power of the federal courts to issue injunctions to stop logging). As a result, hard-fought victories that protect endangered and threatened species are getting axed. In fact, passage of the rider was the primary means by which Congress required the Forest Service to double its planned salvage cut to 4.5 billion board-feet in 1996.[51] This quota in turn virtually guaranteed major sales of this country's most pristine wilderness.

Because such areas are hard to reach and roadless, the rider also is costly to the Forest Service and American taxpayers, who must pay the expense of building roads and providing other infrastructure and services. In fact, salvage logging subsidies to the timber industry will likely cost American taxpayers more than $1 billion.[52] Thus, Clinton and Gore are continuing the Reagan–Bush strategy of promoting lower-cost timber by providing government subsidies and tax breaks, as well as other programs, including accelerated sales of federal timber reserves to logging interests at rates far below going market prices.

The salvage logging rider is but one mechanism by which the cooptive neoliberal strategies of the Clinton–Gore administration have proved more effective in overriding environmental protections than would a confrontational approach. Under the administration's infamous "Option 9" plan, logging in the contentious old-growth federal forests where the spotted owl and other endangered species live will be curtailed.[53] But as part of the deal, in which the administration garnered the support of numerous mainstream environmental organizations located in Washington, DC, on the reasoning that the plan was "better than nothing," timber interests will be able to log remaining forests of equal or greater volume throughout the Pacific Northwest.[54] This plan represents a tremendous defeat for grassroots environmental movements struggling to protect wilderness in the western United States.

Despite these political victories by the timber industry, logging interests are vacating much of the Pacific Northwest in a search for more profitable sources of lumber. Pushed out by lower-cost Canadian competition, logging companies are being pulled into the southern and eastern United States by greater timber inventories, weaker unions and cheaper labor costs (wages and benefits are 50 percent less), lower manufacturing costs, new production technologies, and a variety of more favorable governmental policies.[55] This exodus began in the 1970s, well prior to the infamous spotted owl injunction of 1991 utilized by the environmental movement to protect federal forests, and is accelerating. Seven of the country's largest manufacturers of lumber and wood products—companies such as Louisiana Pacific and Weyerhauser (the country's largest private landowner, which now owns more land in the south than in the west)—have during this time reduced their capacity in the Northwest by 35 percent while simultaneously increasing their capacity in the south by 121 percent.[56] In fact, when Georgia-Pacific, the largest timber company in the country, relocated its corporate headquarters from Oregon back to Georgia in 1982, its earnings subsequently rose dramatically, from $20 million to $661 million seven years later.[57]

Due to this relocation, southern forests (once devastated a hundred years ago but now recovering) are once again being assaulted. The red-cockaded woodpecker, for instance, which lives in the endangered old-growth longleaf pine forests in the southeast, is now closer to extinction than the much publicized spotted owl.[58] In response to such destruction, efforts by Save America's Forests, a nationwide coalition of forest protection organizations, including more than 200 such member groups in the eastern United States, and other community-based organizations are growing. The better known Redwood Summer protests in northern California are now paralleled by similar environmental actions in the south and east, such as the 1990 Hardwood Summer among the 150-year-old hickories, oaks, and tulip trees in the Shawnee National Forest in southern Illinois. Similar battles are also taking place in the 40,000 square miles of woodland comprising the great northern forests of New England. Since 1986, more than 9 percent of Maine (an area the size of the state of Rhode Island) has been logged (including partial clearing of an area three times that size).[59]

Another accelerating trend in the timber industry is to convert diverse forest ecosystems into sterile monocultures. Pine plantations are exemplary, with a short-rotation schedule that yields quicker returns on investments but also eventually wears out the land and poses greater perils for endangered species. (i.e., the overexploitation of renewable resources beyond their ability to regenerate).[60] The increasingly unsustainable destruction of America's woodlands has become so bad that in an uncharacteristically pointed report issued in the spring of 1993 by the usually logger-friendly 18,000-member Society of American Forests called for a dramatic departure from the current destructive practices of the timber industry. Stating that more emphasis must be given to an "ecosystem approach" that would protect wildlife, water quality, and the overall ecological health of America's forests, the report recommended managing efforts aimed at broad landscapes, as large as one million acres at a time. Such an approach would lead to closer regulation of industry practices on private lands, in recognition of the impact private development has on public lands managed by the Forest Service and other Interior Department agencies.

Other sectors of landed capital are also benefiting from the Clinton–Gore policy of give-and-take. In August 1996, before the fall presidential election, the administration brokered a highly public gold mine deal with Crown Butte Mines, a Montana subsidiary of a Toronto-based company, in which the company agreed to give up its interests to mine a site in a national forest just north of Yellowstone National Park. Many environmentalists who had mounted a strong campaign against the project applauded the announcement, arguing that toxic wastes from the site would threaten the region's waterways and wildlife habitat. Such concerns were well founded. In fact, a 1985 EPA report to Congress estimates that hardrock mining creates up to 2 billion metric tons of solid waste every year in the United States—more than twice as much as all the cities and other industries in the Toxic Release Inventory (TRI) data combined. Much of it contains copper, lead, cadmium, zinc, and arsenic, all linked to cancer and brain damage in children.[61]

But in exchange for giving up efforts to mine gold, silver, and copper in the area—an event the courts would most likely not have allowed to occur because of the severe damage it would have caused—Clinton–Gore established a bad precedent by quietly pledging the company tens of millions of dollars worth of other mineral-rich federal property as compensation. Despite decades of court rulings that have found that corporations and private property owners are not entitled to such compensation, the administration embraced the *taking compensation* principles as embodied in the Republican "Contract with America," a huge concession to industry. Furthermore, on August 14, 1996, one day after pleasing environmentalists by blocking the mine project, Clinton signed major legislation that dramatically eased government regulation (i.e., access, rents and royalties, taxes, and other costs) for companies drilling for oil and gas on federal lands.[62]

Coal mining interests are also benefiting. In September 1996, less than two months before his reelection, in a highly publicized event, President Clinton declared 1.7 million acres of southern Utah's red-rock cliffs and canyons a national monument. By creating the Grand Staircase–Escalante National Monument, the move effectively blocked development of Kaiparowits Plateau, the Escalante River Canyons, and the Grand Staircase. The area, 70 miles north of the Grand Canyon, is marked with natural arches and bridges, high cliffs of red, white, and yellow sandstone, and deep canyons, and is one of the largest known coal reserves in the country. However, the legislation was largely modeled on a proposal by Utah Congressman James Hansen, and not only undercuts a citizens' campaign proposal to protect some 5.7 million acres (instead of only 1.7 million acres) but also opens up most of the state's remaining wilderness to development. The Dutch mining company Andalex Resources holds coal leases on the 600,000-acre plateau and had already begun some mining operations. As with the Crown-Buttes case above, the federal government will seek negotiations with the company to trade leases in the area for federal assets elsewhere. These events reveal the cooptive strategy by which Democratic Party neoliberalism grants narrow "preemptive" concessions to the environmental movement on high-profile issues and well-organized campaigns in a way that enables Clinton and Gore to embrace the mantle of "environmentalists," while at the same time the administration quietly grants major concessions to business that contradict and counteract the very victories that much of the environmental movement claims to have won.

Efforts to exploit western coal reserves such as those in Utah are being driven by a number of political–economic forces, including "free-market" environmental policy. In New York, New England, Canada, and the Great Lakes, problems of acid rain stem mainly from the 200-foot-high smokestacks of Midwestern power plants. Under the Clean Air Act, these power plants are required to drastically reduce sulfur dioxide emissions that cause the acid rain (among other major and related health problems).[63] Rather than installing pollution control equipment (such as scrubbers) that costs $100 million or more, the utilities are responding with a less expensive method of blending high-sulfur coal with low-sulfur coal. As a result, the demand for low-sulfur coal is dramatically increasing, while the demand for high-sulfur coal is dropping to such an extent that less efficient producers cannot make profits and are closing the mines. For those companies unable to restructure and become more efficient (i.e., to cut costs by increasing the rates of exploitation of labor and nature), bankruptcy awaits.

Since 1990, the Peabody Holding Company (the nation's largest coal producer) and other companies have shut down nearly 1,000 high-sulfur coal miles—mostly in Appalachia and the East. Some 2,500 are left, many of which are being automated so as to reduce labor and production costs. As a result, coal mine employment in Illinois, Indiana, Ohio, Pennsylvania, western Kentucky, and northern West Virginia has plummeted more than 30 percent since the passage of the Clean Air Act in 1990. Of the mostly 25,906 union workers who remain, the power of the United Mine Workers (UFW) to bargain for better wages, benefits, job security, and working conditions has been greatly diminished.

In contrast, demand for low-sulfur coal from mines in eastern Kentucky, southern West Virginia, and the western United States, in particular, has grown dramatically.[64] In 1996, domestic coal production reached a record 1.05 billion tons. Most of this gain was made in the west, where coal production jumped 26 percent (to 504 billion tons) between 1990 and 1996. In contrast, production fell 14 percent (to 542 billion tons) in the east during this same period.[65] Spurred by the lower cost of labor and the utilization of more highly automated, surface-mining techniques in such states as Kentucky, Colorado, Utah, Wyoming, and Montana, production methods that are not only much cheaper but also much more environmentally destructive and harmful to gains won by union coal miners are increasingly being adopted.

By giving greater power to "free-market" and quasiprivate "environmental mediation" policy approaches, these examples are illustrative of the ways in which neoliberalism is exacerbating social and environmental injustices. Again, much of the mainstream environmental movement has been an accomplice. In 1985, the Acid Rain Roundtable, a corporatist group representing select environmental organizations, utilities, coal companies, and state officials, formulated recommendations for burning more low-sulfur coal (which were then incorporated into the Clean Air Act of 1990). As we can see, this action effectively shifted the environmental protection burden onto western-based environmental movements, Native American reservations, and Latino communities, as well as western and northern Appalachian coal miners, an unsurprising outcome given the fact that these parties were not represented on the Roundtable. Anyone wanting to split environmentalists, labor, and social justice activists (especially Native American movements) could not have designed a better plan for doing so. It is in this "free-market environmentalist" context that the Navajo–Hopi controversy at Big Mountain,[66] the massive Pittstown coal strike, the Grand Staircase–Escalante National Monument signing, and other events must be interpreted. To unite these affected workers and communities into a campaign for ecological democracy is yet another challenge confronting the environmental justice movement.

CAPITAL MOBILITY AND THE EXPORT OF ECOLOGICAL HAZARDS

Last but not least, under policies of "new federalism" and the rhetoric of "states' rights," responsibilities are being shifted from the federal government to the states, many of which are financially strapped. The neoliberal hope is that many states will neglect their responsibilities or engage in bidding wars with other states to attract capital to their home regions by offering more favorable investment conditions, including less worker and environmental regulation and enforcement (i.e., to aid in efforts at cost minimization). One reason that economic problems in the northern "rust belt" are deeper than in most of the rest of the country has been the disproportionate relocation of capital to the "sunbelt" in search of cheaper labor, lower taxes and real estate costs, and less stringent environmental regulations. Increased capital mobility is thus a primary mechanism by which American business is restructuring itself to minimize costs. Hence, the political–economic power base in the 1980 and 1990s for both neoliberal Republicans and Democrats is shifting to the south (through such figures as Carter, Perot, Bush, Clinton, and Gore) and west (Nixon and Reagan—the first presidents ever to come from California).[67]

One such business haven is the state of Arkansas, which has one of the highest low-wage job creation rates in the 1980s–1990s. As governor, Bill Clinton attracted industry to Arkansas by supporting a number of antiunion ("right-to-work") initiatives, giving numerous tax breaks to businesses while raising regressive taxes on the working and middle classes, and selling out on pledges to improve the workers' compensation process, to push "right-to-know" legislation, and clean up the state's environmental problems. One of the worst dioxin waste dumps in the country is located in Jacksonville, Arkansas—a community of 29,000 that suffers from elevated mortality rates due to cancer. Despite the urgings of the National Toxics Campaign (now defunct), Clinton stubbornly refused to stop the planned incineration of tens of thousands of drums of this deadly chemical (one of the most dangerous known to humanity) in the community, or to evaluate Jacksonville for a potential buyout, even though soil contamination is well above the levels that led to the evacuation of Times Beach, Missouri. Furthermore, the Arkansas Pollution Control and Ecology Department is permitting companies to boost profits through the underground injection disposal of toxic chemicals and dangerous industrial wastes. As a result, the health of the huge Sparta aquifer, a supply tapped by community water districts and industry throughout south Arkansas, is now threatened.

Unlike states such as Ohio, which requires extensive well monitoring and seismic testing before granting well permits, Arkansas requires no monitoring. A 1992 EPA evaluation of the Arkansas Department of Health water supply program found that lax enforcement was forcing the EPA to intervene routinely to force state enforcement. In short, the records reveal that under Clinton's tenure the Arkansas Pollution Control and Ecology Department botched inspections, was unwilling to live up to federal standards and lacked the will to crack the legal whip on polluters, and in fact had one of the worst records in the country.

Arkansas is typical of the lax enforcement that the rise of cowboy capitalism has brought to the sunbelt. Fifteen southern states alone account for 33 of the 50 most polluting plant sites in the nation. The state of Texas possesses five of the ten most polluted zip code areas in the country and leads the nation in total air, water, and land releases of carcinogenic pollution. A 1995 report by the Environmental Defense Fund showed that refineries in Texas, Mississippi, West Virginia, and Kansas are the nation's most environmentally inefficient (in terms of pollution releases and waste produced per barrel of oil refined per day). Refineries in northern states such as New Jersey, which have some of the country's toughest pollution laws, are among the best. Furthermore, an "emissions-to-jobs ratio" report by environmental science professor Paul Templet of Louisiana State University in Baton Rouge showed that Louisiana's chemical plants, especially those located in poor African American communities in the corridor between New Orleans and Baton Rouge known as "Cancer Alley," released nearly ten times as much pollution per worker as such plants in New Jersey and California, where law enforcement and industry spending for pollution control and abatement are greater.[68]

Last but not least, the legislative battles waged by labor and environmentalists in the United States brought about regulations in the 1960 and 1970s that, in turn, indirectly contributed to the internationalization of capitalist production and the consequent export of environmental degradation and health problems to the Third World and elsewhere in the 1980 and 1990s. Aided by recent "free trade" initiatives under the Clinton administration, such as the North American Free Trade Agreement (NAFTA), these processes of ecological imperialism include the export of more profitable yet more dangerous production processes and consumer goods, as well as waste disposal methods, to developing countries where environmental standards are lax, unions are weak, and worker health and safety issues are ignored.[69] Along the United States–Mexico border there are more than 2,000 factories or *maquiladoras*, many of them relocated U.S.-based multinational corporations. One study in the border town of Mexicali indicated that stiff environmental regulations in the United States and weaker ones in Mexico were either the main factor or a factor of importance in their decision to leave the United States.[70] In fact, Lawrence Summers, Clinton's recent appointee as Undersecretary of the Treasury for International Affairs, is infamous for writing an December 12, 1991, memo as a chief economist at the World Bank that argued that "the economic logic behind dumping a load of toxic waste in the lowest wage country is impeccable," and that the Bank should be "encouraging more migration of the dirty industries to the LDCs [less developed countries]."[71] Forging links with Third World popular movements combating such abuses is yet another profound challenge confronting the U.S. environmental justice movement.

THE CRISIS OF ENVIRONMENTALISM IN AMERICA: NEW CHALLENGES FOR THE ENVIRONMENTAL JUSTICE MOVEMENT

In the new age of globalization and cost minimization, social and environmental injustices are growing worse. And by embracing ecoliberalist policy approaches—a politics that fails to adequately subordinate short-term processes of capital accumulation and restructuring to longer-term democratic economic planning based on human and environmental needs—the environmental and labor movements are showing themselves incapable of arresting these inequities. In fact, the mainstream environmental movement's single-issue legislative approach is compelling capital to displace costs in different forms from one site to another, and from more affluent sectors to less powerful segments of the popular classes at home and abroad. As a result, ecological racism is causing ever greater damage to oppressed communities of color. A lack of unity with the labor movement has contributed to the worsening of health and safety conditions inside and outside of the factory for working-class Americans. Furthermore, the absence of ecological and labor internationalism in opposition to capital mobility and the globalization of production has proven disastrous in the Third World, where labor and environmental protection is weakest.

The ecology movement's weak analysis of the current crisis of capitalism has therefore helped lead to unintended, adverse effects on the well-being of many working class communities and their environments and has undermined the formation of progressive political alliances with labor, civil rights, and other social movements. Some regional and local movements and coalitions by and large have not looked beyond their own areas to analyze the effects elsewhere of their own local or regional successes in environmental protection. In effect, many of the environmental movement's legislative victories in the 1960 and 1970s have become one source of its failures in the 1980 and 1990s. The result is that environmentalists, labor, and community groups today find themselves unequipped for fighting new forms of ecological degradation and environmental injustice.

The organizational form that the environmental movement has assumed over the past three decades has contributed to today's impasse. The original role of the movement was to conduct studies, assist in drafting legislation, and organize constituencies to support passage of legislation to halt environmentally destructive initiatives. In the 1970s, however, environmentalists became professionalized and institutionalized within the federal government, state and local governments, and in nongovernmental organizations as well. The semi-institutionalization of the mainstream environmental movement as a government watchdog over the implementation and enforcement of environmental legislation, with an emphasis on lobbying rather than expanding grassroots organizing (as well as a strong public perception that environmental problems are in fact being tackled), resulted in a partial demobilization of the membership and centralization of decision-making power in many environmental organizations. By the late 1970s and 1980s, this process of grassroots disempowerment, combined with the corporate counterattack, put the movement on the defensive. After a series of defeats over the course of the "antienvironmental decade" of the 1980s, the movement professionalized itself even more along corporate models. The aim has been to regain legitimacy and expert status in increasingly hostile policy circles. The effect, however, has been to reduce internal democratic practices within some environmental organizations and the new state regulatory agencies as well, and also democratic accountability to the movement by the state.

The focus on technical–rational questions, solutions, and compromises, rather than issues on of political power and decision making, has recently caused a growing disjunction between the mainstream organizations and the grassroots, particularly in the environmental justice movement. The split in the movement since the mid-1980s largely originated with this issue of professionalization, which developed not only because the general emphasis shifted from law making to law implementation but also because of the new hostility to ecological politics in general on the part of business and the federal government. "Professionalized" mainstream movement approaches that neglect social and environmental justice issues, and therefore restrict the scope of policy considerations, are a largely liberalist strategy to reach compromises with, and win concessions from, an "unfriendly" neoliberalist state.

The reaction at the base is a revitalized grassroots politics by community-based movements for environmental justice. Tactics include the use of direct action and civil disobedience against timber companies, corporate polluters, and indifferent state agencies, as well as challenging much of the mainstream movements' racial and class-biased politics. The environmental movement's internal conflicts in the 1980s and 1990s are now between new broadly based direct-action groups that seek to combat environmental inequities (distributional justice) and destruction at their source (productive justice) versus a more pragmatic ecoliberalist politics of institutional consensus, compromise, and professionalization. The problem is that those organizations and labor coalitions that ignore the global and local connections between the political crisis of liberal interest group politics and the combined and interrelated forces of economic and ecological crises feed divisions within the movement. Not only environmental, worker, and community health is at stake; also at issue is the viability of the traditional political strategies, tactics, and vehicles utilized by environmentalists and other social movements.

In order to build itself into a political force capable of addressing the fundamental roots of America's social and ecological crises, a series of challenges now confronts grassroots movements for environmental justice. First, in that the environmental movement as a whole (not to speak of the labor movement) has weakened itself by its failure to revive the struggle to democratize the state and the workplace, the environmental justice movement must build unity with trade unions and the labor movement around issues of productive justice. Only by gaining greater democratic governance over the investment and production processes can both the occupational and environmental interests of all working Americans be harmoniously safeguarded. Second, in that the environmental movement has failed to fight against ecological racism, the environmental justice movement must incorporate and build unity between oppressed people of color and broader segments of the white working class (including the labor movement). Only by gaining greater democratic governance over community planning and national economic development can a potentially divisive "not-in-my-backyard" politics oriented to distributive justice be replaced with a truly transformative "not-in-anyone's-backyard" politics oriented to productive justice.

Finally, in that the environmental movement has been slow to combat the globalization of capital and the export of ecological hazards, the environmental justice movement must develop solidarity with those movements and governments in the Third World that appreciate that capitalist economic development, ecological degradation, and human poverty are different sides of the same general problem. The growing ability of multinational corporations and transnational financial institutions to evade the true issues and dismantle unions, environmental safeguards, and worker/community health and safety regulations in the United States is being achieved by crossing national boundaries into politically repressive and economically oppressive countries, such as in Mexico and Central America generally.[72] As a result, various nationalities and governments are increasingly being pitted against one another in a bid to attract capital investment, leading to one successful assault after another on labor and environmental regulations seen as damaging to profits. And in this context, utilizing the rhetoric of "free trade," "free enterprise zones," and "jobs versus the environment," capital has further weakened and divided America's social movements against one another and has worsened inequities of all kinds. Only by achieving greater social governance over trade and lending institutions can the process that leads different countries to sacrifice human and environmental health in order to compete in the world economy be overcome.

In summary, the challenge confronting the environmental justice movement is to help forge a truly broad-based political movement for ecological democracy. While the traditional environmental movement has played a critical and progressive role in stemming many of the worst threats posed to the health of the planet and its inhabitants, the movement is now rapidly losing power in the face of neoliberalism and the economic restructuring of U.S. and international capitalism. Traditional liberalist environmental strategies are now, at best, limited or ineffective, or, at worst, increasingly self-defeating. Environmentalists must embrace a multi-issue/multimovement approach that emphasizes productive justice for all Americans if this power is ever to be restored. To do so requires developing a more comprehensive understanding of the political–economic terrain on which this approach now struggles. Socialist ecology is essential to this endeavor.

And as the crisis of U.S. capitalism deepens, and global ecological conditions worsen, the need for a mass-based international ecology movement that unites the struggle for both social and ecological justice will become more pressing. Just as in the 1930s, when the labor movement was forced to change from craft to industrial unionism, so today does it appear to many that labor needs to transform itself from industrial unionism into an international conglomerate union (inclusive of women and all racial/ethnic minorities)—just to keep pace with the restructuring of international capital. And just as in the 1960s, when the environmental movement changed from a narrowly based conservation/preservation movement to include the middle class and sections of the white working class, so today does it seem to many that it needs to change from single-issue local and national struggles to a broad-based multiracial international movement. The environmental and other social movements in the United States must be made to realize that they need strong environmentalism and worker health and safety throughout the rest of the world in order to protect local initiatives and gains. This historic task now confronts the environmental justice movement.

ACKNOWLEDGMENTS

This chapter is in part based on a previous work by Daniel Faber and James O'Connor, "The Struggle for Nature: Environmental Crisis and the Crisis of Environmentalism in the United States," *Capitalism, Nature, Socialism,* Issue 2, Summer 1989, pp. 12–39. I am solely responsible for the theoretical modifications, elaborations and updated information contained in this version, but would like to acknowledge James O'Connor for his earlier contributions.

NOTES

1. For concise discussions of the history of the environmental, including the manner in which the movement has transformed attitudes and the ecological consciousness of the American public, see Mark Dowie, *Losing Ground: American Environmentalism at the Close of the Twentieth Century* (Cambridge, MA: MIT Press, 1996); Kirkpatrick Sale, *The Green Revolution: The American Environmental*

Movement 1962–1992 (New York: Hill and Wang, 1993); Robert Gottlieb, *Forcing the Spring: The Transformation of the American Environmental Movement* (Washington, DC: Island Press, 1993); and Victor B. Scheffer, *The Shaping of Environmentalism in America* (Seattle: University of Washington Press, 1991).

2. One study, conducted by researchers at the Harvard School of Public Health, Brigham Young University, and the American Cancer Society, which was released on March 10, 1995, and appeared in the *American Journal of Respiratory and Critical Care Medicine,* estimated some 60,000 annual air pollution deaths. Another Natural Resources Defense Council study (released on May 8, 1996) on 239 cities across the country found some 64,000 Americans to be dying each year from air pollution, even at levels which the federal government considers safe.

3. According to a 1993 report by the American Lung Association, some 66 percent of U.S. citizens live in areas that violate standards for ozone (which causes lung tissue to become inflamed and impedes breathing); carbon dioxide (which impedes the blood's ability to carry oxygen to the heart); and lead (which causes brain and organ damage). Because their lungs are particularly sensitive, at highest risk are the 31 million children and 19 million elderly who live in these polluted areas.

4. For an analysis of Environmental Protection Agency data that demonstrate the growing severity of the ecological crisis, see Barry Commoner, *Making Peace with the Planet* (New York: Pantheon, 1987).

5. See A. J. McMichael, *Planetary Overload: Global Environmental Change and the Health of the Human Species* (Cambridge, UK: Cambridge University Press, 1993), p. 103.

6. See Nancy Sokol Green, *Poisoning Our Children: Surviving in a Toxic World* (Chicago: The Noble Press, 1991).

7. See National Research Council, *Environmental Epidemiology: Public Health and Hazardous Wastes* (Washington, DC: National Academy Press, 1991).

8. For a review, see Environmental Research Foundation, *Rachel's Hazardous Waste News,* 332, April 8, 1993, pp. 1–2.

9. For a discussion, see Eric J. Krieg, "Toxic Wastes, Race, and Class: A Historical Interpretation of Greater Boston" (PhD dissertation, Northeastern University, 1995), pp. 1–26.

10. See Sandra Steingraber, *Living Downstream: An Ecologist Looks at Cancer and the Environment* (New York: Addison-Wesley, 1997).

11. See Terry L. Anderson and Donald R. Leal, *Free Market Environmentalism* (San Francisco: Pacific Institute for Public Policy, 1991).

12. Since 1982, the stock market has experienced one of its greatest long-term bull runs in history. However, in terms of the broader economy, this does not mean rising real investment and more prosperity for the majority of Americans. In fact, virtually no investment funds are currently being raised through new stock offerings. That is, since the bull run began, nonfinancial firms have actually retired $733 billion more stock than they've issued, mainly due to corporate takeovers, buybacks, and elimination of excessive debt. In fact, it is likely that low rates of return on new capital in industry are spurring capital to increasingly seek speculative profit possibilities—by conjecturing that profits will eventually rise to a sufficient rate at a later date. Thus, it is speculation (the hope of realizing windfall profits), that, along with the influx of baby boom generation retirement money, that is in large part fueling the stock market.

13. See John Miller, "Is the Boom and Bust Cycle Over?," *Dollars and Sense, 211*, May/June 1997, pp. 10–12.

14. While many political commentators equate the term "crisis" with a total breakdown or collapse of the economy, I, like Joyce Kolko, use the term in a much broader sense to signify "a critical turning point to separate periods, the future from the past. For capitalism, crisis is a condition when and where the problem of accumulation and the form it takes is no longer one of productive expansion—a crisis of accumulation." Thus, while "a depression is always a crisis, a crisis in the operations of the capitalist economy is not always a depression." See Joyce Kolko, *Restructuring the World Economy* (New York: Pantheon, 1988), pp. 3, 20.

15. See Robert J. S. Ross and Kent C. Trachte, *Global Capitalism: The New Leviathan* (Albany: State University of New York Press, 1990); and Peter Dicken, *Global Shift: The Internationalization of Economic Activity*, 2nd ed. (New York: Guilford Press, 1992).

16. Labor productivity is one of the most important means for determining the economic health of industry. And during the expansion of the 1990s, labor productivity has grown only 1 percent a year, far below the 2.9 percent annual increases in productivity during the boom years from 1960 through 1973. See Miller, "Is the Boom and Bust Cycle Over?," p. 11.

17. See James O'Connor, "20th Century Limited: Capital, Labor, and Bureaucracy in the Age of Nationalism," *Capitalism, Nature, Socialism, 5*(3), September 1994, pp. 1–34.

18. See Roger H. Bezdek, "The Net Impact of Environmental Protection on Jobs and the Economy," in Bunyan Bryant, ed., *Environmental Justice: Issues, Policies, and Solutions* (Washington, DC: Island Press, 1995), pp. 86–106. Bezdek, however, argues that environmental regulations have a positive economic impact.

19. In the 1970s, nearly one-half of all capital investment was made in polluting industries—e.g., oil, petrochemicals, electrical power, strip mining, metalworking—that supply inputs to other industries. For instance, construction costs of the 116 coal-fired electric plants built from 1971 through 1978 increased 68 percent in real terms, with 90 percent of this attributable to antipollution improvements required by the EPA. See "Clean Coal—What Will It Cost?" Interview with Charles Komanoff, *Dollars and Sense,* March 1980, p. 2.

20. As Chairman of the Presidential Task Force on Regulatory Relief in the early 1980s, then Vice-President George Bush and key members of the Reagan administration compiled a hit list of burdensome regulations that it had solicited from trade associations, state and local governments, and corporations. The largest number of requests received concerned EPA regulations, particularly in the automobile and petrochemical industries. The recommendations of this task force outlined the attack on liberal environmental policy that has continued to this day. See Richard Andrews, "Deregulation: The Failure at EPA," in Norman J. Vig and Michael E. Kraft, eds., *Environmental Policy in the 1980s* (Washington, DC: Congressional Quarterly, 1984), pp. 161–180.

21. See Richard Kazis and Richard Grossman, *Fear at Work: Job Blackmail, Labor and the Environment* (New York: Pilgrim Press, 1982).

22. See Carl Deal, *The Greenpeace Guide to Anti-Environmental Organizations* (Berkeley, CA: Odoniam Press, 1993).

23. Within weeks of taking power, the Republican-dominated 104th Congress tried to close national parks and cut spending for environmental protection by 25

percent. By late 1995, stinging under the heavy fire of incensed constituents, even House Speaker Newt Gingrich admitted that the Republican Party was "strategically out of position on the environment." Along with then Senate Majority Leader Robert Dole, the party promptly changed their confrontational rhetoric. Instead, a new plan of attack designed to limit public attention to and debate over their antienvironmental measures was adopted. The party began quietly placing measures intended to roll back environmental policy into unrelated legislation, in particular, a series of confusing budget and appropriations bills needed to keep the U.S. government running. When President Clinton objected, this tactic eventually precipitated two federal government shutdowns. This strategy backfired badly for the Republicans. Climaxing in a presidential veto of a budget bill that would have allowed oil exploration in a north Alaskan wildlife refuge, the Republican party's confrontational approach enabled President Clinton to seize the public mantle of nature's defender well before the 1996 election.

24. In the words of DuPont Corporation's Chairman Edgard Woolard, "We do not seek a wholesale dismantling of the regulatory framework, . . . just smarter, more effective regulation where we need it." See Ronald Begley, "Deregulation: Too Much of a Good Thing?," *Chemical Week,* April 19, 1997, p. 68.

25. Under Clinton, EPA Administrator Carol Browner has launched at least ten initiatives aimed at improving efficiency, cutting paperwork, allowing industry more flexibility in solving problems, and requiring the EPA to prove that the benefits of major rules outweigh the costs. Many of the new rules are also preemptive in nature. On August 8, 1996, President Clinton signed a new version of the Safe Drinking Water Act, with a number of strong provisions, including a requirement that water companies give their customers an annual report listing the health-threatening contaminants found in their drinking water. However, several of the major health protections in the Act, including requirements to track down water contamination sources, were not funded as part of the law.

26. See Brian Tokar, "Trading Away the Earth: Pollution Credits and the Perils of Free Market Environmentalism," *Dollars and Sense,* 204, March/April 1996, p. 24–29.

27. In Marxist terminology, the most important mechanism by which higher rates of labor exploitation are achieved is by increasing the productivity of labor power per unit of time via the introduction of new technologies and production processes (many of which are damaging to environmental and worker health).

28. See David Gordon, *The Fat and Mean: The Corporate Squeeze of Working Americans and the Myth of Managerial "Downsizing"* (New York: Free Press, 1996).

29. Miller, "Is the Boom and Bust Cycle Over?"

30. A 1993 survey by the National Law Journal and Arthur Anderson Environmental Services found that two-thirds of the corporate lawyers representing manufacturers, mining companies, insurance and real estate firms, and other industries acknowledged that their companies had violated environmental laws during the preceding year and believed that full compliance was impossible.

31. This annual EPA report has been criticized at length by environmental activists for the manner in which the actual amount of toxic chemicals released into the environment is grossly understated.

32. Some 45 million tons of this hazardous waste is considered to be "toxic" by the federal government. See William Ophuls and A. Stephen Boyan, Jr., *Ecology*

and the Politics of Scarcity Revisited (New York: Freeman, 1992), p. 151; cited in the Environmental Research Foundation, *Rachel's Hazardous Waste News, 332*, April 8, 1993, pp. 1–2.

33. See Robert D. Bullard, *Dumping in Dixie: Race, Class, and Environmental Quality* (Boulder, CO: Westview Press, 1990); Robert D. Bullard, ed., *Unequal Protection: Environmental Justice and Communities of Color* (San Francisco: Sierra Club Books, 1994); and Rodger C. Field, "Risk and Justice: Capitalist Production and the Environment," Chapter 3 in this volume.

34. In the words of Lewis Regenstein, "the long history of inadequate enforcement of the RCRA [Resource Conservation and Recovery Act] helps defeat the purpose of not only this statute, but of other environmental laws as well." See Lewis Regenstein, *How to Survive in America the Poisoned* (Washington, DC: Acropolis Books, 1986), p. 160

35. Florida, Illinois, Arizona, California, and Colorado also allow class-B sludge to be recycled as fertilizer. See Weston Kosova, "Sludge on the Range," *Audubon, 98*(6), November/December 1996, pp. 70–76.

36. See Donald Rebovich, *Dangerous Ground: The World of Hazardous Waste Crime* (New Brunswick, NJ: Transaction Publishers, 1992); Russel Mokhiber, *Corporate Crime and Violence: Big Business Power and the Abuse of the Public Trust* (San Francisco: Sierra Club Books, 1989); and Alan A. Block and Frank R. Scarpitti, *Poisoning for Profit: The Mafia and Toxic Waste in America* (New York: William Morrow, 1982).

37. The May 20, 1997, EPA report stated that because more wastes are being recycled or sent offsite for treatment and disposal, the amount of pollution getting into the environment continues to decline. Although such a claim is itself highly dubious, it also neglects the issues of inequity and ecological racism that are accompanying this process.

38. See Benjamin Goldman and L. Fitton, *Toxic Waste and Race Revisited: An Update of the 1987 Report on the Racial and Socioeconomic Characteristics of Communities with Hazardous Waste Sites* (New York: Center for Policy Alternatives, National Association for the Advancement of Colored People, and the United Church of Christ Commission for Racial Justice, 1994).

39. According to the 1995 *Workers Compensation Yearbook*, the insurance system that compensates workers who have been injured on the job cost American business $60 billion in 1992. These costs, which had been escalating in previous years, are now only beginning to stabilize in response to the cost-cutting techniques of capital, including the practice of bringing injured workers back to work more quickly and better corporate oversight and promotion of less costly medical treatments afforded by managed health care programs and insurance reform.

40. These figures are drawn from Charles Levenstein and John Wooding, "Dying for a Living: Workers, Production, and the Environment," Chapter 2 in this volume.

41. According to the report "Accident Facts" by the National Safety Council, deaths resulting from workplace accidents declined from 13,000 to 8,500 a year between 1981 and 1993. However, some of this decline can be attributed to a change in statistical methods that undercounts actual numbers, as well as a decline in the overall size of the industrial work force (by downsizing, relocation, etc.), which had some of the highest accident rates.

42. In a government raid in August 1995, 72 Latino workers in a garment

factory in El Monte, California, were found to be toiling in what Reich called "slave labor" conditions for 17 hours a day at 60 cents an hour. The three dominant factors behind this trend toward sweatshops are: (1) increased competition on U.S. producers from the highly exploitable, low-wage work forces of Asia and Latin America; (2) the growing dominance of the retail clothing market by huge chain stores, who leverage legitimate manufacturers to outsource to illegitimate lower-cost producers; and (3) the growth of poor Latin America and Asian immigrants into the United States. Cited in Fred Kaplan, "The Fruit of Their Labor Is Misery," *Boston Globe*, Friday, July 12, 1996, pp. 1, 12.

43. See the Community Environmental Health Program, *Environment and Development in the USA: A Grassroots Report for UNCED* (New Market, TN: Highlander Research and Education Center, 1992).

44. See Eyal Press, "Almighty Oil," *The Nation*, May 9, 1994, pp. 617; and Jack Doyle, *Crude Awakening* (Washington, DC: Friends of the Earth, 1994).

45. Exxon cut almost 20 percent of its labor force between 1981 and 1985 alone, while Mobil lopped off one-third of its employees between 1982 and 1986. See Michael Tanzer, "Growing Instability in the International Oil Industry," in Arthur MacEwan and William K. Tabb, eds., *Instability and Change in the World Economy* (New York: Monthly Review Press, 1989), pp. 225–240.

46. See Nicholas A. Ashford et al., *The Encouragement of Technological Change for Preventing Chemical Accidents: Moving Firms from Secondary Prevention and Mitigation to Primary Prevention* (Cambridge, MA: MIT Press, 1993), pp. iii–vi.

47. According to the National Environmental Law Center (NELC), from 1988 through 1992 there were 34,500 "accidents" involving toxic chemicals—an average of 19 accidents per day. However, the NELC study, which was based on the Emergency Response and Notification System (ERNS) database maintained by the EPA, collected data on only 800 of the 66,000 chemicals commonly in use, and further omitted data on accidents involving only petroleum products (fuel oil, gasoline, etc.) that comprise 52 percent of the ERNS data. Thus, the NELC figures are greatly understated. For instance, the attorney general's office in New York State conducted a study that found a total of 3,496 toxic chemical accidents in the state during an overlapping three-year period (compared to 496 accidents recorded on the ERNS database during this same time period). For these reasons, the Environmental Research Foundation has a much higher estimate. See *Rachel's Environmental & Health Weekly, 408*, September, 22, 1994, pp. 1–2.

48. Overall, it is estimated that the oil industry wastes the energy equivalent of 1,000 Valdez oil spills every year, costing the country $10 billion annually in oil replacement, pollution and property damage, and public health and other costs. See Press, 1994, p. 617.

49. See William R. Freudenburg and Robert Gramling, *Oil in Troubled Waters: Perceptions, Politics, and the Battle Over Offshore Drilling* (Albany: State University of New York Press, 1993).

50. See Judi Bari, *Timber Wars* (Monroe, ME: Common Courage Press, 1995).

51. While the rider applies only to "salvage logging," traditionally understood as the harvest of dead or dying trees, industry lobbyists have broadened the definition to include diseased or "associated" trees. In effect, the rider lets loggers cut whatever they want to cut. See Reed McManus, "Logging without Looking," *Sierra: The Magazine of the Sierra Club, 81*(4), July/August 1996, pp. 30–31.

52. The U.S. Forest Service adds thousands of miles each year (at an annual cost of $95 million) to the 360,000 miles of logging roads in public woodlands, which is eight times more than the entire interstate highway system. The Forest Service also grossly undervalues trees sold to timber companies at $2.85 for a thousand board feet of lumber, which is about 1 percent of the normal commercial rate. Prices are so artificially low that the value of many trees cut down by timber companies is less than the cost of the roads built into the forests, representing an estimated $427 million of Federal government wealth annually transferred to these companies (not counting tax loopholes of about $520 million a year). Of the 120 forests the USFS managed in 1994, 87 percent lost money. In fact, between 1985 and 1994, the USFS lost more than $5.6 billion. The rider will greatly increase these costs. See Mark Zepezauer and Arthur Naiman, *Take the Rich Off Welfare* (Tucson, AZ: Odoniam Press, 1996), pp. 107–110.

53. On June 29, 1995, the Supreme Court upheld the government's right to restrict development on private land to protect the habitat of endangered plants and animals. Lawyers on both sides of the "spotted owl" case called it one of the most significant environmental decisions in years, because the 6–3 majority upheld government regulation in a direct confrontation over property rights. Despite enjoying the support of environmentalists, Interior Secretary Bruce Babbitt immediately stated after the finding that "we will continue to aggressively pursue a variety of reforms to make the act less onerous on private landowners."

54. While "Option 9" does curtail annual logging from an average of 4 billion board feet to 1.2 billion per year over the next ten years—a figure that even the Congressional Research Service says is unsustainable—it also opens up 2.3 million (or 40 percent) of the 5 million acres of old-growth forests east of the Cascades to logging. This represents up to 2 billion board feet per year in additional cuts. Forest "reserves" and other areas comprising the remaining 60 percent will also be subject to salvaging and thinning. The agreement also includes no permanent and inviolate reserves; no prohibition on clearcutting in roadless watersheds; no ban on raw log exports; fewer restrictions on private land; and more intensive cutting in the forest matrix (i.e., the land between reserves). See Alexander Cockburn, "Beat the Devil: Munich in the Redwoods," *The Nation,* August 23–30, 1994, pp. 199–200.

55. New technologies employed in the gluing process in the southern plywood industry, or in production processes that utilize hardwood chips for the first time (which are then mixed with softwood pines to produce high-quality coated papers for fax machines, computers, and glossy magazines) are allowing a growing number of pulp mills and chip mills in the south to exploit the younger and less expensive stands of trees that predominate in the region. Until these technological innovations appeared, these smaller sized trees served as a major barrier to the fuller development of these industries. on the other hand, costs are increasing in the Northwest, as big timber has cleared the trees off the more easily accessible areas and has moved into tougher, steeper terrain.

56. In 1988, the total timber volume cut in the Northwest was 13 billion board feet (bbf), compared to 7.2 bbf in 1992. In the south, the total volume grew from 12.7 bbf in 1988 to some 14.4 bbf in 1992. See Jay Letto, "Go East Young Timberman!," *E: The Environmental Magazine,* 5(1), February 1994, pp. 27–32.

57. See Ray Raphael, *More Tree Talk: The People, Politics, and Economics of Timber* (Covelo, CA: Island Press, 1994).

58. To save the woodpecker, environmentalists want to preserve the remaining

longleaf-grassland forests, which thrived in colonial times on the coastal plain from Virginia to Texas. Only 1.8 million acres of the original 92 million acres of longleaf remains, as foresters have replaced it with other species, primarily loblolly and slash pine, which grow quickly to maturity and are ideal for pulpwood production. See John Tibbetts, "A Spotted Owl by Any Other Name," *E: The Environmental Magazine, 5*(1), February 1994, pp. 15–16.

59. As in the Pacific Northwest, much of the improved profitability of the timber industry in New England has been brought about by labor-saving technologies. Equipment such as a forwarder, a long-armed machine that can carry 50 trunks from the woods in one trip, and a mechanized tree harvester called the feller-buncher, which can clip a tree every 30 seconds, much like a lawnmower. Each of these pieces of equipment is operated by one person and can clear two acres of timber a day. In the past, this work would have required 15 oldtime loggers.

60. In 1952, only 660,000 acres of the 42 million acres of total industry forest land in the South were in pine plantations. By 1985, the figure had increased to more than 13 million acres. See Letto, 1994, "Go East Young Timberman!," p. 33.

61. "Exact numbers of mining pollution are hard to come by, but in the first year of TRI reporting—released in 1989 for the year 1987—Kennecott Copper and its Bingham Canyon Mine in Utah mistakenly thought they were required to participate, and so dutifully disclosed that they produced over 158 million pounds of waste, including 130 million pounds of copper residue. That ranked Kennecott, among 18,000 facilities, as the fourth-largest polluter in the nation." Cited in Bruce Selcraig, "Minding the Mines," *Sierra,* January/February 1997, p. 42.

62. The law is designed to address the complaints of oil and gas producers that the president heard during his first vacation at Jackson Hole in 1995, including the argument that the 1982 royalty law and conflicting court interpretations that followed had resulted in overpayment of royalties, uncertainty about financial liabilities, burdensome paperwork, and unnecessary costs. Royalty money collected from mineral leases on federal land, including oil and gas royalties, represent about $4.2 billion a year in governmental revenues.

63. Excessive mercury contamination of water and fish by these emissions is one such problem. According to an early 1995 draft of a study ordered by Congress as part of the Clean Air Act of 1990, the EPA estimates that some 80,000 to 85,000 pregnant women are exposed to mercury levels high enough to produce risk in their children. But in order to placate the interests of coal-burning power companies, the leading industrial source of mercury emissions, and prevent the adoption of more stringent regulations, the Clinton administration has repeatedly delayed the release of the EPA study on this problem. By calling for a review by an outside science advisory board, the administration will likely be able to prolong the report's release until 1999 or possibly into the next century.

64. Since 1990, coal-mining employment in the West has grown 2 percent, and now stands at 5,514.

65. Production figures cited in Peter T. Kilborn, "East's Coal Towns Wither in the Name of Cleaner Air," *New York Times,* February 15, 1996, p. A16.

66. See Kathy Hall, "Changing Woman, Tukunavi and Coal: Impacts of the Energy Industry on the Navajo and Hopi Reservations," *Capitalism, Nature, Socialism, 3*(1), March 1992, pp. 49–78.

67. For an excellent discussion, see Mike Davis, *Prisoners of the American Dream* (London: Verso, 1986).

68. Fifteen of the twenty facilities reporting the largest total releases of toxic chemicals in 1994 were located in the southern United States. None was located in New England or along the east coast, and only two were located in the Midwestern states. Two DuPont plants, in Pass Christian, Mississippi, and New Johnsonville, Tennessee, led the nation in total toxic releases, with 117 million pounds between them. All but a small portion were injected underground. See Bruce Selcraig, "What You Don't Know Can Hurt You," *Sierra,* January/February 1997, pp. 38–43.

69. See Barry Castleman and Vicente Navarro, "International Mobility of Hazardous Products, Industries, and Wastes," *Annual Review of Public Health, 8,* 1987, pp. 1–19; Ralph Nader et al., eds., *The Case against Free Trade: GATT, NAFTA, and the Globalization of Corporate Power* (San Francisco: Earth Island Press, 1993); and Tim Lang and Colin Hines, *The New Protectionism: Protecting the Future against Free Trade* (New York: The New Press, 1993).

70. See Roberto A. Sanchez, "Health and Environmental Risks of the Maquiladora in Mexicali," *Natural Resources Journal, 30,* Winter 1990, pp. 163–170.

71. In his new capacity, Summers oversees the operations of the World Bank, the African Development Bank, the Asian Development Bank, the Inter-American Development Bank, the European Bank for Reconstruction and Development, the International Monetary Fund, all G-7 negotiations, and all U.S. economic aid to Russia.

72. See Daniel Faber, *Environment Under Fire: Imperialism and the Ecological Crisis in Central America* (New York: Monthly Review Press, 1993).

Chapter 2

Dying for a Living

Workers, Production, and the Environment

Charles Levenstein
John Wooding

> What place, working people are asking themselves, are we going to have in the brave new, green world of tomorrow? The environmental movement's failure to answer this question is the principal obstacle to meaningful labor–environmental coalition.[1]

In a major effort to improve the quality of water in Boston Harbor and in response to environmental concerns, the city of Boston is constructing a nine-mile-long sewage outfall tunnel to get sewage away from inland waters. As a result, 400 feet beneath the harbor, hundreds of workers are tunneling through layers of bedrock. The working conditions are horrendous: ventilation is supplied by a solitary plastic tube; two feet of water covers the ground in which they work; slime oozes from the walls; water drips constantly so that all must wear rain gear. The workers are bent double as they labor, and toilet facilities consist of a few Port-O-Johns located two miles from the tunnel face (workers must walk to these, bent over at the waist). Time pressures on the job do not permit workers to travel all that distance to go to the bathroom. The consequences are obvious: urine and excrement fill the tunnel floor. Recently, the vent pipe fell into this mess and was only partially cleaned, contaminating the air supply with bacteria.

Laboring in these conditions, over a hundred workers have experienced pneumonia, acne, difficulty breathing, and other unspecified illnesses. Several have been killed by falls or by equipment. Electrical cables sit in pools of water, posing a serious hazard of electric shock. The workers work in three shifts and never see daylight during their working hours. This is the working life for these people—a mole-like existence, full of danger.[2]

The danger is very real for other working Americans as well. Workplace injury and illness is an unacknowledged epidemic: some 16,000 workers are injured on the job every day, of which about 17 will die (crushed by machines, falling to their deaths from scaffolding, run over by trucks or forklifts, electrocuted, or shot). Another 135 will die every day from diseases caused by exposure to toxins and chemicals in the workplace. This toll is the equivalent of a major airplane crash every day in which all passengers and crew are killed.[3] The economic costs are equally staggering: $173.9 billion in direct and indirect costs, or 3 percent of the Gross Domestic Product.[4]

As new forms of work come into being, new types of dangers become evident. Ergonomic problems are increasing, and repetitive-motion disorders, such as carpal tunnel syndrome, are becoming ubiquitous: 27,000 repetitive-motion injuries were recorded by the Bureau of Labor Statistics in 1983, 224,000 by 1991, and this problem continues to worsen.[5]

The Occupational Safety and Health Administration's (OSHA) 1993 budget was $289 million, the Food Safety and Inspection Service (the federal agency that oversees the nation's food quality) had nearly twice that allocation ($490 million), and the Environmental Protection Agency (EPA) $7 billion. To police the workplace OSHA has fewer than 1,000 active inspectors. With 6.5 million workplaces in the United States this means that OSHA can inspect each workplace once every 87 years.[6] The federal government, through OSHA, will spend just $3 annually per worker to deal with this problem.[7]

But it is not just the health threats or the economic costs that such conditions pose (serious as they are). It is the fact that workers like those in the Boston harbor sewage tunnel, and millions of others labor under intolerable and degrading conditions. Few of us fully understand the degrading and awful conditions of work as experienced by many Americans—tunnels may be among the worst, but mines, meat-packing plants, small factories, warehouses, and thousands of print works, auto repair shops, machine tool plants, and maintenance rooms are too often places full of dirt and grime, where toilets are filthy, fresh air virtually nonexistent, and eating facilities either absent or not fit even for animals.[8] Imagine how a business executive might react to conditions such as these. What outrage there would be if these were the typical work environments in corporate corridors and government offices!

The larger irony here, of course, is that the Boston workers are building a tunnel to improve the environmental quality of Boston harbor. Few would argue with that goal, but how many think about what kind of life results for those who labor underground? Do most environmentalists consider the role of work and workers when they demand a cleaner environment? Are they conscious of the impact on jobs of environmental regulation, the demands made of workers in the plants, the suffering caused by chemicals and toxics in the workplace? Too often, for many in the labor movement, the answer has seemed to be "no."

On the other hand, environmentalists complain (with some justification) that workers and unions tend to side with management over environmental questions—viewing workers as reactionary and uninformed, labor unions as bureaucratic and hierarchical, and the labor movement as committed to jobs and growth without much thought for the impact this has on the environment. There are very real conflicts here, ones that make alliances between workers and environmentalists difficult and worker–environmental coalitions problematic.[9]

These attitudes and antagonisms spring from a complex set of values that divides not only "workers" from "environmentalists" but also defines work and production in hierarchical and nondemocratic ways—in short, the divisions are based on the way market capitalism has structured our view of the nature and purpose of production. It is, therefore, difficult for any of us to conceptualize what democratic and socialized forms of production would look like or how work might be safer, as much as it is for us to think of what might constitute a "green" future.

Clearly the conflict between the environmental movement and workers and unions arises from a number of intersecting (often contradictory) political, economic, and cultural divisions. Indeed, the extent to which cooperation and conflict exist among workers, unions, and the environmental movement varies widely, depending on the given location, level of industrial development, the particular issues at hand in various industries (e.g., timber cutting, oil refineries), the strength of the labor movement, and the class, race, and gender composition of particular groups. This tension is playing out in the United States against the backdrop of profound changes in American politics and international capitalism.

Despite these many conflicts there is also some clear evidence of very successful coalitions. Labor and community/environmental groups have created a significant counterhegemonic force to capital's control and dominance of both the point of production and the environment.[10] While these examples are not yet common, there are lessons to be learned from their activities and efforts to bridge the gap between worker health and safety struggles and movements for environmental and ecological health.

In what follows we present the issue of occupational safety and health in the United States and its connection to environmental degradation and crisis. We examine the link between safety and health and environmental degradation in the context of attempts to forge links between workers and environmentalists around the issues of toxic exposure, worker safety, and the destruction of the environment.

UNDERSTANDING THE HISTORY OF OCCUPATIONAL SAFETY AND HEALTH STRUGGLES

Why is the workplace so dangerous? Why have so many lives been lost, health ruined, and families destroyed when most of the causes of these problems are well known and the technology to prevent them readily available? These deaths and injuries are clearly avoidable. The understanding of what causes most of these problems exists, as does (for the most part) the technological mechanisms to mitigate exposure to dangerous machinery and chemicals.

We will argue that the magnitude and pattern of occupational disease and injury in a particular society is strongly affected by the level of economic and technological development, by the societal distribution of power, and by the dominant ideology of a particular social and political system. All of these factors (and more) bear on the way in which diseases and injuries are "produced," the recognition and prevention of these problems, and the extent to which workers receive compensation for them.

The question is: who controls the decisions about what and how to produce goods and services? Clearly, this is about power. Capital still exploits labor. It may be more subtle than in Marx's day and the class distinctions more complex, but, for the most part—across the globe—workers have very little control over what is produced, the conditions under which they work, and the rewards they gain for their labor. Capital exploits both workers and the environment, and it does so simultaneously. This is nowhere more evident than in the connection between threats to worker health and safety at the point of production and the threat to the environment posed by these production activities. Thus it is critical that the environmental movement be joined with labor in a struggle for economic democracy and control over production decisions, since a safe and healthful workplace and a healthy environment cannot be achieved by one movement alone.

In the United States, occupational health and safety has rarely received much attention (as in most societies). Historically, capital's commitment to economic growth through technological innovation has been blind to the toll on workers' health, as it frequently has to the ecological costs. Workers themselves have been engaged in the more pressing task of supporting their families and often ignore widespread occupational safety and health problems. The labor movement in the United States has not been strong enough to force continuous public concern about these issues. As a result, relatively little attention (with some notable exceptions in organized labor and among a number of social reformers and activists) was given to the problem of occupational illness and injury in the United States until the late 1960s.[11] There is still little emphasis on the link between the health problems of the work environment and the crisis of environmental and ecological health.

In most countries the process of industrialization that resulted in the creation of the factory system radically changed people's experience of work. Forced by economic necessity in the nineteenth century into the newly created factories of the machine age, workers found themselves controlled by bosses whose sole concern was the maximization of profit. Working in large-scale plants and using the new technology of modern industry, workers confronted a whole new set of conditions: powerless and tied to the speed of the machine they served, facing the ever present dangers of physical injury from conveyor belts and speeding looms and exposed to a range of dyes, bleaches, and gases, the workplace became a source of injury, disease, disability, and death.

With the help of social reformers and professionals, workers—newly organized into unions—fought back against these conditions. Laws restricting working hours and the employment of women and children, as well as promoting protection against safety hazards and some hazardous chemical exposures, increased. A system of factory inspections was established in Great Britain by the mid-nineteenth century, and Germany moved to control working conditions by the beginning of the twentieth century. In Europe these efforts built on an earlier tradition of occupational medicine, an acceptance of government intervention and paternalism, and a relatively powerful workers' movement. By the twentieth century workers and unions had achieved political representation in the form of labor, socialist, or social democratic parties. This development gave workers the power to demand reform and was a major factor in using laws to improve working conditions.

In the nineteenth century the Industrial Revolution also brought to the United States—as it had to Europe—a host of safety problems and some public concern about these problems. Massachusetts created the first factory inspection department in the United States in 1867 and in subsequent years enacted the first job safety laws in the textile industry. The Knights of Labor, one of the earliest labor unions, agitated for safety laws in the 1870s and 1880s. Social reformers and growing union power did gain, by 1900, minimal workplace health and safety legislation in the most heavily industrialized states. In the United States the regulations and the systems of inspection were, however, inadequate. The absence of a powerful and radicalized union movement and the lack of social democratic parties that had any hope of gaining political power thwarted the possibility that health and safety regulation would equal that in most of western Europe. Those states that had some legislated protections rarely enforced them and focused largely on safety issues; little was done to protect workers from exposure to the growing number of chemicals in the workplace.

During the "Progressive" era bludgeoning Taylorist production systems and the intensification of manufacturing created enormous worker, urban, and environmental health problems. In response, workers were joined by professional public and worker health activists, unions, and worker advocacy groups in efforts to ameliorate working conditions and urban environmental problems. Much of their activity focused on demands for a state-sponsored regulatory system, interposing government authority between labor and capital.[12]

After 1900, the rising tide of industrial accidents resulted in passage of state workers' compensation laws, so that by the mid-1920s virtually all states had adopted these no-fault insurance programs. But this system simply provided very minimal support to injured workers and allowed employers and companies to avoid internalizing the real economic and social costs, displacing them onto the working class.

The growth of Fordism in the 1920s and 1930s and the domination of big capital with intensified mass production and mass consumerism resulted in the professionalization and cooptation of health, safety and emerging environmental movements.[13] In the workplace, business unionism was spearheaded by the American Federation of Labor, and, as the Great Depression took hold, unions struggled to maintain wages and jobs. The creation of the Congress of Industrial Organizations (CIO) and the workplace struggles of the 1930s provided the impetus for coalitions of solidarity around workplace, community, and environmental issues, but the problems of unemployment and economic crisis pushed public health issues in general, and workplace health and safety and environmental concerns in particular, to the back burner.[14]

Throughout the 1920s in the United States, the rise of company paternalism and Fordism was accompanied by the development of occupational medicine programs. Much attention was paid to pre-employment physical examinations rather than to industrial hygiene and accident prevention. Occasional scandals reached the public eye, like cancer in young radium watch dial painters. But it was not until the resurgence of the labor movement in the 1930s that there was some national legislation to control working conditions.[15]

The mobilization for war required the U.S. government to become deeply involved in the organization of production. Concern for the workers' health increased since a healthy work force was considered indispensable to the war effort. After World War II, however, health and safety receded from public attention (one exception to this general neglect was passage of the Atomic Energy Act in 1954, which included provision for radiation safety standards) as postwar economic growth proceeded apace and unions and management entered into a period of collaboration that emphasized continued growth in jobs, wages and profits—not concerns about working conditions.

As Fordism gained ground in the war years it brought with it a vast increase in mass production and the introduction of thousands of new chemicals, processes, and technologies, most of whose health impacts were unknown. The dynamic of mass production and mass consumption explicitly contained in Fordism successfully hid many of the negative aspects of production (particularly worker health and safety problems and environmental degradation). In the immediate postwar years, with U.S. dominance of the global economy and a quiescent labor movement, little attention was paid to health and safety issues and even less to environmental problems. Diverted by the anticommunism of the Cold War, divided by surburbanization and race and bought off by rising standards of living, the labor movement, environmentalists, and community groups rarely sought collective action against the growing environmental crisis. With the state mediating labor–management conflicts and providing symbolic regulatory relief, it was only in the emerging civil rights, public health, and antinuclear movements that the seeds for new labor/environmental coalitions were sown. Nursing these seeds to life required direction and changing economic and social conditions:

> Under Fordism, labor/community collaboration was initially hampered by political repression, the capital labor-accord, the fragmentation of working class communities through surburbanization, and attendant political conservatism within the working class. Whereas communities of solidarity once developed more readily, they now had to be cultivated more strategically.[16]

The strategic efforts of the 1960s were never very clearly articulated, and the institutionalization of protest movements into state-led programs such as the Great Society divided the working class into narrowly defined specific interest groups. The antiwar and urban rebellion movements were extensive and angry but failed ultimately to garner true democratic reforms or loosen the control of capital over the American economic and social landscape. It was only in the late 1960s that these movements developed into a resurgence of worker, environmentalist, and community collaboration.

Not until the 1960s, when labor regained some political clout under the Democratic administrations of Presidents Johnson and Kennedy, did the worker health and safety movement reemerge as significant. Injury rates rose 29 percent during the 1960s, prompting union concern, but it was a major mine disaster in 1968 in Farmington, West Virginia, when 78 miners were killed, that captured public sympathy. In 1969 the Coal Mine Health and Safety Act was passed, and, finally, the first comprehensive federal legislation to protect workers was created when the Occupational Safety and Health Act (OSHAct) became law in 1970.[17] This was a time of increasing public concern with environmental problems. The publication of *Silent Spring* in the late 1950s, the spread of suburban middle-class life, and a new concern with urban and suburban environmental conditions led to growing environmentalism in the 1960s and to much of the clean air, water, and environmental laws of the 1970s.[18]

In the past two decades, although many countries have provided regulatory protection for workers and unions (often demanding safe working conditions through collective bargaining agreements), the occupational health and safety problems facing workers have increased. The prevalence of new and exotic chemicals in the workplace, cutbacks in regulatory enforcement, and the demands of an increasingly competitive global economy all exacerbate the need to maintain and improve working conditions.

The globalization of production, trade, and consumption has resulted in occupational—and environmental—safety and health problems becoming ubiquitous. Workers in developing and newly industrialized countries now face a range of workplace hazards. Stricter environmental regulations in the industrialized countries make it attractive for companies to use Latin America, Asia, and Africa as dumping grounds for toxic waste and as places to export highly toxic substances and hazardous industries. Most of these countries are ill equipped to deal with these hazards.

In North America, the development of continental free trade may threaten the comparatively advanced work environment standards of Canada and the United States while bringing many new hazards to Mexico. In Europe, integration has made the movement of capital and labor across borders much easier—industries can move to countries with less strict occupational and environmental standards. In some cases this intrusion threatens worker and environmental health; in others the more advanced standards of some countries are being imposed on the less advanced, improving working and living conditions. In both situations, conflicts over standards have arisen. The export of hazardous technologies, hazardous products, and hazardous wastes represents increasing challenges to public health worldwide.

Despite these pressing global problems, our understanding of the nature of health hazards to workers and communities has been improving. The restructuring of the world economy may undercut the political will to control these hazards, however.

TOWARD A NEW POLITICS OF WORKER HEALTH AND SAFETY

The most telling and significant events of the past twenty years—that is, changes in the national and international economic order—such matters as the growth of new markets and the disappearance of old ones, new technologies, new competitors, demographic shifts, and shifts in investment—all of these directly affect the structure of production and work. More importantly, they have conditioned and altered the political structures, modes of political interaction, and the politics of change in the United States and elsewhere. In the past decade or so, both the American and European political landscapes have witnessed a profound rightward shift, resulting in the rise of neoconservative forces in the mainstream political system, an ideological movement against state interference and regulation, political attacks on workers, unions, and the poor. The degradation of the environment, cutting of wages, decline in real incomes for the majority of Americans and the speedup of work all indicate a seeming success for capital in the class struggles of the New World Order. The end of the historic conflict with the former Soviet Union and triumph of the market over planned economies, the development of capitalism in Eastern Europe and southeast Asia, the success of GATT, NAFTA, and the World Trade Organization in promoting free trade, and the rapid and free movement of capital all seem to indicate a world cast firmly in capitalism's mold.

Through all of this an essential truth remains: the system still produces goods and services and, in fact, is doing so at an unprecedented rate. Production still takes place. It may not be in the mine or the blast furnace, but it remains in the factory, the office, the warehouse, and in vast transportation systems across the globe. In many countries, ores are still mined, fields tilled and factories hum. Workers work as they always have, the products may be very different, the conditions of work infinitely variable, and the workers themselves more diverse than at any time in history—but work they do. Harder than at any time, for less real income, and in conditions too often like those in the Boston harbor sewage tunnel.

That this is so concerns us. We believe that the rapid pace of change conceals an unchanging truth: production still matters. What goes on at the point of production still conditions much else in the world. Changes in the national and international economic order—the growth of new markets and the disappearance of old ones, new technologies, new competitors, demographic shifts, and shifts in investment—all directly affect the structure of production and work. But production remains an essential feature. It is production that creates a work environment. And what is created in that environment also creates what is consumed, what pollutes, what kills, what becomes waste and trash.

All of these features are well known. They provide the political–economic setting in which struggles over workplace and environmental contamination take place. Over the past twenty years the fate of American capitalism has hinged on its ability to become a world capitalism (exploiting and expropriating workers and the environment on a global scale) while, at the same time, squeezing the domestic economy by pushing down wages, reducing welfare, decreasing government regulation and breaking union power. Such a context makes any progressive struggles against the economic and ideological hegemony of capitalism problematic.

One way, however, to increase the social and political control over workplace and environmental hazards and to reassert that political will is to build movements that link workers with environmentalists, community organizations, and antitoxic movements.

Unions provide workers a voice in determining the rules and conditions of work, wage rates, and benefits. They are the collective strength that provides a counterweight to management power and prerogatives. Some unions have been deeply involved in health and safety issues, but for most unions the health and safety issues are only a few among many. In the United States, given the weakness of unions and American capital's historic antagonism to them, unions have not always been able to mobilize the necessary resources to protect their members from workplace hazards.[19] In the United States, however, with its relatively weak labor movement and the absence of social-democratic or labor parties, most unions do offer some minimal protection against arbitrary exercise of authority and some provision for injured workers.

The strength of a labor movement determines a host of issues that directly influence worker health, including what information is generated about workplace hazards, who has access to it, what workplace standards are set and who enforced them, the options open to workers encountering a hazard, and the effectiveness of workers' compensation.[20] Unionized workers are more likely to be informed about the presence of health and safety hazards than nonunion members in the same jobs.[21] In addition to union-sponsored education programs, the union provides a shield against employer discrimination. This shield is extremely important for health and safety because employers may fire a worker for raising concerns about health and safety problems.

Unions in the United States and elsewhere have fought to create legislation requiring employers to cleanup the workplace, control the exploitation of women and children, as well as hours of work, and set and enforce industrial hygiene standards. In the United States, where OSHA requires that workers be informed about the hazards associated with the chemicals they work with, unions have pushed to make sure that capital complies with these "right-to-know" regulations. When there was no federal right-to-know law, some unions negotiated this right, as well as the right to refuse unusually hazardous work.[22]

The fight to create state-sponsored legislation is an important one, and laws to protect workers and the environment are significant forces for change. They are, however, a necessary but not sufficient condition for protecting the workplace and the environment. What is needed is collective action by workers and their communities to pursue control over what, how, and when production takes place. The point of production still determines much of how workers and the environmental movement confront health and safety threats. There is, therefore, an obvious identity of interest between workers and their unions and the activists who fight for the environment. But what may be obvious is not always realizable, and the effort to link workers and their unions into progressive coalitions with environmental and community groups has been no easy task. The success of labor, community, and environmentalist collaboration depends heavily on the recognition of an identity of interests among these groups. It also depends on the ways in which divisions are defined and made manifest. As

Estabrook notes, these divisions are class-based, ideological, racial, gendered, and spatial.[23] For example, the environmental and workers movements are divided by class and class interests, while unions have not always accepted people of color and women as "equal" workers. Further, these movements are structured by the historical specificity of collaboration and the stages of economic, political, and social development manifested by American capitalism:

> Labor/community collaboration has occurred most notably during the major social reform periods: the Progressive Era, the New Deal, and the 1960s Civil Rights and Great Society programs. . . . Labor/environmental relations, as well, experienced high points in the Progressive Era, the late 1960s and the early 1970s, and in the late 1980s and 1990s.[24]

LABOR AND THE ENVIRONMENTAL MOVEMENT: NEED FOR A NEW COALITION

For many years now the progressive health and safety movement has recognized the link between workplace health and safety and the environment and the need for organized labor and environmentalists to work together to defend the health of workers and the community. In turn, a number of environmental groups and activists, particularly those in the emerging environmental justice movement, have made serious efforts to form coalitions with labor and support labor's perspective.[25]

As American capitalism enters a new crisis phase in the 1990s, confronting increased competition from Europe, Japan, and the Newly Industrialized Countries (NICs) of the Pacific Rim, new forms of flexible production are replacing Fordism. American capital is also politically challenging the successes of institutionalized liberal reform as manifest in the environmental and health and safety laws of the 1970s.[26] Thus, efforts to build labor, environmental, and community coalitions in the 1990s confront a new and well-armed corporate capitalism intent on destroying the modest gains of the 1970s. Since the Reagan–Bush era, corporate downsizing and loss of union power have left most workers ineffective in challenging capital's hegemonic control of the economy and a dominant public discourse emphasizing the benefits of a subordinated state in a market economy, the freedoms of private property, and the value of free-market liberalism.

In the 1970s some significant progress on building a worker–environment alliance was evident, especially as the result of efforts by the Oil, Chemical and Atomic Workers union (OCAW), the Steelworkers, the International Chemical Workers Union (ICWU), and activists in a number of local and national environmental organizations—all of whom understood the absolute imperative of building this kind of coalition.[27] In addition to such initiatives as OCAW's and the Steelworkers', environmentalists have also attempted to link their struggles to unions. The creation of Environmentalists For Full Employment (EFFE), the Labor Committee for Full Employment, and the relationship EFFE sought with the AFL-CIO's National Committee for Full Employment in the mid-1970s are instructive examples.[28] In 1978 the United Electrical and Machine Workers joined environmentalists to protest General Electric's occupational health and environmental policies and to save jobs. In the early 1980s labor and environmental groups worked together around issues related to nuclear power. But the forces operating against these efforts have not been idle. Local company bosses and big industry have undertaken wherever possible to pit workers against communities and communities against workers, seeking to prevent any alliance between environmentalists or the community and labor. Today this is especially evident in health and safety struggles, where companies use intimidation, the fear of layoffs, of closing down, of "goin' South" (or, more likely, leaving the country altogether) to split and defeat such an alliance. All of these tactics (and the very real dilemmas they engender) have been ably documented and discussed elsewhere, but while the forces arrayed against progressive coalition building are formidable, they are only part of the story.[29] Powerful as these forces may be, there are signs of progressive actions that link labor and communities together in common and often *successful* struggle against a variety of repressive and antiprogressive institutions and individuals.

LABOR AND THE ENVIRONMENT: THE EXAMPLE OF OCAW

A union, such as the Oil, Chemical and Atomic Workers (OCAW), in industries producing toxic substances cannot avoid dealing with environmental issues. Due to its particular history and its position representing workers in especially dangerous and toxic industries, OCAW has been at the forefront in connecting labor to the environmental movement. The difficulties of such coalitions are legendary inasmuch as conflicts about jobs, investment, class differences all tend to mitigate a commonality of interests emerging between workers and environmentalists. While such problems are large, OCAW's strategies have done much to connect the two forces and build an important base within the American environmental movement. These strategies have focused on a simple truth, namely, that hazards in and outside of the factory impact the lives of both workers and community members, who are quite often one and the same.

In the 1950s, a local of the Gas, Coke and Chemical Workers (a forerunner of OCAW) on Long Island in New York collected thousands of baby teeth for a campaign against Strontium 90 contamination (Strontium 90 builds up in teeth over time). The local worked with other groups to pressure the Atomic Energy Commission into taking action to reduce the level of contamination and to get support for the 1963 Nuclear Test Ban Treaty.[30] In another example of union concern for environmental issues, OCAW President Grospiron, when he joined the AFL-CIO Executive Council in 1969, voted against the supersonic transport on environmental grounds.[31] In 1970, Tony Mazzocchi (a leading figure in the occupational safety and health movement and then secretary-treasurer of OCAW) represented labor as a speaker at the first Earth Day celebration.[32]

The 1973 strike by OCAW against Shell Oil pushed union demands for health and safety and, with Mazzochi's efforts, gained the support of environmental and public interest groups. In May 1973 25 scientists published a statement in *The New York Times* supporting OCAW's strike and emphasizing the connection between workplace and environmental contamination.[33]

In a further effort at building a labor–environment coalition, OCAW prepared a study of 21 plants that closed between 1970 and 1976, causing some 1,700 members to lose their jobs. The study concluded that it was not environmental regulation that occasioned the plants' demise:

> It is our experience that environmental considerations are generally not the overriding factor in the decision to close a plant facility. Too many other factors seem to be involved. At best, environmental regulations assume the character of the straw that broke the camel's back.[34]

The study, like much of the union's activity in this period, was an attempt to reduce the impact of environmental regulation on the jobs of union members and to diminish the effectiveness of corporate "environment versus jobs" rhetoric. It represented an important step in combating the ideological hegemony of "science" and "economics" in the service of capital.

In 1976, OCAW members working for American Cyanamid in Willow Island, West Virginia, were pressured into having themselves sterilized in order to prevent being transferred to lower-paying jobs.[35] Mazzocchi, then Vice President of OCAW, took up the issue: "American Cyanamid, and other companies of like persuasion, are attempting to alter their workers instead of their plants in order to comply with government regulations."[36] The American Civil Liberties Union (ACLU) was drawn into the struggle, supporting the workers. Eula Bingham, then head of OSHA, charged American Cyanamid with violation of the "General Duty" Clause, which requires employers to maintain healthy and safe workplaces. Her action, however, was overturned by the Occupational Safety and Health Review Commission, and the ACLU sued American Cyanamid for the five women in civil court. They were later awarded about $25,000 each.

More important, however, than the victory was that OCAW and the ACLU were joined by women's groups and other industrial unions in this struggle. While the coalition fizzled out during the Reagan years because of lack of funding, the action provided a profound lesson in the efficacy of a union's reaching out to other like-minded forces and the potential such a coalition might have.[37]

Again, in 1979, the union sought the support of the environmental movement when it struck Goodyear Atomic over radiation hazards in the plants:

> After the strike had lasted seven months, a number of Washington-based environmental and anti-nuclear groups—SANE, the Urban Environment Conference, Environmentalist for Full Employment, Environmental Policy Center, Environmental Action Foundation, Mobilization for Survival, and Friends of the Earth—formed a Strike Support Group. The Group helped the striking workers generate publicity and political pressure, arranging for about 100 workers to come to Washington, hold a press conference with Ohio Senators John Glenn and Howard Metzenbaum, and meet with Department of Energy officials.[38]

A few years later, OCAW moved to another level of political activity with its BASF corporate campaign. In 1984, BASF, a German owned conglomerate, locked out OCAW members in its plant in Geismar, Louisiana. The campaign that developed out of this situation exemplifies OCAW's broad coalition-building strategy. BASF accused the OCAW of: "retrieving data from the company's computer network, acquiring and threatening to disclose the secret formula for its product Allugard 340-2 . . . , petitioning the Justice Department to block a $400 million acquisition, and publishing personal financial information on the company's top executives."[39] The union proudly listed further company allegations: "causing a Congressional investigation into BASF's safety record, publicly embarrassing the company in international forums through filing of complaints with the Organization for Economic Cooperation and Development."[40]

OCAW also commissioned a study of BASF activities in South Africa and succeeded in getting the National Rainbow Coalition and Jesse Jackson's support for the strike, including a speech by Jackson at the Louisiana state capitol in Baton Rouge.[41] Supporting the action, the Sierra Club produced a report on chemical industry dumping in Louisiana, five Louisiana environmental organizations condemned chemical industry polluters, including BASF, and the Natural Resources Defense Council reviewed an OCAW report on safety at the BASF plant, drawing analogies to Bhopal.[42] In addition, a resolution in support of OCAW was passed by the Occupational Health Section of the American Public Health Association, and the German chemical workers union, IG Chemie, actively backed OCAW by sending a delegation and a strike support donation to the United States.

This corporate campaign brought together a wide variety of progressive activists including people of color, environmentalists, and unionists (both domestic and international), providing yet another lesson in the effectiveness of coalition building in supporting union initiatives in the struggle against the prerogatives of capital.[43]

By the mid-1980s, however, the period of coalition building went into decline as a result of pressures on jobs stemming from corporate downsizing, capital flight, and increased international competition. This first era of labor–environment links reflected a largely liberal cast, as the mainstream environmental movement (based primarily in Washington, DC) sought links with organized labor at the national level. More recently, two new strategies have emerged: the building of coalitions with grassroots environmentalists (antitoxic and hazardous waste campaigns and the environmental justice coalitions) and the development of the Toxics Use Reduction strategy.[44]

OCAW worked very closely with the National Toxics Campaign until its demise in the early 1990s. The union brought Lois Gibbs of Love Canal fame and the Citizen's Clearinghouse for Hazardous Waste (now the Center for Health, Environment, and Justice) to its first training session for educators who were developing courses on health and safety for the union rank and file in the 1980s.

Common goals provided the basis for cooperation, but there were also some issues that were difficult to deal with—principally "jobs vs. the environment."

> Those of us in the trade union movement, who represent people employed in industries that either use or produce hazardous materials, now find ourselves caught on the horns of a difficult dilemma: On the one hand, our first concern is to protect the jobs, incomes, and working conditions of our members. That is what they elected us to do, and if we can't do it, they are within their rights to replace us. Which they do. On the other hand, people who work in hazardous industries, as most OCAW members do, want safe jobs and a healthy environment, just like everyone else. And they expect us to do everything we can to ensure that their employers provide them with a workplace free from all recognized hazards, as the Occupational Safety and Health Act requires. The problem is that these two goals often seem to conflict with one another. Ask workers at home if they support government efforts to clean up the air or the water, and they will most likely say yes. Ask them at work, after the company has told them that new OSHA or EPA regulations threaten their jobs, and they will most likely say no. . . . Both the trade unions and the environmental movement need to find a way to help working people escape from this cruel choice. . . . This is not an easy task. It requires, first of all, that we each—community activist as well as trade unionist—know as much as we can about the scientific, medical, and economic issues involved.[45]

Perhaps the most fundamental challenge that OCAW put to the environmental movement was its call for a "Superfund for Workers."[46]

OCAW made plain its willingness to ally with environmentalists and community activists if they in turn would commit themselves to the economic security of affected workers—attacking the income/job security issue head-on. OCAW suggested that the GI Bill, providing educational benefits to war veterans, was a useful model for dealing with the problem of workers displaced by the closing of polluting plants and that a labor–environment coalition should include this in its program.

This Superfund is clearly controversial and has led to heated debates among both workers and environmentalists. Some national environmental groups viewed efforts to build in adjustment assistance for affected workers in the Clean Air Act Reauthorization in 1992 as an attempt to derail environmental legislation. Despite these worries, the (now defunct) National Toxics Campaign and other grassroots community environmental groups, such as the Labor and Community Strategy Center Los Angeles, were receptive to the Superfund for Workers concept.

OCAW continues to pursue the Superfund strategy and, in further efforts to overcome the tension between jobs and environmental protection, has developed a "Jobs and Environment" training course, offered jointly to rank-and-file chemical members of the union and community activists.[47] These are powerful stories and important initiatives, but quite often OCAW and a few other unions stand as rather isolated examples. The tension over jobs and the environment and the class, racial, gender, and spatial distinctions that it exemplifies clearly pose formidable barriers to further coalition building.

"No work, no food—eat an environmentalist"—a slogan heard a lot during the early 1970s and again during the struggles around environmental policies in the timber-producing regions seems to embody all the conflicts between the environmental and labor movements.[48] Despite the efforts of unions such as OCAW, the perceived material interests of workers are often at odds with those of the environmental movement. The class divisions between workers and environmentalists frequently exacerbate the divisions. Social, cultural, and ideological connections are difficult to establish when workers and environmentalists live in different areas, share very different values and lifestyles, and lack a common vocabulary.

CONCLUSIONS

Occupational health and safety problems and environmental hazards emerge from the nature of production. We have argued here that worker health and safety and environmental problems are deeply linked. Environmental and worker advocates must join forces to cleanup industry, but doing so is difficult. The U.S. economy, like the labor and environmental movements, is highly fragmented. Trade unions cannot rightly be accused of being monolithic and in fact are frequently incapable of working in unison. This is especially true on such contentious issues as occupational and environmental health and safety—the interests of the Farmworkers Union do not frequently coincide with those of the United Steelworkers of America; the Carpenters Union may be at odds with the Laborers' over jobs and access to work; and the International Chemical Workers' Union may be competing with the Oil, Chemical and Atomic Workers for members. Even within individual unions there may be much diversity of opinion. OCAW members in the nuclear weapons industry have very different material and cultural goals than members, say, in the petrochemicals industry.

Similar divisions have split the environmental movement. In fact, it may be impossible to speak meaningfully of a collective environmental movement. Diverse in approach, actions, and identity, environmentalists have worked to pass federal, state, and local legislation to protect the environment. They have defended existing laws, protested plant sitings, demanded the cleanup of hazardous waste sites, critiqued consumerism, and advocated recycling initiatives. Environmentalists have taken radical actions to defend specific ecological systems and habitats. Some work through Washington, others build community coalitions. The movement is increasingly diverse. In recent years it has failed to develop a national collective voice—class, racial, ethnic, and gender divisions still abound, and often political ideologies are unspecified and inchoate. Many of the problems confronting the current environmental movement are mirrored by the trade unions.

Yet, like the labor movement, environmentalism has begun to be strengthened by the infusion of activists representing people of color and women and by reforming pressure from rank-and-file community members and workers. The development of the environmental justice movement in the late 1980s, the reform of the larger unions, the success of Jobs with Justice, and a host of coalition groups provide the basis for some optimism.[49]

Other strategies, such as efforts to promote toxics use reduction (TUR) by eliminating pollutants before they are produced or used, will not only save the lives and health of workers but also reduce the emissions of toxics into the wider environment. Twenty-five years of "environmental protection" have made it clear that cleanup and "end-of-pipe" control strategies are not long-run solutions to the degradation of the environment. These failed strategies are based on the "externalities" arguments of free market economists who believe in the fundamental wisdom of the entrepreneur and the stupidity and malevolence of government. An alternative notion, that society has a right to intervene directly in decision making about the uses of hazardous technologies, is gaining support among the critics of environmental protection. Pollution prevention and cleaner production have become the rubric of a new generation of environmentalists. The TUR strategy is meeting with some success.[50] But all these strategies must recognize the voices and needs of workers—stopping the production of dangerous chemicals is important, but workers also need to know what place they have in "the brave, new, green world of tomorrow."

The problem of "sustainability" is central to the new environmental agenda. But to define, discuss, and debate "sustainability" in a political and economic environment dominated by free market economists and their most retrograde political allies seems abstract at best and, on good days, an exercise in futility. For clean production to be a rallying cry of a broader and more effective movement, concrete examples are required: hope depends on even small successes in the transition from our productivist economy to something more "sustainable." Even as we need to build political coalitions for social change, we need to demonstrate that sustainable production can mean environmental improvement for communities, good and safe jobs for workers, and economic viability for managers. Model projects of this sort also become the basis for progressive criticism of recalcitrant companies and the basis for a green future for all.

ACKNOWLEDGMENT

We would like to thank Daniel Faber and the members of the *Capitalism, Nature, Socialism* Boston Editorial Collective for their many helpful comments on an earlier draft of this chapter.

NOTES

1. Michael Merrill, "Accepting the Challenge: A Response to Lin Kaatz Chary," *New Solutions, 3*(1), 1992, p. 15.
2. Interview with Joan Parker, Director of the Office of Safety, Massachusetts Attorney General's Office, January 1995.
3. Michael Silverstein, "Remembering the Past, Acting on the Future," *New Solutions, 5*(4), 1995, p. 80.
4. J. Paul Leigh et al., *Cost of Occupational Injuries and Illness in 1992*, Final NIOSH Report for Cooperation Agreement with ERC, UGO/CCU902886, 1996.
5. Robert Reich, speech on the twenty-second anniversary of the signing of the OSHA Act, Press Associates, May 3, 1993.
6. Ibid.
7. Silverstein, "Remembering the Past, Acting on the Future."
8. One recent example: a small lead smelter, itself a story of the link between occupational and environmental health, provided no lunch room for its workers. Instead, a microwave was perched on top of the toilet tank in the one lavatory. The smelter was completely contaminated by lead dust on all work surfaces. All of the workers were immigrants, and there was no union. See Joan Parker and Gina Solomon, "Decades of Deceit: The History of Bay State Smelting," *New Solutions, 5*(3), Spring 1995.
9. For a full discussion of these problems, see Richard Kazis and Richard

Grossman, *Fear at Work: Job Blackmail, Labor and the Environment* (Philadelphia: New Society Publishers, 1991).

10. There are examples of major successes and innovative strategies: the Calumet Project, begun in the mid-1980s; the Labor/Community Strategy Center in Los Angeles, founded in 1989; the Green Works Alliance in Toronto, founded in 1991; the Citizens Clearing House for Hazardous Waste; and the Jobs with Justice campaign, among others.

11. Unions such as the Steelworkers, the Oil, Chemical and Atomic Workers, the United Mineworkers, and the United Autoworkers were very active in pushing for passage of OSHA and have done much to keep attention on the issue over the past 25 years. The Coalitions/Committees on Occupational Safety and Health are also extremely important in the movement for workplace health and safety.

12. See D. Rosner and G. Markowitz, eds., *Dying for Work* (Bloomington: Indiana University Press, 1987).

13. Here we use the term "Fordism" to denote assembly-line manufacturing using unskilled mass labor. We also define it (following Gramsci) in terms of a rationalized social and economic system involving not only mass production but also mass consumption.

14. One exception was Saul Alinsky's leadership of the Back of the Yards campaign in the late 1930s. See Stanley Horwitt, *Let Them Call Me Rebel: Saul Alinsky, His Life and Legacy* (New York: Vintage Press, 1992); cited in Thomas Estabrook, "Labor/Community/Environment: The Spatial Politics of Collective Identity in Louisiana," unpublished PhD dissertation, Clark University, 1996, p. 108.

15. The Walsh–Healey Public Contracts Acts of 1936 required federal contractors to comply with health and safety standards, and the Social Security Act of 1935 provided funds for state industrial hygiene programs. For the history of this period see Rosner and Markowitz, eds., *Dying for Work*; L. Teleky, *A History of Factory and Mine Hygiene* (New York: Columbia University Press, 1948); Bennett Judkins, *We Offer Ourselves as Evidence: Towards Workers' Control of Occupational Health* (New York: Greenwood Press, 1970); Paul Weindling, ed., *The Social History of Occupational Health* (London: Croom Helm, The Society for the Social History of Medicine, 1985); and M. J. Fox and G. S. Nelson, "A Brief History of Safety Legislation and Institutions in the U.S. and Texas," *Business Studies*, North Texas State University, *11*(2), Fall 1972.

16. Estabrook, "Labor/Community/Environment," p. 119.

17. For a discussion of the founding of OSHA, see J. MacLaury, "The Job Safety Law of 1970: Its Passage Was Perilous," *Monthly Labor Review, 104*(3), March 1981; Charles Noble, *Liberalism at Work: The Rise and Fall of OSHA* (Philadelphia: Temple University Press, 1986); Daniel Berman, *Death on the Job: Occupational Health and Safety Struggles in the United States* (New York: Monthly Review Press, 1978); P. G. Donnelly, "The Origins of the Occupational Safety and Health Act of 1970," *Social Problems, 30*(1), October 1982; Bureau of National Affairs, *The Job Safety and Health Act of 1970: Text, Analysis and Legislative History* (Washington, DC: Bureau of National Affairs, 1971); and M. Page and Mary-Win O'Brien, *Bitter Wages* (New York: Grossman, 1973).

18. For a history, see Robert Gottlieb, *Forcing the Spring: The Transformation of the American Environmental Movement* (Washington, DC: Island Press, 1993).

19. At the time of writing, union membership in the private sector in the

United States is about 12 percent of the total work force. There has been some increase in membership in unions in the public sector during the past couple of years. The total percentage of workers who are members of unions is about 16 percent.

20. See Ray Elling, *The Struggle for Workers' Health: A Study of Six Industrialized Countries* (New York: Baywood Press, 1986).

21. David Weil, "Reforming OSHA: Modest Proposals for Major Change," *New Solutions, 2*(4), Summer 1992, p. 27.

22. Parts of this section were previously published in C. Levenstein, J. Wooding, and B. Rosenberg, "The Social Context of Occupational Health," in B. Levy and D. Wegman, eds., *Occupational Health: Recognizing and Preventing Work-Related Diseases* (Boston: Little, Brown, 1994).

23. Estabrook, "Labor/Community/Environment," pp. 92–97.

24. Ibid., p. 93.

25. Richard Hofrichter, ed., *Toxic Struggles: The Theory and Practice of Environmental Justice* (Philadelphia: New Society Publishers, 1993).

26. This was particularly evident in the attacks on OSHA. See Charles Noble, *Liberalism at Work.*

27. For a discussion of the many problems this creates see Richard Grossman, "Environmentalists and the Labor Movement," *Socialist Review, 15*(4 & 5), July–October 1985, pp. 63–87; and Robert Gottlieb, "A Question of Class: The Workplace Experience," *Socialist Review, 22*(4), 1992.

28. A full discussion of these problems and the formation and impact of EFFE can be found in, Grossman, "Environmentalists and the Labor Movement."

29. By far the best source is Kazis and Grossman, *Fear at Work*. See also J. Brecher and T. Costello, eds., *Building Bridges: The Emerging Grassroots Coalition of Labor and Community* (New York: Monthly Review Press, 1990).

30. Kazis and Grossman, *Fear at Work*, p. 243.

31. Ray Davidson, *Peril on the Job: A Study of Hazards in the Chemical Industry* (Washington, DC: Public Affairs Press, 1970), p. 333.

32. Anthony Mazzocchi, "Finding Common Ground: Our Commitment to Confront the Issues," *New Solutions, 1*(1), Spring, 1990.

33. Kazis and Grossman, *Fear at Work*, p. 236.

34. Ibid., p. 20

35. "Women Charge American Cyanamid Pressured Them to Become Sterilized," *OCAW News, 34*(12), February 1979.

36. Ibid.

37. Sylvia Kieding, former Director OCAW Health and Safety department, telephone interview, August 11, 1995.

38. Kazis and Grossman, *Fear at Work*, p. 253.

39. *OCAW Reporter*, November–December 1986, p. 9.

40. Ibid.

41. Ibid.

42. Ibid.

43. For an excellent detailed study of these events, see Thomas Estabrook, "Labor/Community/Environment".

44. See Robert D. Bullard, *Dumping in Dixie: Race, Class, and Environmental Quality* (Boulder, CO: Westview Press, 1990); and Ken Geiser, "Toxics Use Reduction and Pollution Prevention," *New Solutions, 1*(1), pp. 43–50.

45. Anthony Mazzocchi, "Finding Common Ground."

46. Michael Merrill, "Accepting the Challenge."

47. This section borrows from John Wooding, Charles Levenstein, and B. Rosenberg, "The Oil, Chemical and Atomic Workers International Union: Refining Strategies for Labor," *International Journal of Health Services*, 27(1), 1997.

48. Cited in Robert Gottlieb, "Forcing the Spring," p. 145.

49. For a discussion of the environmental justice movement, see Richard Hofrichter, ed., *Toxic Struggles: The Theory and Practice of Environmental Justice* (Philadelphia: New Society Publishers, 1994).

50. See Ken Geiser, "Toxics Use Reduction and Pollution Prevention."

Chapter 3

Risk and Justice

Capitalist Production and the Environment

Rodger C. Field

Tip O'Neill said that all politics is local, and the same may be said of environmental risk. Whether it's the risk from breathing polluted air or from the consumption of contaminated fish, abstract risk manifests itself in real harm to real persons in particular places. A growing number of studies have shown convincingly that these risks are not evenly distributed, but are concentrated disproportionately in poor and minority communities, often in urban areas. Using a variety of methodologies—the location of pollution sources, ambient environmental quality data, or health information—these studies confirm what Friedrich Engels observed about 1840s Manchester: that dirty air, dirty water, and dirty industries are invariably located in poor neighborhoods far removed from middle-class suburbs.[1]

As in Dylan Thomas's *A Child's Christmas in Wales,*[2] which told "everything about the wasp, except why," these studies too often describe the phenomena of uneven environmental risk without discussing the national and global economic forces that produce and distribute pollution burdens in the first place. This omission has led to a debate in the United States focusing on the political institutions that control and distribute pollution burdens rather than the prior issue of how capitalist forces and relations of production create pollution and shape how it is distributed. Only by examining such production practices can we understand the processes that determine where pollution sources come to be located, what kinds of pollution are produced, and how pollution increasingly moves through commerce to poor and minority communities.

Existing liberalist environmental regulations, aimed as they are at *pollution control* rather than *pollution prevention,* fail to address issues of production to any significant degree. Indeed, by "capturing" waste, pollution control efforts have had the effect of commodifying it into a form that can be transported across local, state, and national boundaries far from the initial point of generation. This ever increasing mobility of pollution (which also occurs through pollution trading and similar mechanisms) is thus creating a "waste circuit" whose end point is increasingly within poor and minority communities. The resulting explosion of new landfills and incinerators in the 1980s to accommodate this national commerce in waste was a significant factor in the rise of the environmental justice movement. Understanding this larger dynamic is crucial if we are to understand why, in the words of one environmental justice advocate, residents in southeast Chicago live within the center of a "toxic donut."[3] The way forward, while probably counter to current political trends, is toward pollution prevention rather than pollution control and greater democratization of the administrative process that regulates pollution.

This chapter begins with the background of the environmental justice movement and provides a brief theoretical overview of the ways in which the unequal distribution of pollution can be viewed. It then proceeds to consider how local risks are produced and distributed and concludes by examining the failure of the present environmental system to regulate production practices with some provisional responses to this situation.

THE ENVIRONMENTAL JUSTICE MOVEMENT

The grassroots organizations that collectively constitute what is now called the environmental justice movement are heirs to earlier environmental movements in the United States dating back at least to Jane Addams and the urban progressives of the nineteenth century. What is distinctive about the present movement is the explicit alliance between environmental and civil rights organizations, first evidenced by a 1982 protest against the disposal of PCB-contaminated waste in poor, predominantly African American Warren County, North Carolina. Since then, local environmentalists and civil rights leaders have increasingly found common ground in focusing on local environmental issues.

A common feature of these grassroots organizations as opposed to mainstream environmental groups is a shared belief that environmental concerns are not separate from other social issues. The issues addressed by these groups are usually local and often involve efforts to oppose unwanted facilities within a community. Lois Gibbs, whose experience as a housewife near the Love Canal Superfund site eventually led her to organize the influential Citizen's Clearinghouse for Hazardous Waste (now the Center for Health, Environment, and Justice), exemplifies this grassroots environmental movement. Unlike mainstream groups, these organizations eschew lobbying in favor of direct pressure on local authorities and pollution sources. Disparagingly dismissed as NIMBY (not-in-my-backyard) groups by conservatives and hostile business interests, they evoke a forgotten history of class-based urban struggle exemplified by Jane Addams and her fellow urban reformers (some of whom were labeled "sewer socialists"), even down to the fact that many of the leaders of these organizations are women, like Addams, Florence Kelley, Alice Hamilton, and other Hull House activists.[4]

Their tactics are a rejection of what Habermas terms "the rational–technical discourse of politics" and a belief in what Benjamin Barber calls strong, that is, participatory democracy.[5] The strong grounding of the movement in local, often urban, communities entails a rethinking of more reified conceptions of nature taken for granted by mainstream environmental organizations. Communities with their own experiential and historical realities become the "idea, the place and the relations and practices" that generate new ways of configuring the relationship between the human and nonhuman worlds.[6] To put it another way, environmental justice activists have exploded the traditional notion of "the environment" as the static physical media (i.e., air, water, and land) in which we exist. Instead, in the words of Dana Alston at the First People of Color Environmental Leadership Summit in Washington, DC, in 1991: "The environment, for us, is where we live, where we work, and where we play."[7]

These tactics and the local orientation of these grassroots organizations should be contrasted with the large mainstream organizations, such as the Environmental Defense Fund and the Natural Resources Defense Council, established after Earth Day 1970, which emphasize global and national issues. Through lobbying and legal action, these organizations were (and still are) instrumental in shaping the basic environmental laws of this country. With the exception of the Sierra Club, however, the mainstream organizations do not have a strong system of local chapters. They are national, not local, organizations. From the beginning, the belief that environmentalism was a middle-class "amenities" movement far removed from social justice issues led to an uneasy relationship between environmentalists and civil rights leaders, who believed the environmental movement would likely deflect political energy from the issues of poverty and race. Mayor Richard Hatcher of Gary, Indiana, is reported to have said in the early 1970s that the environmental movement had accomplished what George Wallace was unable to do, namely, distract the nation from the human problems of race in America.

The environmental justice movement has benefited from the active participation of a number of important academic researchers, most notably Robert D. Bullard and Bunyan Bryant, whose writings form a kind of primer for the issues within the movement.[8] Not surprisingly, then, the environmental justice movement has given rise to a number of studies designed to explore whether pollution burdens are concentrated in poor and/or minority areas. The most influential of these was a report issued by the United Church of Christ Commission for Racial Justice (then headed by Rev. Benjamin Chavis) in 1987. Studying commercial hazardous waste facilities and uncontrolled toxic waste sites, it found that three of every five African American and Latino citizens in the United States lived in communities with one or more uncontrolled hazardous waste sites and that 40 percent of total hazardous waste disposal capacity is located in communities of people of color. It found that people of color were twice as likely as whites to live in communities with a commercial hazardous waste facility and three times as likely to live near a large landfill or multiple waste facilities.[9] There have been a number of studies since 1987 that have used a similar methodology, i.e., comparing the number of pollution sources within poor or minority communities with the number in other communities. With few exceptions, these studies have concluded that the distribution of pollution sources is not random, but falls more heavily on poor and/or minority communities.[10]

Other studies have examined actual pollution levels in different areas of the country. Most of these studies compared ambient air quality in various geographic areas against the health-based standards established by the U.S. Environmental Protection Agency (EPA). Again, they conclude that poor air quality affects poor and minority communities to a significantly greater degree than other communities. A recent study, for example, found that African Americans are 40 percent more likely to live in an area which does not attain EPA standards for air quality, and Latino citizens are almost 90 percent more likely to live in such areas compared to whites.[11] The EPA itself has concluded that minorities have greater exposure to environmental pollutants, although, citing lack of data, it deferred from concluding that such exposure is the cause for disproportionate health effects (except for lead poisoning which, as discussed below, is known to affect African Americans disproportionately).[12]

It is well established that health correlates to race and income. For the first time this century, for example, African American life expectancy actually declined during the 1980s. Age-specific death rates are higher for African Americans in all age groups, 0–84, compared to whites, and death rates from cancer are 33 percent greater for African American males and 16 percent greater for African American females.[13] Two examples are particularly illustrative. Asthma is on the rise in the United States among all children, but African American children have the greatest impairments from asthma and the most frequent hospitalizations. Death rates from asthma are far greater for African Americans than whites.[14] Similarly, lead poisoning in children, an environmental hazard of major proportions, is estimated to affect three to four million children under the age of six. One out every six children in the United States has a blood lead level higher than the trigger levels established by the Centers for Disease Control and Prevention, but African American children consistently have higher levels than whites, and among inner-city low-income African American children the rate approaches an astounding 70 percent.[15]

PERSPECTIVES ON ENVIRONMENTAL JUSTICE RESEARCH

These studies, using different methodologies, present a consistent picture of uneven distribution of environmental burdens. Whether the studies examined the number of pollution sources or ambient environmental conditions or health effects, the results suggest conclusively that the burdens of pollution are inequitably imposed on poor and minority residents. Yet, these research efforts fail to offer an adequate political vision to account for this situation. I suggest that there are three possible perspectives from which to view this information: (1) from the perspective of risk, in which case the issue is whether poor and minority communities are in fact at greater risk from pollution; (2) from the perspective of fairness, in which case the issue is whether the political system fairly distributes risks (for example through the siting process); or (3) from the perspective of production, in which case the issue is whether local citizens have adequate control over the economic decisions that affect their communities. These viewpoints obviously overlap, but it is important to mark the implications of each because they lead to different political outcomes.

The current system of environmental regulation in the United States and most other industrialized nations is for the most part based upon a logic of risk that, in turn, has its roots in the state's inherent "police power" to regulate matters pertaining to health and safety. Health and safety issues are recognized as legitimate governmental concerns, although bounded by the prohibition against undue interference with the use of private property.[16] In the environmental context, risk is the concept that delineates the boundary between the legitimate authority of the government to act in matters of health and safety and illegitimate interference with private property (similar to the legal concept of "nuisance," which historically mediated between the right to exploit property for profit and the right to "quiet enjoyment" of property). Since the burden of proving risk is always on the regulator, such laws are fundamentally compatible with traditional property concepts and the limited role of government.

A system of environmental regulation based upon the logic of risk asks two questions: what is the acceptable level of risk, and what controls can be imposed to keep pollution within such limits? Applying this logic to environmental justice issues means that the movement's proponents must demonstrate with reasonable scientific rigor that poor and minority communities are at greater risk. This they attempt to do by showing, as with the United Church of Christ report, that poor and minority communities contain a disproportionate number of pollution sites. Opponents, however, counter that numbers alone say nothing of actual risk. A proper understanding of risk, they argue, would require a detailed examination of the actual releases from each pollution source, the likelihood of exposure to residents, and a judgment about the health consequences of such exposure. This would, in other words, require a community-wide cumulative risk assessment. Opponents also argue that the causes for disproportionate health effects are not known and have more to do with individual "lifestyle" than actual exposure to pollutants.

The "science" of risk assessment for even a single chemical or facility is controversial enough.[17] Beyond that, there is simply no accepted protocol for assessing community-wide risks. It is obvious, however, that the logic of risk can lead to a highly professionalized debate as to both the extent of risk and its causes. Clearly, it is desirable to reduce risks in burdened communities. However, to the extent that the "logic of risk" is relied upon exclusively or even fundamentally to legitimize the goals of the environmental justice movement, there is danger that the debate will become mired in a highly technical discourse over the extent of risk and will lose sight of the equally profound issue of democratic control over the economic aspects of community life that is also presented by this movement.[18]

Similarly, exclusive reliance upon the logic of fairness can, as Robert Lake has pointed out, lead to a tendency to locate environmental justice issues exclusively in the political process, which mediates how and where sources come to be located.[19] The debate breaks down into two camps: those who look at the data from such studies as the United Church of Christ report and argue that siting laws and other institutions are racist and class-biased[20] and those who argue that the siting of pollution sources is due at least as much to market forces as the institutions of government. There are a number of subissues in considering whether the system is unfair,[21] but the fundamental implication of the argument over fairness is that the political system that locates pollution sources in communities should more adequately represent the interests of disempowered poor and minority communities.

The logic of risk and the logic of fairness are strongly related since the political system that sites pollution sources will be considered unfair only to the extent that siting produces greater risk. I do not deny the validity of either of these positions. Indeed, there is strong reason to believe both that poor and minority communities do bear greater risk and that institutional racism and class bias exist within the systems that distribute these risks. In this debate, however, there has been insufficient attention given to the role of capitalist production in producing these risks in the first place. Concerns about the risks of pollution or fairness in siting already assume the existence of pollution as a natural part of industrial production. In fact, the question of who bears pollution risks is historically specific, and, as argued here, is located fundamentally in the forces and relations of production in our economy that not only produce risks but also shape the way they are distributed.

THE PRODUCTION AND DISTRIBUTION OF ENVIRONMENTAL HARMS

Industrial production occurs through decisions made by individual firms within an existing economic, geographic, and regulatory context. These decisions are crucial to the distribution of environmental risks in determining where production will occur, what kind of pollution is produced, and how pollution moves across borders to the most vulnerable communities, each of which is discussed below.

Where Pollution Is Produced

Environmental risks are borne to the greatest extent by those in physical proximity to pollution sources. The decision to locate a production facility depends upon a variety of factors. In the case of industries that extract natural resources, this will depend on where these resources are located. While it may seem so obvious as to not require discussion, capital extracts wealth by means of environmental exploitation as well as exploitation of labor. Although Marx did not develop a theory of nature as he did a theory of labor, this proposition is clearly implicit in his work. This is nicely captured in the following passage in which Marx writes about robbing the soil, but which can be equally applied to robbing the earth's plant, animal, and mineral wealth:

> All progress in capitalist agriculture is a progress in the art, not only of robbing the worker, but of robbing the soil; all progress in increasing the fertility of the soil for a given time is progress towards ruining the long-lasting sources of that fertility. . . . Capitalist production, therefore, only develops the techniques and the degree of combination of the social process of production by simultaneously undermining the original sources of all wealth—the soil and the workers.[22]

James O'Connor has elaborated on this point in a more thorough examination of "ecological Marxism" that extends traditional Marxist theory to include a focus on "the *conditions* of capitalist production" in addition to an examination of the forces and relations of production.[23] External physical conditions, including natural wealth and resources, form one aspect of the conditions under which production occurs. (The social/geographic infrastructure discussed elsewhere in this chapter is another.) O'Connor argues that in addition to the contradictions inherent in the relations of production under capitalism, which are the point of departure for traditional Marxist analysis, "ecological Marxism" requires consideration of the ways in which capitalism also impairs and destroys the very conditions (natural and societal) of capitalist production.

Because there is a locational or spatial element to where valuable natural resources are discovered, the health and environmental burdens associated with their extraction will fall on local communities near such areas. This extraction, primarily by large mining and timber corporations, requires the massive application of machinery and extensive processing that produces toxic by-products and scars the earth.[24] Governments have long encouraged the large-scale extraction of natural resources as a source of national wealth through subsidies and other incentives, thus making the extraction of virgin natural resources more profitable than environmentally more desirable alternatives such as recycling, reuse, or reduction in material usage. In the United States, for example, the depletion allowance allows mining companies to deduct 5–22 percent of their income from taxation. This tax break, which is in addition to normal depreciation for capital expenditures, is based upon the now outdated assumption that companies that extract supposedly finite natural resources, such as minerals or oil, need incentives to invest capital resources in a depleting resource. Another throwback to an earlier century is the still applicable General Mining Act of 1872 that allows anyone who finds hard-rock minerals in public territory to buy the land for $12 per acre.[25]

The national policy of encouraging the extraction of virgin natural resources has a direct impact on particular communities. One of the great ironies of the history of repression against Native Americans in this country is that the unwanted lands to which they were forcibly relocated contain some of the richest mineral deposits and other natural resources in the country. Two-thirds of all uranium resources in the United States, for example, lie under Native reservations, as does a third of all low-sulfur coal in the United States.[26] The quest for natural resources, then, imposes specific environmental risks on peoples such as Native Americans who reside near, and are dependent on, natural resources.[27]

For nonextractive industries, the decision where to locate production facilities (and hence where pollution risks will be borne) is more complex and occurs within a particular geographic landscape that itself is shaped over time by forces within the capitalist economy. This (primarily) urban landscape significantly influences how environmental burdens will be distributed. Any discussion of local environmental risks would be incomplete without an examination of how this landscape has been prepared.[28]

Here, environmentalists have much to learn from geographers who have studied the spatial landscape of the industrial city and the political and economic factors that shape it. Until the nineteenth century, the urban landscape was remarkably stable. As David Harvey has pointed out:

> Up until the sixteenth or seventeenth centuries, urbanization was limited by a very specific metabolic relation between cities and their productive hinterlands coupled with the surplus extraction possibilities (grounded in specific class relations) that sustained them. . . . The basic provisioning

> (feeding, watering, and energy supply) of the city was always limited by the restricted productive capacity of a relatively confined hinterland. Cities were forced to be "sustainable" to use a currently much favored word, because they had to be.[29]

Succeeding innovations in transportation and communications—notably railroads in the nineteenth century and the automobile in the twentieth—altered this "metabolic relation." William Cronon's history of Chicago, *Nature's Metropolis,*[30] records the ways in which technological advances in transportation allowed Chicago to bring within its market sphere timber, livestock, agricultural products, and other resources far from its immediate hinterlands. As more and more territory was sucked into Chicago's market sphere, local industries became national ones. When Armour and Swift, for example, located their meatpacking operations in Chicago in the 1870s, residents in and around Packingtown were forced to live with the blood and guts of the hogs and cattle that fed not only Chicago but a good part of the nation as well.[31]

Modern industrial production is characterized, among other things, by the growth of the factory system, expansion of wage labor, increased use of machine production, and the rise of the industrial city.[32] The ways in which newer cities, such as Chicago and Pittsburgh, accommodated to the needs of industrial production are starkly obvious. In the older cities of Europe and the eastern seaboard, development was more complicated, as the forces of production were superimposed upon a preindustrial landscape. In either case, however, the geography of the city becomes an adjunct for industrial production. Waterways became a convenient method for waste disposal; the air, a sink for smoke; land, a commodity that could be created (as with the extensive landfilling of the nineteenth century) and used with no restrictions except traditional legal notions of nuisance and trespass that mediate between conflicting rights of property owners.[33] Historically, industries were located near waterways just outside of the commercial core. Typically, the commercial core was first encircled by a band of industrial activity, surrounded by worker housing and finally, at the outer core, more wealthy residences, thus creating the "concentric circle" model of urban development. In the industrial city, the workers, who were often recent immigrants, were most likely to live closest to industrial activity and bore the brunt of industrial pollution.

The ability of local government to regulate the manner of this development or to place controls on these industries was minimal at best. Beginning with Chicago and Cincinnati in 1881, for example, many cities, especially those in the Midwest which relied on high-sulfur bituminous coal, passed local "smoke laws." These proved ineffectual for a number of reasons: penalties were low (Justice Gibbons, who handled smoke cases in Chicago, is said to have regarded a $100 fine as cruel and unusual punishment), and enforcement was lax.[34] The political power of industry over local government was overwhelming in an age where faith in economic growth and big business swept aside other concerns. Without adequate means to control industrial production, urban environmental reform in the nineteenth century consisted primarily of public works projects to improve sanitation and public health. By the end of the century, for example, Chicago had built almost 1,500 miles of sewers, had reversed the flow of the polluted Chicago River away from Lake Michigan, and had constructed a system of tunnels to convey pure drinking water from underground water supplies beneath Lake Michigan. In major cities across the country, streets were paved to cut down on dust, and eventually cities began providing street cleaning and trash collection services. Led by urban landscapers such as Frederick Law Olmstead, land was set aside for parks and playgrounds. Taken together, these projects addressed some of the most serious health problems such as cholera and typhoid and undoubtedly improved the conditions of city dwellers,[35] but progressive urban reformers made little headway in asserting effective legal or regulatory control over polluting industries.[36] That is to say, the traditional prerogative of capital to control production continued unabated during this period of great urban reform. Accordingly, the response of many people (at least those who could afford it) to urban pollution was to place as much distance as possible between them and the sources of pollution. This was social policy by default, but no less significant therefore in shaping urban space.

Technological changes meant that workers were no longer tied to housing located within walking distance from the workplace. Beginning with the first mass transportation systems and continuing most dramatically with post–World War II suburbanization, the work force began a steady march away from areas of industrial activity. This trend essentially became this nation's urban policy as it was embedded in a series of political decisions during the postwar period. A number of governmental activities—easy credit for the purchase of single-family homes, mortgage interest tax deductions, highway construction, mass transit policy—all came together to create the urban/suburban landscape as we know see it.[37] The "concentric circle" model of urban development broke down into today's patchwork of industrial and residential neighborhoods intersected by large corridors to the suburbs.

This trend has had a number of spatial implications that bear on the distribution of environmental harms. First, those who still live in the older industrial areas (which are those, by and large, who are too poor to move) bear the historical contamination that is a legacy of an earlier industrial stage (soil contamination, old dump sites, etc.). Second, it also means that new polluting industries, such as waste disposal facilities that are part of the newly emerging waste industry, searching opportunistically for areas of low property values and communities with less political power, locate in the hollowed-out areas of former industrial activity. Third, because this flight from urban industrial areas was shaped by historic patterns of racial discrimination, minorities who are prevented from relocating outside the inner city were and are often forced to live in formerly industrialized areas.[38] These factors tend to ensure that industrial pollution will fall most heavily upon poor and minorities.

Overlaid on this landscape is an additional phenomena. As David Harvey among others has noted, a new era of post-Fordist production has seen the emergence of "industrial ensembles" (to use Harvey's term) that are at once centralized administratively and financially and decentralized for purposes of production. This entails both horizontal decentralization, where companies will relocate operations to take advantage of cheap labor or lax labor and environmental laws, as well as vertical decentralization, where more and more activities are being outsourced. Surrounding the industrial center is a web of suppliers, fabricators, and the like who exist in a symbiotic relationship with the center. The monoliths like IBM survive on circuit boards, chips, and other computer components produced in countless sweatshops in the "valleys and glens" that surround major cities. Economies of scale have given way to economies of scope. Harvey notes for example that 75 percent of machine parts are produced in batches of 50 or less.[39]

This phenomenon of industrial production has a spatial, therefore distributional, aspect. In addition to the large, old-style factories (which still exist and still generate environmental risks), new, smaller pollution sources have also become more widespread. There are, for example, some 120,000 facilities in this country that produce or distribute chemical products. It is no longer sufficient to assume, as did the writers of our environmental laws, that environmental risks come only from major sources. Instead, in the same way that textile and shoe companies have contracted out to independent sweatshop operators, major industrial sectors increasingly contract out the dirty and toxic operations to small contractors who are often least able to handle these toxic chemicals safely. Electroplating companies or computer component manufacturers that often use extremely toxic chemicals can exist literally in someone's garage. These are often undercapitalized, low-budget operations that operate in the poorest sections of the industrial landscape, in the interstices of an older, Fordist landscape. Large production facilities continue to impose burdens on poor and minority communities, as demonstrated, for example, by the large petrochemical facilities concentrated in Cancer Alley in Louisiana. Together with the newer "post-Fordist" industries, they present a double hit to the poor and minority residents who live in the hollowed-out areas of our industrial centers. As in the case of extractive industries, the consequences of industrial production continue to be felt most profoundly by those who live in proximity to these sources.

What Kind of Pollution Is Produced

It is not only the location of pollution sources that generates environmental risks but also the production process that determines the amount and kind of pollution that is produced. Modern methods of mass production have created more pollution now than at any other time in history: with Fordism[40] came mass production, and with mass production came increased pollution. Industrial production in the United States, which had its start in the last century, saw its full realization after World War II. Almost every sector experienced great increases in production and consumption. Between 1945 and 1985, for example, the annual production of organic chemicals rose 15-fold, from 6.7 million metric tons to 102 million.[41] Since 1950, per capita energy use has climbed 60 percent, car travel has more than doubled, plastic use has multiplied 20-fold, and air travel has jumped 25-fold.[42]

At one level, the simple fact of a mass production economy is relevant to the distribution of environmental harms. The more waste and pollution, the more likelihood of exposure—and the more smokestacks, landfills, incinerators, transfer stations, and waste handling sites. These are precisely the facilities that are often targeted for poor and minority areas.

As important as the overall increase in production is the fact that the manner of production has changed in two important respects that are relevant to the way harms are distributed. First, as environmentalist Barry Commoner has shown, since the end of World War II, the use of new and often toxic chemicals has increased dramatically, and, at the same time, synthetic materials have been substituted for natural materials.[43] In the context of the present economic crisis, where capital seeks to increase productivity and decrease costs, industry attempts to substitute capital (in the form of technical innovation and chemicals) for labor. In other words, economic restructuring means increased use of toxics. This can most dramatically be seen in the area of agriculture, where some 50,000 pesticides are registered for use. Between 1950 and 1987, total U.S. agricultural output increased by 80 percent; at the same time, labor input decreased by 71 percent, mechanization was constant, but the use of chemical pesticides increased 484 percent. The same trend holds true in other sectors as well, resulting in, as Commoner writes, "a trinity of problems: less demand for labor (a factor in unemployment), a greater demand for capital . . . , and much more environmental pollution."[44]

There are some 70,000 chemical substances in commercial use, with about 1,500 new chemicals introduced annually, most of which have come into use since the end of World War II. We live in world of synthetic chemicals, as evidenced by the dramatic growth in the petrochemical industry. The widespread use of such chemicals means that their potential toxic effects can crop up anywhere, most commonly in the farm workers and other workers who are engaged in the production or application of such chemicals, but also in consumers and others who are exposed to such substances. Such ubiquitous use also means an increase in the amount of hazardous waste (a subset of all industrial waste) to be disposed of in hazardous waste landfills or incinerators. While some argue that we all benefit from having so many synthetic consumables, the fact remains that the full impact of the 200–300 million tons of hazardous waste generated in the process is borne by the 3,000 or so local communities that host the facilities that treat, store, and dispose of the most toxic waste.

How Pollution Moves

Because pollution itself has become a movable commodity as a result of changes in the economic and legal landscape over the past two decades, capital increasingly moves pollution from the point of production across state and national borders. Ironically, this development is directly related to the passage of new environmental laws in the early 1970s. Well into the twentieth century, the full cost of getting rid of waste was avoided because companies used the commons (i.e., air, land and waterways not privately owned or controlled) as a sink for waste disposal. Beginning with the Clean Air Act in 1970 and the Clean Water Act in 1972, industry was obligated to install pollution control equipment to capture pollution, which had the effect of transforming the problem of air and water pollution into a problem of land pollution. It also meant that pollution became more mobile and less susceptible to a private, local resolution. As Andrew Hurley has described with respect to Gary, Indiana, companies began filling up every available hole in the ground with captured waste in an attempt to comply with new Clean Air Act and Clean Water Act regulations. Of course, these holes in the ground were not randomly located, but were concentrated in industrial and quasi-industrial areas near communities that were too poor to join in the exodus from the inner city to the suburbs or were excluded from doing so by racial discrimination. Under a system of pollution control (as opposed to prevention), the generalized benefit of clean air was achieved in Gary at a cost of increased industrial waste concentrated in specific communities.[45]

As companies filled up the readily available "holes in the ground" at or near their facilities, they increasingly arranged with outside companies for waste disposal. The number of these companies rose dramatically during the 1970s, as anyone with access to a suitable site could make a great deal of money by accepting industrial waste. It is not surprising that many of the sites that are on EPA's Superfund list date back to this period. Initially, the outside waste disposers were small, marginal operations, and waste disposal was virtually unregulated. However, about the time that hazardous waste regulations were promulgated in 1980, a few large companies (Waste Management, Browning-Ferris, Laidlaw, etc.), sensing huge possibilities for expansion in the waste disposal area, started acquiring smaller operations, so that the industry is now dominated by a handful of large corporations.

Waste Management, Inc. (now WMX), for example, expanded from a company with $76 million in revenues in 1971 to a Fortune 500 company with 6 billion in revenues in 1990. Waste Management is a "fully integrated" waste disposal company that, in addition to ordinary trash collection, has subsidiaries specializing in hazardous waste (Chemical Waste Management), nuclear waste (Chem-Nuclear), asbestos removal (Brand Industries), trash incinerators (Wheelabrator Technologies), and hazardous waste cleanups (ENRAC).[46]

Presently, to the extent pollution is no longer released to the commons, it moves in commerce, with responsibility for waste disposal increasingly given over to independent companies such as Waste Management. Commodified waste is disassociated from the particular plant or region that produced it and placed in commerce for handling by the now huge waste disposal industry. This directly impacts the spatial distribution of pollution. Since there is no tie between a given industry and the companies that handle waste, the treatment and disposal of waste can occur anywhere. Prior to the rise of the waste industry, a company would generally arrange for disposal at or near its plant. To this extent, the beneficiaries of the plant who generally lived nearest to it were also those who bore the burdens. Andrew Hurley, for example, in a study of Gary, Indiana, reports that both workers and managers at U.S. Steel's Gary works lived next to the plant (although in different neighborhoods) and both suffered the black smoke from factory operations.[47] Yet, as waste is captured and then handled by outside industries, it can travel across state lines and even beyond national borders in search of a final resting ground. Looking only at what EPA defines as hazardous waste (a subset of all industrial waste), almost 8 million tons were shipped offsite, over half of that to another state.[48] The interstate transport of nonhazardous waste has increased steadily as well. Not surprisingly, the ultimate disposal sites are often located in poor communities that derive no benefits from the industrial operations, where property values are low, and where community political power is weak.

As a commodity, waste becomes abstract, no longer identified with a particular plant or source. We can now talk of the "waste crisis" without reference to the source of waste. The obstacle to solving the crisis becomes the "selfishness" of NIMBY activists. The state (not capital) becomes responsible for ensuring sufficient disposal capacity one way or the other. Indeed, local, state, and federal governments have supported the waste disposal industry in a variety of ways. The methods range from direct subsidies, such as the (recently repealed) Retail Rate Law by which the state of Illinois subsidized "waste-to-energy" incinerators, to using state authorities to override local opposition to siting decisions, to actually constructing disposal facilities as the federal government is attempting to do in constructing a high-level radioactive waste disposal area in the Yucca Mountains of Nevada (a region, not surprisingly, claimed by the western Shoshone Indians). Using the "waste crisis" as the justification, waste disposal becomes a locational problem for the state rather than a production problem for capital, as Lake and Disch have pointed out.[49] Recent Supreme Court decisions have facilitated the economic transformation of the waste disposal industry by holding that local governments cannot restrict waste disposal facilities within their jurisdiction from receiving outside waste since this would violate the constitutional prohibition against restraints on interstate commerce.[50] We have come full circle, then, from a society where waste disposal is local to full and free commerce in pollution.

The movement of pollution is likely to increase. The widely touted provisions of the 1990 amendments to the Clean Air Act (enacted with the support of some environmental groups such as the Environmental Defense Fund), for example, encourage the movement of pollution in commerce by permitting companies to buy and sell the right to emit pollution. Power companies that are permitted to emit certain quantities of sulfur dioxide can now sell the right to pollute to another company, and trading of these credits now occurs regularly at the Chicago Board of Trade. This policy of tradable pollution rights is rapidly gaining currency as a way of regulating all manner of pollutants. Again, however, this has a distinctly spatial aspect since it distributes pollution burdens to communities far removed from the plant or factory that generated them.

All of these aspects of industrial production are related and feed back on one another. The fact that, with very few exceptions, the environmental laws do not prevent pollution, but attempt to capture and control it, means, as discussed previously, that the current system often ends up generating wastes that are redistributed to poor and minority communities by large waste disposal companies with the state (in the form of environmental regulators) responsible for the distribution of pollution burdens. Only in the depopulated, hollowed-out, politically weak sections of the older industrial areas is it possible to locate dirty businesses such as waste disposal facilities. At the same time, the existence of such facilities contributes to the desire to leave such areas. It is this dynamic relationship between the forces that produce waste and pollution and the urban infrastructure in which they operate that shapes the distribution of past and present environmental burdens. The common thread is the failure to assert effective control over the processes of production.

LOCAL RISKS AND NATIONAL ENVIRONMENTAL POLICIES

The preceding discussion has attempted to demonstrate the ways in which production decisions produce and distribute pollution burdens. Current environmental policy, however, fails to address the forces that produce local environmental risks in at least two important respects. First, the present system's emphasis on capture and control is inadequate for a number of reasons. As argued above, these practices increase the burdens placed on poor and minority communities. Moreover, even on its own terms, pollution control often fails to regulate or underregulates contaminants. The assumption behind the environmental laws enacted in the 1970s was that pollution was primarily caused by a handful of pollutants (referred to by regulators as "conventional" pollutants) coming from major sources. This has meant that the existing scheme does not regulate various classes of smaller sources, although, as mentioned above, these sources are increasingly part of the post-Fordist landscape. It is now apparent, moreover, that the handful of "conventional" pollutants that so preoccupied the early years of environmental regulation have now been overwhelmed by the flood of new synthetic chemicals in use today. Precisely because of the constant introduction of new chemicals, it is impossible to answer the question demanded by a risk-based statutory scheme: what pollutants are harmful and at what levels? The risk assessment process by which "safe levels" are established is expensive and time consuming. It is also problematic and fraught with scientific uncertainty and hidden policy assumptions. Each of these assumptions, while usually embedded in highly technical documents, has enormous consequences. The use of the apparently reasonable scientific concept of average risk, for example, means that data from the most sensitive individuals, such as children, will not be the basis for regulation, but rather data from the statistically average person. As Commoner has noted, this is not so much a scientific decision as it is a moral judgment.[51]

Because the current system attempts to control pollution by capturing waste by-products, important areas of contamination that are not introduced into the environment in the form of "waste" are simply not addressed through these controls. Lead contamination is an excellent example of this. As mentioned earlier, lead poisoning is one of the most serious problems facing persons living in the inner city, with an astounding 70 percent of inner-city, low-income African American children affected. People are exposed to lead from multiple sources. Lead smelters and industrial sources regulated by the Clean Air Act are only a small factor in lead exposure. The primary sources of lead exposure are: (1) lead in soil (from residues from the combustion of leaded gasoline); (2) lead in drinking water (from lead in plumbing fixtures); and (3) lead in the home (from lead paint). The most important observation to be made about lead contamination is that lead does not enter the environment simply as a waste emitted from a smokestack or discharge pipe, but in the form of commercial products: gasoline, paint, and plumbing fixtures. Traditional pollution control methods are ineffective in addressing this kind of pollution burden. The only effective response has been to mandate a different way to formulate paint and gasoline and different plumbing construction techniques.

The second way in which the current system fails to address the uneven distribution of risks is the limited way in which local communities can participate in the environmental decision-making process. While the law requires that the public be given notice and an opportunity to comment on most environmental decisions (such as regulations or permit decisions), this form of participation has been criticized as occurring too late. Consider an example cited by Lake. In McDowell County, West Virginia, one of the poorest counties in the state, where most of the coal mining jobs have been eliminated through automation, a developer proposed building a 6,000-acre waste disposal facility. It would have been the state's largest repository of out-of-state nontoxic waste delivered on 150–200 railroad cars per day. In return, the developer would have paid for a county sewer system, paid taxes, and created 300 jobs. This places the citizens of McDowell County who are asked to approve or disapprove the facility in an untenable situation. As Lake points out, this right of approval most certainly "does not constitute self-determination for the County's residents if they were systematically excluded from prior decisions that determined the choices available for consideration."[52]

The siting of new facilities, the preeminent fairness issue for many environmental justice groups, is another example of the limited ability of citizens to assert control over facilities to be operated within their community. Federal law establishes a variety of technical requirements that may apply to a facility seeking an operating permit, but the location of the facility is not considered to any significant degree, in effect divorcing technical issues from fairness issues. Assuming a facility complies with national pollution control requirements, it generally matters little whether it will be located in a rural area or in an inner-city community already the location of multiple facilities.[53] Consequently, the overall effect of multiple facilities in a community is generally not considered anywhere in the permit process. Real democratic choice would involve not simply the location of the waste facility but the whole set of prior issues about how the options available to a community are to be developed. Existing statutes, based as they are on the logic of risk, are ill equipped to incorporate democratic principles in the administrative process.

CONCLUSION

I have argued here that environmental risks originate in the production decisions of capital and that the existing system of environmental controls fails to assert control over production practices. A political response will involve moving toward a system based upon pollution prevention (with its inherent involvement in the production processes). Congress, it should be noted, has enacted the Pollution Prevention Act of 1990, which declares that "pollution should be prevented or reduced at the source whenever feasible." Taken literally, this would provide a clear alternative to the pollution control approach established in existing environmental laws. It is, however, entirely voluntary at present, with no direct obligations on industry other than certain reporting requirements. Mandatory provisions were rejected earlier when Congress enacted the current hazardous waste law (Resource Conservation and Recovery Act) in 1976. At that time, Congress failed to enact a much more comprehensive bill that would have included the authority to review industrial processes.[54] Industry has been vigilant in lobbying against any attempts to control the manufacturing process to the extent of attempting to prevent EPA from developing an "eco label" program on grounds that setting standards for such labels would be an opening wedge for EPA to interfere with the production process.[55]

A second response is to demand greater participation in the administrative process. Two conditions are necessary for meaningful citizen participation. The first is information. Here, the national policies contained in the Freedom of Information Act enacted in the 1960s and more recently the Community-Right-to-Know Act enacted in 1986 (which requires that companies describe toxic substances located at their facilities) contribute positively to citizen participation. Community activists have had some success in negotiating "good neighbor agreements" with companies using information obtained under the Community-Right-to-Know Act. The second condition is having a defined and legitimate role in the decision-making process. A strong version of this would include citizen participation in the regulatory and research agenda, as AIDS activists have achieved to some extent,[56] and an even stronger version would include formal (legal) rights over at least the most significant economic decisions that may affect a community (such as the siting of pollution sources).[57]

Consider again the three ways of viewing environmental justice issues discussed earlier, that is, from the perspectives of risk, fairness, and capitalist production.[58] A compete political response will be informed by each of these perspectives. Since the current system of environmental laws is built on a managerial model of preventing undue risk, it is ill equipped to address issues of production or democratization. Indeed, reformers of the Progressive era who established administrative agencies as a vehicle for clean government self-consciously desired to place the administration of government in the hands of persons who were not dependent on partisan politics. The source of legitimacy for administrative agencies, then, has never rested on a claim to democratic principles, but rather on nonpartisan expertise. The challenge, therefore, is to develop a new theory of legitimacy for the administrative state that can incorporate class and pluralist considerations such as those presented by the environmental justice movement. That is to say, like any political actor, administrative agencies must have a constituency to which they are accountable since they can no longer rely upon a mystique of expertise to remain above the fray.

The environmental justice movement has made an important contribution to progressive politics by demonstrating the unequal distribution of environmental burdens. The implications of its findings go well beyond seeking improvements in traditional risk control measures or redesigning state and federal environmental agencies that apportion risks through siting decisions, as important as these efforts may be. Barry Commoner is emphatic in reminding us that the broad questions of socializing the economic sphere go far beyond narrow questions of reducing environmental risk, a desirable but subsidiary benefit to the broader goal of pollution prevention. As he states:

> The large-scale transformations of the national system of production that must implement pollution prevention, however powerfully motivated by the need for environmental improvement, transcend even that urgent but singular goal. They call for national policies to govern the relation between production and nearly every sector of public life. These are not simply environmental policies, but *production* policies. . . . Simply stated, the solutions to environmental problems are not in the realm of science, but public policy—or, more plainly, politics.[59]

ACKNOWLEDGMENTS

I would like to thank Daniel Faber and the *Capitalism, Nature, Socialism* Boston Editorial Group for their helpful comments on an earlier draft of this chapter. I would like to dedicate this chapter to the memory of Sharon Stephens, with thanks for her support and suggestions.

NOTES

1. As Engels wrote: "The upper classes enjoy healthy country air and live in luxurious and comfortable dwellings which are linked to the centre of Manchester by omnibuses which run every fifteen or thirty minutes. . . . These plutocrats can travel from their houses to their places of business, which run entirely through working-class districts, without even realizing how close they are to the misery and filth which lie on both sides of the road." See Friedrich Engels, *Conditions of the Working Class in England* (Stanford, CA: Stanford University Press, 1968), p. 55.

2. Dylan Thomas, *A Child's Christmas in Wales* (Boston: David Godine, 1980), p. 21.

3. This is the image often used by Hazel Johnson, longtime environmental justice activist, who is the head of People for Community Recovery, an environmental organization located in one of Chicago's oldest public housing projects.

4. Marcy Darnovsky, "Stories Less Told: Histories of U.S. Environmentalism," *Socialist Review,* 22(4), 1992, pp. 11–56; and Robert Gottlieb, *Forcing the Spring: The Transformation of the American Environmental Movement* (Washington, DC: Island Press, 1993), pp. 47–80.

5. Robert Lake, "Negotiating Local Autonomy," *Political Geography,* 13(5), 1994, pp. 423–442.

6. Giovanna Di Chiro, "Nature as Community: The Convergence of Envi-

ronment and Social Justice," in William Cronon, ed., *Uncommon Ground: Toward Reinventing Nature* (New York: Norton, 1995), p. 310.

7. Gottlieb, *Forcing the Spring,* p. 5.

8. Bunyan Bryant, ed., *Environmental Justice: Issues, Policies, and Solutions* (Washington, DC: Island Press, 1995); Robert D. Bullard, ed., *Confronting Environmental Racism: Voices from the Grassroots* (Boston: South End Press, 1993); and Robert D. Bullard, *Dumping in Dixie: Race, Class, and Environmental Quality* (Boulder, CO: Westview Press, 1990).

9. United Church of Christ Commission for Racial Justice, *Toxic Wastes and Race in the United States: A National Report on the Racial and Socioeconomic Characteristics of Communities with Hazardous Waste Sites* (New York: Author, 1987).

10. Benjamin Goldman, *Not Just Prosperity: Achieving Sustainability with Environmental Justice* (National Wildlife Federation, December 1993), p. 10.

11. Ibid., p. 15.

12. The EPA report, for example, specifically cites the fact that a large proportion of racial minorities reside in metropolitan areas with poor air quality and reside in physical proximity to potential pollution sources such as hazardous waste sites. (It also noted that fish consumption was greater among Native Americans and minority anglers and that minority farm workers had greater exposure to pesticides.) See U.S. EPA, *Environmental Equity: Reducing Risk for All Communities,* EPA 230-R-92-008 A, Vol. 2 (Washington, DC: EPA, June 1992), pp. 7–15.

13. Ibid., pp. 4–6.

14. Richard Evans, "Asthma among Minority Children: A Growing Problem," *Chest, 101*(6), 1992, pp. 368–371.

15. U.S. EPA, "Environmental Equity," pp. 8–9; and Karen Florini, George Krumbhaar, Jr., and Ellen Silbergeld, *Legacy of Lead: America's Continuing Epidemic of Childhood Lead Poisoning* (Washington, DC: Environmental Defense Fund, March 1990).

16. Jennifer Nedelsky, *Private Property and the Limits of American Constitutionalism* (Chicago: University of Chicago Press, 1990).

17. Barry Commoner, "The Hazards of Risk Assessment," *Columbia Journal of Environmental Law, 14*(2), 1989, pp. 365–378; Howard Latin, "Good Science, Bad Regulation, and Toxic Risk Assessment," *Yale Journal of Regulation, 5,* 1988, pp. 89–148. It should be noted that technical risk assessments are distinct from community-directed health surveys, which are an important way to gain information about actual health effects (not risks) and which have been an important tool for community organizing. See Patrick Novotny, "Popular Epidemiology and the Struggle for Community Health," *Capitalism, Nature, Socialism, 5*(2), 1994, pp. 29–42.

18. Barry Commoner, "Pollution Prevention: Putting Comparative Risk Assessment in Its Place," paper presented to conference on "Setting National Environmental Priorities: The EPA Risk-Based Paradigm and Its Alternatives," Annapolis, MD, November 15–17, 1992.

19. Robert Lake, "Volunteers, NIMBYs, and Environmental Justice: Dilemmas of Democratic Practice," *Antipode, 28*(2), April 1996, pp. 160–174.

20. Bullard, ed., *Confronting Environmental Racism,* p. 11; Richard Hofrichter, ed., *Toxic Struggles: The Theory and Practice of Environmental Justice* (Philadelphia: New Society Publishers, 1993).

21. Among them are (1) whether disproportionate siting is "really" based upon race or upon class; (2) whether, based upon the historical record, poor people and minorities "come to" a pollution source after it has been sited for nondiscriminatory economic reasons (Vicki Been, "Locally Undesirable Land Uses in Minority Neighborhoods: Disproportionate Siting or Market Dynamics?" *Yale Law Journal, 103,* 1994, pp. 1383–1415); and (3) whether the responsibility to address disproportionate siting arises from a showing of intentional discrimination or whether evidence of a discriminatory impact is sufficient (Laura Pulido, "A Critical Review of the Methodology of Environmental Racism Research," *Antipode, 28*(2), April 1996, pp. 142–159.

22. Karl Marx, *Capital,* Vol. 1 (New York: Vintage, 1976), pp. 637–638, as quoted in John Bellamy Foster, *The Vulnerable Planet: A Short Economic History of the Environment* (New York: Monthly Review Press, 1994), p. 63. See also John Bellamy Foster, "The Limits of Environmentalism without Class: Lessons from the Ancient Forest Struggle of the Pacific Northwest," *CNS, 4*(1), 1993, pp. 11–41.

23. James O'Connor, "Capitalism, Nature, Socialism: A Theoretical Introduction," *Capitalism, Nature, Socialism, 1,* 1988, pp. 11–38 (emphasis in original).

24. Mining, for example, involves the removal of large amounts of overburden to uncover the desired metal. Once extracted, the metal is generally concentrated to remove impurities, leaving large quantities of "tailings," which are disposed of in and around the mining site. The concentrated ore is reduced to crude metal through a smelting process and later heated and refined again. The whole process is quintessentially dirty and produces high levels of contamination at every stage. Disturbed soil and tailings permit minerals that were once inert to be released into the environment, resulting in sediment and water contamination. (Sulfuric acid, which is created by exposure of disturbed sulfur in soil to rainwater, is a common example.) The smelting process, which is generally done at or near the site of extraction, produces air pollution, often creating "dead zones" where no vegetation survives. Forty-eight of the 1,189 Superfund sites in the United States, including the largest (a 136-mile stretch of the Silver Bow Creek and the Clark Fork River in Montana), are former mineral operation sites. See John Young, "Mining the Earth," in Lester Brown, ed., *State of the World: 1992* (New York: Norton, 1992), pp. 100–118.

25. There is a segment of mainstream environmentalists who believe that removing subsidies and internalizing the costs of pollution are all that is necessary to address pollution problems, that is, the market will weed out heavy polluters and provide incentives to clean industries. The clear limitations to this approach are set forth in J. Martínez-Alier, "Political Ecology, Distributional Conflicts, and Economic Incommensurability," *New Left Review, 211,* May/June 1995, pp. 70–88, and Martin O'Connor, "On the Misadventures of Capitalist Nature," in Martin O'Connor, ed., *Is Capitalism Sustainable: Political Economy and the Politics of Ecology* (New York: Guilford Press, 1994), pp. 125–151.

26. Winona LaDuke, "A Society Based on Conquest Cannot Be Sustained: Native Peoples and the Environmental Crisis," in R. Hofrichter, *Toxic Struggles,* pp. 98–106.

27. Al Gedicks, "Racism and Resource Colonization," *Capitalism, Nature, Socialism, 5*(1), 1994, pp. 55–76.

28. J. O'Connor, 1988, "Capitalism, Nature, Socialism," p. 17.

29. David Harvey, *Justice, Nature and the Geography of Difference* (Cambridge, UK: Blackwell, 1996), pp. 410–411.

30. William Cronon, *Nature's Metropolis: Chicago and the Great West* (New York: Norton, 1991).

31. Louis Wade, *Chicago's Pride: The Stockyards, Packingtown and Environs in the Nineteenth Century* (Champaign: University of Illinois Press, 1987).

32. Foster, 1994, *The Vulnerable Planet,* p. 53.

33. Morton Horwitz, *The Transformation of American Law: 1780–1860* (Cambridge, MA: Harvard University Press, 1977), pp. 74–78.

34. R. Dale Grinder, "The Battle for Clean Air: The Smoke Problem in Post–Civil War America," in Martin Melosi, ed., *Pollution and Reform in American Cities, 1870–1930* (Austin: University of Texas Press, 1980), p. 91.

35. See, for example, Stuart Galishoff, "Triumph and Failure: The American Response to the Urban Water Supply Problem, 1860–1923," and Joel A. Tarr, James McCurley, and Terry F. Yosie, "The Development and Impact of Urban Wastewater Technology: Changing Concepts of Water Quality Control, 1850–1930," and other articles in M. Melosi, *Pollution and Reform in American Cities.*

36. Wade, *Chicago's Pride.*

37. Michael Smith, *City, State, and Market: The Political Economy of Urban Society* (Oxford, UK: Basil Blackwell, 1988); and Kenneth Jackson, *Crabgrass Frontier: The Suburbanization of the United States* (New York: Oxford University Press, 1985).

38. For an excellent account of how these forces affected one community, see Andrew Hurley, *Environmental Inequalities: Class, Race, and Industrial Pollution in Gary, Indiana, 1945–1980* (Chapel Hill: University of North Carolina Press, 1995).

39. David Harvey, *Conditions of Postmodernity* (Oxford, UK: Basil Blackwell, 1989), pp. 147–155.

40. "Fordism" has a number of characteristics, but I use the term here to refer to a method of production that mass produces goods in a factory-like setting, often involving an assembly-line type of process.

41. Sandra Postel, "Controlling Toxic Chemicals," in Lester Brown, ed., *State of the World 1988* (New York: Norton, 1988), pp. 118–136.

42. Alan Durning, "How Much Is Enough?," *World Watch, 3*(6), 1990, pp. 12–19.

43. Barry Commoner, *Making Peace with the Planet* (New York: Pantheon, 1990), pp. 44–45.

44. Ibid., p. 83.

45. Hurley, *Environmental Inequalities.*

46. Charles Cray, *Trash into Cash: Waste Management Inc.'s Environmental Crimes and Misdeeds* (Washington, DC: Greenpeace, 1991).

47. Hurley, *Environmental Inequalities,* p. 30–31.

48. U.S. EPA, *National Biennial RCRA Hazardous Waste Report,* EPA 530-R-92-027 (Washington, DC: EPA, February 1993).

49. Robert Lake and L. Disch, "Structural Constraints and Pluralist Contradictions in Hazardous Waste Regulation," *Environmental Planning, 24,* 1992, pp. 663–681.

50. Kirsten Engel, "Reconsidering the National Market in Solid Waste: Trade-Offs in Equity, Efficiency, Environmental Protection and State Autonomy," *North Carolina Law Review, 73,* 1995, pp. 1481–1560.

51. Barry Commoner, "The Hazards of Risk Assessment," *Columbia Journal of Environmental Law, 14*(2), 1989, pp. 365–378.

52. Lake, "Volunteers, NIMBYs, and Environmental Justice," pp. 166–167.

53. The Clean Air Act does restrict some operations if the area in which a facility is proposed has not attained the health-based standards in the Act. In Los Angeles, for example, which does not meet the ozone standards, a new facility may not be constructed unless certain reductions from other sources are available. This is an attempt to vary environmental regulation based upon the environmental conditions in the community. However, since, as mentioned above, the Clean Air Act has established these limits for only a limited number of "conventional" pollutants, this in no way restricts the construction of minor sources or other sources besides ozone producers.

54. Andrew Szasz, *EcoPopulism: Toxic Waste and the Movement for Environmental Justice* (Minneapolis: University of Minnesota Press, 1994), pp. 17–22.

55. There is some precedent, however, for moving toward pollution prevention. The state of Massachusetts, for example, has enacted a Toxic Use Reduction Act that has served as a model for other states. Under its provisions, a company must develop a comprehensive toxic use reduction plan. Technical assistance is provided by the state, but eventually the state is empowered to establish standards that must be achieved by industrial sectors. These steps, however small, move in the direction of greater socialization of at least some aspects of the production process. See Paulette Stenzel, "Toxics Use Reduction Legislation: An Important 'Next Step' after Right to Know," *Utah Law Review*, 1991, pp. 707–748.

56. Steven Epstein, "Democratic Science? AIDS Activism and the Contested Construction of Knowledge," *Socialist Review*, *21*(2), 1991, pp. 35–64.

57. There are, however, significant obstacles to rectifying the present deficiencies in the current scheme of environmental laws that can only be alluded to here. The U.S. political structure is extremely fluid, with different governmental functions being carried out at the local, state, and federal levels. Fragmentation of government both vertically (local, state, and federal) and horizontally (judicial, executive, and legislative) generally works to favor capital, which is well equipped to take advantage of any and all opportunities; see Charles Noble, *Liberalism at Work: The Rise and Fall of OSHA* (Philadelphia: Temple University Press, 1986). Moreover, in a capitalist economy, the authority of the government is always constrained at some level by the threat of capital flight and loss of business confidence, as has been recognized by state theorists; see Martin Carnoy, *The State and Political Theory* (Princeton, NJ: Princeton University Press, 1984).

58. These perspectives, it may be noted, parallel those identified by Alford and Friedland to describe the ways in which the state acts; that is, from the pluralist perspective, which highlights a logic of democracy (parallel to the "logic of fairness"); from a managerial perspective, which highlights a logic of bureaucracy (parallel to the "logic or risk"); and from a class perspective, which highlights a logic of capitalism (which parallels the "logic of capitalist production"). See Robert Alford and Roger Friedland, *Powers of Theory* (Cambridge, UK: Cambridge University Press, 1985).

59. Commoner, "Pollution Prevention," 1992, pp. 25–26.

Chapter 4

Environmental Justice from the Grassroots

Reflections on History, Gender, and Expertise

Giovanna Di Chiro

The environmental justice movement (EJM) has emerged within the past decade as a social movement in the United States that both challenges dominant discourses of environmentalism and produces new constructs of environmental theory and action. The language used by activists in the movement abounds with political conviction, dynamism, and hope. Phrases such as "transforming a movement," "reclaiming the landscape," "empowering ourselves," and "reshaping our communities" speak to the power and promise that motivate this historical moment of social movement building. The term "environmental justice," which first appeared in the United States sometime in the mid-1980s, problematizes popular notions of "environment" and "social justice" and discursively produces something different.

The vast majority of activists in the EJM are low-income women and predominantly women of color.[1] From the start, the gender, race, and class composition of movement activists distinguish it from that of the "mainstream"[2] environmental movement, whose constituents have historically been white and middle class. Women activists in the EJM, together with men in their communities, have begun to produce the conditions for social and environmental change, locally and nationwide, by reinventing socioenvironmental terms and definitions; standard narratives of social movement history; constructs of gender, race, and class politics; strategies for coalitions; and questions regarding the validity of scientific knowledge and expertise. Drawing heavily on personal interviews with women in leadership roles and a variety of movement literatures, I will discuss some of the critical issues and innovations articulated through the voices of the EJM.

"ENVIRONMENT" AND SOCIAL JUSTICE: RETHINKING THE TERMS OF DEBATE

The history of mainstream environmentalism locates its adherents within an ideological position that constructs a separation between humans and the "natural" world. Environmentalists, therefore, are often identified as being obsessed with preserving and protecting those "wild and natural" areas defined as the place where humans are not and *should* not be in large numbers. Social movement historians have occasionally referred to environmental justice activists as the "new environmentalists,"[3] a term that I find misleading. Many of the grassroots activists I have spoken to are reluctant to call themselves environmentalists at all, much less newly converted ones. In part, this is due to the dominance of the mainly white, middle-class, and uncritically "preservationist" political culture from which much of the mainstream environmental thinking has developed.[4] Again, in these mainstream terms, what gets to count as environment-related is limited to such issues as wildland preservation and endangered species protection. Issues pertaining to human health and survival, community and workplace poisoning, and economic sustainability are generally not considered to be an important part of the "environmental" agenda. Additionally, many activists perceive much of mainstream environmentalism to be either fixated on antiurban development campaigns (read as "no jobs for city-dwelling people") or utterly indifferent to the concerns of urban communities altogether. Many of the community organizations that comprise the EJM are located in low-income and working-class communities in and around industrialized urban centers throughout the country. Crucial issues in these communities include lead and asbestos poisoning in substandard housing, toxic waste incineration and dumping, and widespread unemployment. Until relatively recently, these are problems that the mainstream organizations have located outside the domain of "environment."[5]

Environmental justice activists define environment as "the place you work, the place you live, the place you play." Moreover, Dana Alston, director of the environment program at the Public Welfare Foundation, argues that environmental justice must be "seen through an overall framework of social, racial and economic justice, and the environment is just one piece in a whole linkage. . . . It calls for a total redefinition of terms and language to describe the conditions that people are facing and to come up with solutions."[6]

Many mainstream environmentalists would find this formulation incomprehensible, even ethically indefensible, because of its apparent anthropocentrism. Putting humans at the center of environmental discourse is erroneous, they argue, because humans are the perpetrators of environmental problems in the first place. Environmental justice activists maintain that some humans, especially the poor, are also the victims of environmental destruction and pollution, and that, furthermore, some human cultures live in relatively ecologically sound ways. Pam Tau Lee, the labor coordinator for the Labor and Occupational Health Program at the University of California, Berkeley, and a board member of the Southwest Organizing Network and the now defunct National Toxics Campaign Fund, describes environmental justice as being

> able to bring together different issues that used to be separate. If you're talking about lead and where people live, it used to be a housing struggle, if you're talking about poisoning on the job it used to be a labor struggle, people being sick from TB or occupational exposures used to be separate health issues, so environmental justice is able to bring together all of these different issues to create one movement that can really address what actually causes all of these phenomena to happen and gets to the root of the problems.[7]

The merging of social justice and environmental interests, therefore, assumes that *people* are an integral part of what should be understood as the "environment." Numerous studies have demonstrated that it is primarily low-income communities of color that are often targeted for industrial and toxic waste disposal sites.[8] Activists argued that this fact is a reflection of the people/nature dichotomy linked to the histories of racism and colonialism that associate poor and dark-skinned people with a "lowly" nature. Alston discusses how the EJM's redefinition of "nature" and "environment" to account for the presence of people reflects one of the primary discrepancies between it and the mainstream movement.

> The Nature Conservancy defines itself as the "real estate" arm of the environmental movement and as being about saving nature, pristine areas, sensitive ecosystems, endangered species, and rainforests. But the reality of the situation is that there is hardly anywhere in the world where there aren't people living, no matter how remote you get, and the most vulnerable cultures are in the areas that are most remote, whether you are talking about here in the U.S. or in Latin America or wherever, so immediately, it puts us in confrontation with the Nature Conservancy. We continue to raise these issues not only in the international arena but here as the Nature Conservancy goes to buy large tracts of land in New Mexico or out west where indigenous and Chicano people have lived for decades and have sovereignty or land-grant rights . . . with total disregard for how these real estate dealings affect the social, political, and economic life of our communities. We feel that many of these communities are just as much endangered species as any animal species.[9]

Consequently, activists in the EJM are unlikely to identify themselves as the "new environmentalists" because they do not view themselves as an outgrowth of the "old" environmental movement together with its "save the whales and rainforests" sloganeering. It would be more accurate to regard environmental justice activists as the "new" civil rights or "new" social justice activists, since many of the prominent organizers affirm their roots in and political continuities with the social justice movements of the 1960s, including the civil rights, welfare rights, and labor and farm worker movements. Moreover, the term "new environmentalists" suggests that the members of these emerging grassroots organizations, who predominantly come from African American, Latino, Native American, and Asian American communities, are only recently becoming interested in or aware of the importance of "environmental" concerns. Numerous histories of activism by people of color on environmental issues exist but often are not classified by mainstream groups as authentic "environmental history" because of these crucial questions of definition.[10]

According to Alston, the results of an exit poll conducted the day of the 1992 presidential election by one of the "Big Ten" environmental organizations were that

> in the white population, 28 percent said that the environment was one of the key factors that helped them determine who to vote for, but when it came to African Americans or Latino Americans it was close to 48 percent. Before, when you asked a black person what were the important issues they would say, "oh the air smells terrible, the water tastes terrible, my child has asthma, there's lead poisoning in the apartments," but they hadn't defined those as "environment" because how it was being defined by others was whales and ancient forests and national parks, etc. Now that gap in definition is being closed, but we always said that people of color were much more interested and invested in these issues than what was being said about us, that is, that "we're not interested, we're too busy surviving" even though the Black Congressional Caucus has had the best voting record on the environment in the past 20 years.[11]

The notion that these grassroots, community-based, social and racial justice-driven organizations are composed of "new environmentalists" is contested terrain. Questions of the importance of self-representation, definitional clarity, and the agency inherent in "speaking for ourselves" are key issues for movement activists.[12] What is "new" about the EJM is not the "elevated environmental consciousness" of its members but the ways that it is transforming the possibilities for fundamental social and environmental change through processes of redefinition, reinvention, and construction of innovative political and cultural discourses and practices. This includes, among other things, the articulation of the concepts of "environmental justice" and "environmental racism" and the forging of new forms of grassroots political organization. I will illustrate some of these new conceptual inventions by examining a few key historical moments that have defined the EJM.

REVISIONING ENVIRONMENTAL HISTORY: THE POLITICS OF ENVIRONMENTAL RACISM

Two landmark events are represented by activists and movement historians as crucial defining moments of the EJM. One event was the highly publicized struggle in the late 1970s at Love Canal in upstate New York, where community residents fought a successful battle against a large chemical company, forcing it to be accountable for its role in contaminating the local environment with hazardous wastes.[13] The Love Canal struggle is often described as the origin of the "antitoxics" movement in the United States and as instigating the formation by Lois Gibbs of the national grassroots organization, the Citizen's Clearinghouse for Hazardous Waste (CCHW, now the Center for Health, Environment, and Justice). The Love Canal episode also sent the message to potential polluters that low-income communities would no longer passively bear the brunt of toxic waste disposal.

The EJM is defined as different from, or, in some cases, as having evolved from, antitoxics groups.[14] Antitoxics groups, often called NIMBYs (not in my backyard), are usually characterized as undertaking a single issue and engaging in very provincial actions that are meant to begin and end with the immediate, local crisis (often they don't end up remaining local, however, as Lois Gibbs and the Love Canal episode demonstrate). Antitoxics groups do not produce a well-developed critique of the status quo and usually do not appeal to a long history of oppression. They may identify themselves as the unfair targets of big business or corrupt government activities but do not engage in identity politics and so do not articulate themselves as a "class" of oppressed people. Although they tend not to trust local government bureaucracy, antitoxics groups also tend not to have a "question authority" outlook and are usually utterly shocked to learn that their public officials and other experts can be so indifferent to the health and well-being of their neighborhoods. EJM groups, on the other hand, are seen as forging coalitions among diverse disenfranchised communities and engaging in discursive politics, that is, in making claims for the symbolic power of the meanings of such ideas as "nature," "environment," "justice," and "expertise." They also develop strategies and methodologies for changing consciousness and the material realities of their lives by constructing a political culture that is committed to staying locally based, yet always making connections to broader issues. EJM activists often sustain an explicit critique of capitalism and its destructive effects on nonhuman and human systems. EJM activists utilize powerful historical narratives and see themselves as being the descendants of the civil rights, workers' rights, and indigenous rights movements of the past. Many activists I interviewed, if they had not been politically active before, were now strongly committed to a long-term struggle and situated themselves within an "imagined community" or "imagined world" of previously voiceless groups of people making historical changes.

A second major event in the history of the EJM occurred in Warren County, North Carolina, in 1982 and marked the emergence of the organized struggle against racist practices in hazardous waste management in the United States. At this large-scale demonstration of civil disobedience, hundreds of predominantly African American women and children, and also local white residents, used their bodies to prevent trucks loaded with poisonous, PCB-laced dirt from being dumped in their community. The mainly African American, working-class, rural communities of Warren County had been targeted as the dumping site for a toxic waste landfill that would serve industries throughout North Carolina. This demonstration of nonviolent civil disobedience opened up the gates for a series of subsequent actions by people of color and poor people throughout the country. Unlike social activism against toxic contamination that predated this particular event, such as the struggle against Hooker Chemical Company at Love Canal, this action began to forge the connections between race, poverty, and the environmental consequences of capitalism's industrial waste problems. The Warren County episode succeeded in racializing the antitoxics agenda and catalyzed a series of scientific studies that would document the historical patterns of the disproportionate targeting of racial minority communities for toxic waste contamination.

A year after the Warren County protest, Washington, DC, delegate to the House of Representatives Walter Fauntroy pursued an investigation, under the auspices of the U.S. General Accounting Office, into the siting of hazardous waste facilities in eight southern states. The results of this federal study produced strong evidence indicating that African American communities are disproportionately burdened with the dangers of industrial pollution in the southern United States.[15] It was not, however, until the publication of a landmark report sponsored by the United Church of Christ Commission for Racial Justice (UCC-CRJ) in 1987 that an awareness of the widespread existence of "environmental racism" entered into the mainstream political consciousness.

The UCC-CRJ report, *Toxic Wastes and Race in the United States: A National Report on the Racial and Socioeconomic Characteristics of Communities with Hazardous Waste Sites*, compiled the results of a national study that indicates race as the leading factor in the location of commercial hazardous waste facilities. The study, presented to the National Press Club in Washington, DC, that same year, found that people of color suffered a "disproportionate risk" to the health of their families and their environments, with 60 percent of African American and Latino communities and over 50 percent of Asian/Pacific Islanders and Native Americans living in areas with one or more uncontrolled toxic waste sites. The report also disclosed that 40 percent of the nation's toxic landfill capacity is concentrated in three communities—in Emelle, Alabama (with a 78.9 percent African American population), Scotlandville, Louisiana (93 percent African American), and Kettleman City, California (78.4 percent Latino).[16]

The term "environmental racism" entered into political discussion on the environment in 1987 when the Rev. Benjamin Chavis, then the commission's executive director, formulated the term. According to Chavis, environmental racism is "racial discrimination in environmental policy-making and the enforcement of regulations and laws, the deliberate targeting of people of color communities for toxic waste facilities, the official sanctioning of the life-threatening presence of poisons and pollutants in our communities, and history of excluding people of color from leadership in the environmental movement."[17] In the mid- to late 1980s, this process of naming and researching the material realities of environmental racism made possible a significant transformation in what would count as properly "environmental" concerns. This new political concept also provided an organizing tool that could function to galvanize into action the diverse communities and constituencies for whom "environmental racism" was a painful reality.

How did the appearance of the UCC-CRJ report on toxics and race and public acknowledgment of environmental racism affect the national environmental agenda? By 1990 a variety of coalitions of people of color environmental justice organizations had emerged, including the extremely dynamic Southwest Network for Economic and Environmental Justice (SNEEJ).[18] In January and March of that year, representatives from many of these grassroots coalitions composed two recriminating letters to the "Group of Ten" national environmental organizations, "calling on them to dialogue on the environmental crisis impacting communities of color, and to hire people of color on their staffs and boards of directors."[19] The letters presented an analysis of environmental racism and defined the ways that the primarily white, mainstream organizations had complicitly supported it.

> There is a clear lack of accountability by the Group of Ten environmental organizations towards Third World communities in the Southwest, in the U.S. as a whole and internationally. Your organizations continue to support and promote policies which emphasize the clean-up and preservation of the environment on the backs of working people in general and people of color in particular. In the name of eliminating environmental hazards at any cost, across the country industrial and other economic activities which employ us are being shut down, curtailed or prevented while our survival needs and cultures are ignored. We suffer from the results of these actions, but are never full participants in the decision-making which leads to them.[20]

According to the activists with whom I have spoken, responses to these challenges have been varied. At worst, some of the Big Ten have expressed outrage and denial and have all but ignored the invitation to "come to the table as equals." On the other hand, some have begun to enter into discussions about building "multicultural and multiracial organizations," to share resources such as technical expertise, legal assistance, and funding, and to modify their organization's structure and mission in significant ways.

Some promising signs of multiracial coalition building include the series of "Great Louisiana Toxics Marches" initiated in 1988 by the Gulf Coast Tenant's Project in New Orleans. These massive demonstrations (some lasting up to 11 days and spanning a distance of 100 miles), sent marchers through Louisiana's "Cancer Alley" and were organized by a diverse group of grassroots civil rights, church, labor, and tenants' organizations, together with a number of national bodies such as the Sierra Club and Greenpeace. "The 'toxics marches' have helped put race on the environmental agenda and to put the environment on Third World communities' agenda."[21]

In the San Francisco Bay area, another coalition between mainstream environmentalists and organizations of people of color has recently formed. EDGE, the Alliance of Ethnic and Environmental Organizations, provides a forum for multiracial dialogue around issues of sound economic development in light of California's rapidly changing demographics and ethnic composition. EDGE's first conference, "Redefining the California Dream: Growth, Justice and Sustainability," convened in early 1993 and brought together groups as varied as the Japanese American Citizens League, the Environmental Defense Fund, the Latino Issues Forum, and Citizens for a Better Environment.[22]

Further south, in Los Angeles, a multiracial coalition of women community activists and scholars has recently organized the Center for Community Action and Environmental Justice (CCAEJ). This organization's mission is to aid in the creation of "partnerships" among diverse communities interested in linking issues of environmental justice, social justice, and economic development. Members work with local women's health groups, forging connections between the high incidence of breast cancer in African American women and environmental toxins, with the L.A.-based Philippine Action Group for the Environment, and various Third World struggles against the international hazardous waste trade in Asia and the Pacific Islands, and with racially diverse rural communities on building strong organizational skills to combat issues such as unsafe pesticide use and unwanted sewage sludge dumping.[23]

The Highlander Center in Newmarket, Tennessee, has for a number of years instituted STP (Stop the Pollution) Schools, which bring together grassroots activists from poor and racially diverse communities to share resources and strategies aimed at confronting environmental racism in struggles against toxic dumping, lead poisoning, and the destruction of local forests in the paper mill industry. Their project has recently expanded to include "grassroots exchanges" for environmental justice on an international scale. Activists from Kenya, Indonesia, Nicaragua, the Czech Republic,

Mexico, and India have met with activists from around the United States to discuss the global aspects of community environmental issues.

Such examples provide evidence that the EJM has intervened, at least to some extent, in establishing the importance of race and class in organizing for truly effective environmental change. Most of the women activists I interviewed, however, were still cautiously watching to see what may come of these changes. They are particularly suspicious and critical of the ways that some Big Ten organizations are choosing to respond to the challenges put forth by the movement. Alston explains that

> some now see that there are a lot of organizations that have made [environmental justice] a priority, and you see these same organizations going to set up programs. . . . Like the National Wildlife Federation has a grant proposal out for 1.4 million dollars to work with people of color . . . and they are notorious for having terrible relationships with people of color. They have Waste Management, Inc., on their board of directors which engages in supreme environmental racism. They will dump some of the most hazardous materials known in people of color's communities. To watch NWF raising all this money to deal with people of color is very difficult to see.[24]

Similarly, Lois Gibbs, director of the Citizen's Clearinghouse for Hazardous Waste (now the Center for Health, Environment, and Justice), argues that

> some of these "Big Ten" groups have put in proposals to the tune of 1.5 million dollars to do community organizing and outreach to low-income community groups, and people of color groups. I look at that and I know that these people are adopting the rhetoric for foundation purposes. They use it in their newsletters and their PR pieces, but all they're doing is taking away from support groups that could help them, whether it's the Southwest Research Organization or the Southern Organizing Committee, or Greenpeace, these are organizations that could help and there's only a small pool of money. Now these big bluechip organizations are coming in and foundations are saying, "Oh, good they've been around for a hundred years and now we don't have to take the risk." If media people have a choice of tracking down the Southern Organizing Committee or the Environmental Defense Fund, you know who's in their Rolodex. I find it very frustrating because we've built this wonderful movement and it's almost like being victimized one more time.[25]

Diane Takvorian, director of San Diego's Environmental Health Coalition, discusses her attempt to apply for a grant to work on technology transfer of "Right to Know" legislation for an international collaborative project with environmental justice groups in Tijuana, Mexico:

> So this funder says to me, "we just gave this grant (a huge one) to the WWF, they have this great international program and they're trying to

> do that kind of work too, and so maybe you could give them a call and help them out," and I thought, "maybe you could give us a grant." I think there is the level of trust of the bigger, established organizations. Maybe they could have a broader impact than we could, but who are they really working with and for? Both the NWF and the WWF have gone on record in favor of NAFTA, so how do you square that with trying to help communities in Mexico that are inundated with companies that are spewing toxic chemicals in their rivers, groundwater, and the air?[26]

These activists' criticisms of the unacknowledged paternalism inherent in the "Big Ten's" approach to dealing with the issues of environmental justice illustrate the importance of the movement's insistence on self-representation and self-definition, best signified by the movement phrase "We speak for ourselves."[27] This unequivocal rejection of a "partnership based on paternalism" with the mainstream environmental movement explains, in part, the overwhelming enthusiasm surrounding the First National People of Color Environmental Leadership Summit, which convened in Washington, DC, in October 1991.

PRACTICING ENVIRONMENTAL JUSTICE: LEADERSHIP AND DIVERSITY FROM THE GRASSROOTS

The First National People of Color Environmental Leadership Summit signified a watershed moment in the history of the movement. According to conference participants, this event helped to establish the importance of people-of-color environmental groups' insistence on self-representation and "speaking for ourselves."[28] It also represented an unequivocal rejection of a "partnership based on paternalism" with the mainstream environmental movement.

The summit brought together 300 African American, Native American, Latino American, and Asian American delegates from the United States and a number of conferees from Canada, Central and South America, Puerto Rico, and the Marshall Islands to frame the contours of a "multiracial movement for change" founded on the political ideology of working from the grassroots. Conference participants heard testimonies and reports on the local effects of environmental racism, including the extensive poisoning of air, water, and land that disproportionately devastates their environments and health. These discussions also provided a supportive context for people of color to "reaffirm their traditional connection to and respect for the natural world," which was collectively understood as "including all aspects of daily life." Environment so defined expands the terms of environmental problem definition and so includes issues such as "militarism and defense, religious freedom and cultural survival, energy and sustainable development, transportation and housing, land and sovereignty rights, self-determination and employment."[29] Dana Alston describes how the Leadership Summit helped to bring people of color together in a spirit of political solidarity.

> The most important thing that came out of the Summit was the bonding. Many people might think that because they're nonwhite, that they're going to come together, but the society is built on keeping people divided and we all know about the tensions between African Americans and Asian Americans and Latinos and Native Americans but it's the history, the culture, the society that's keeping us divided . . . because that's how the power structure stays in power, by keeping us separate, so we had to from the very beginning put together a set of principles from which we were going to relate to each other.[30]

The composition and program of the second day of the Leadership Summit shifted with the arrival of another 250 participants from a variety of environmental and social change organizations together with a sampling of "professionals" including lawyers, academics, and policy makers. Engaging in critical discussions and debates, the conferees articulated key issues of EJM building, including the definition of environment and environmental problems, leadership and organizational strategy, and the formation of coalitions and partnerships. Working by consensus, the Leadership Summit drew up a set of 17 organizational principles that would guide the emergent political process. These "Principles of Environmental Justice" describe a broad and deep political project to pursue environmental justice in order to "secure our political, economic and cultural liberation that has been denied for over 500 years of colonization and oppression, resulting in the poisoning of our communities and land and the genocide of our peoples."

All of the activists with whom I have spoken maintain that the most significant achievement of the Leadership Summit was its commitment to the construction of diverse egalitarian and nonhierarchical leadership and organizational processes and structures. In contrast to the technocratic rationality and top-down managerialism that the mainstream environmental organizations have adopted by mimicking the decision-making approaches of the very corporations that they are opposing, the participants at the Leadership Summit insisted on something different. As grassroots activists working in direct response to the threats of pollution, resource exploitation, and land-use decisions in their communities, they contend that the decision-making process is itself a primary issue in the debate over environmental problems. They reject the top-down approach as disempowering, paternalistic, and exclusive and instead are committed to developing a more democratic, locally and regionally based, decentralized organizational culture. A commitment to such values, they argue, will build an environmental movement that truly works. Alston remarks:

> I think that those of us who study the history of social movements have learned so much from other movements that we made a commitment to spend the next two years building local and inter-regional structures and to strengthen those, and to then come back together and see. . . . We didn't want one person to emerge as the "spokesperson" because we have worked too long and too hard to have the bonds between us destroyed. The media and the EPA were pushing for this spokesperson, so it's been a real struggle. I'm really glad that we had made that decision.[31]

An interactive environmental justice *network* has emerged as the most identifiable structure organizing the movement. Pam Tau Lee describes this strategy:

> What I see is a wonderful phenomenon . . . the development of networks. Those networks are based on actual work that is coming out of the grassroots. You've got the Southwest Network, forty organizations that are doing real live work, who have joined a network to create "a net that works," as Richard Moore puts it. And you've got another one in the southeast, as the New Orleans Conference on Labor and the Environment showed, real live grassroots organizations coming in and starting from a strong foundation and building up. There was just a meeting in Chicago of the midwest organizations, so the approach is, I feel, one that's going to win. It's a winning strategy, the top-down strategy is not a winning strategy.[32]

Penny Newman, codirector of the Center for Community Action and Environmental Justice (CCAEJ), agrees with the "network" formulation used to represent the organizational development of the EJM. However, she prefers terms such as "fragile coalition" or "diffuse" in her descriptions—phrases that connote much more fluidity and ambiguity.

> It's really a very diverse grouping of people, amazingly so. I think that on many other issues they wouldn't all agree, so this is a very fragile coalition. Right now, there's not a real strong, formal cohesion in the movement, but there's a network that's very much there and it can act together, which is what I think really defines a movement—it's ability to be able to act together. I think it's still developing, more and more people are understanding the issues, but it frightens me that it's also getting pushed a bit.[33]

For Newman, and other activists, the EJM's strength is located in its community-centered, face-to-face organizing practices, and in its history of "breaking the traditional organizing rules" that the mainstream groups have inherited. The loosely aggregated network that remains grounded at the community level is, for many EJM activists, the political culture that they would like to see encouraged. In addition to SNEEJ in the southwest, and the Southern Organizing Committee in the south, both well-established regional bodies, other networks are in the process of formation. For example, the Indigenous Environmental Network and the Asian Pacific Environmental Network have assembled since the early 1990s, and the Northeast region convened its first organizational meeting in 1994.[34]

The activists I interviewed noted the tension between the desire to remain faithful to the imperatives of direct democracy by working locally and on a small scale, versus the inclination to centralize and "go national" in order to gain more clout or to look more like a movement. The parochialism of many grassroots groups and the political limitations of failing to link local struggles to national and global ones have circulated as internal criticisms of the EJM.[35] This tension, together with the historical problems of political coalition building across class, racial, and ethnic differences will, according to some movement activists, be the primary challenges to the ongoing effectiveness of the movement.

A significant issue relating to organizational strategy and composition, often ignored or relegated to secondary status, is that of gender relations and the position of women in the movement. As in many environmental and social justice movements in the United States and internationally, women comprise the majority of "people taking substantive action" in their communities and local organizations. According to numerous accounts by activists, women make up the vast majority of the active participants in the EJM, yet they are not well represented in highly visible leadership roles.[36] This situation indicates one of the key challenges for the EJM that has emerged from the summit's principle of democratic and diverse leadership—namely, how simultaneously to address issues of gender, race, and class in shaping new forms of leadership and organization in this new movement for social and environmental change.

WOMEN AND ENVIRONMENTAL JUSTICE: CHALLENGING THE OLD AND SHAPING THE "NEW" ENVIRONMENTAL EXPERTISE

The apparent historical pervasiveness of the high involvement of women in movements and struggles for social and environmental change has inspired a spate of academic and feminist theorizing. Much of this discussion revolves around the question of the extent to which women's observed connection to nature and to environmental concerns has essentially biological or social origins. The biological explanation contends that women have an innate knowledge of the interconnectedness and value of all life on earth owing to their reproductive capacities. Consequently, their experiences of pregnancy and childbirth compel them to fight for life-affirming policies and clean, healthy environments as mothers in defense of their children and families. This position is criticized as being overly "essentialist"[37] and deterministic by those theorists who argue that women's heroic struggles for social and environmental justice must be understood as emerging from the gendered sociocultural roles they have been expected to fulfill as a result of patriarchal oppression. Since these gendered roles of "mother" and "nurturer" have included child-rearing, food production, and the overall responsibility for the health and survival of the community, women's social location affords them specific knowledge and investment in issues of the "environment." Both of these arguments, and multiple combinations of the two, are represented in numerous ecofeminist writings and actions that have appeared since the late 1970s.[38] How are women activists in the EJM situated in these "ecofeminist" debates around gender and the environment?

AN "UNMARKED" WOMEN'S MOVEMENT

Many ecofeminist writings construct theories as to why women would organize *as women* in their struggles for socioenvironmental change. Such theories, as I mentioned earlier, suggest that women possess unique knowledge about the connections between human health and survival, the environment, and their ever-increasing destruction by the "capitalist–militarist–patriarchal complex." These theorists claim that by virtue of this "innate" or experiential knowledge, women come together in political solidarity.[39] "Womanist" or "motherist" organization around issues of militarism and the material conditions of survival is not something new in the United States. However, the late 1970s and early 1980s marked a particular historical moment in which many, predominantly white, middle-class women explicitly linked feminist and environmentalist concerns. The proliferation of the nuclear industry, specifically the Three Mile Island crisis, prompted women in the United States to come together in 1980 at a conference in Amherst, Massachusetts, entitled "Women and Life on Earth: Ecofeminism in the 1980's." Ynestra King, one of the organizers of the conference and a prominent theorist and activist in the movement, declared in the opening address: "We're here to say the word 'ecology' and announce that for us as feminists it's a political word—that it stands against the economics of the destroyers and the pathology of racist hatred. It's a way of being, which understands that there are connections between all living things and that indeed we women are the fact and flesh of connectedness."[40] The conference catalyzed a series of now famous nonviolent direct actions, including the Women's Pentagon Actions of 1980 and 1981, the Seneca Women's Peace Encampment in 1983, and the Mother's and Others' Day action at the Nevada Test Site in 1987.

During those same years, other women in the United States, not specifically identifying themselves as feminists or even activists per se, organized around issues of the environment and the survival of their communities. At the same time that "ecofeminists" convened the "Women and Life on Earth" conference, Lois Gibbs was waging her now famous battle against the Hooker Chemical Company (Occidental Petroleum Corporation), which had dumped 20,000 tons of toxic waste on neighborhoods near the Love Canal in upstate New York. It was also the same year that Penny Newman was leading the battle against the state of California and a group of private corporations, including McDonnell Douglas and Rockwell International, for siting and dumping over 34 million tons of carcinogenic hazardous chemicals in the Stringfellow Acid Pits, a state-licensed disposal site (now California's top priority Superfund site) overlooking the rural community of Glen Avon in southern California. In the early 1980s, while thousands of women, predominantly white and middle-class, were encircling the Pentagon, weaving symbolic webs of containment made of yarn, flowers, and children's photos through its barbed wire fences, Cora Tucker, a longtime civil rights activist and resident of rural Halifax County, Virginia, was organizing against a local uranium mining project and the proposed siting of a high-level nuclear waste repository near her community.

These stories of women's "environmental" activism represent very different types of organization and strategy, ranging from national, large-scale, often symbolic demonstrations and direct action all the way to local, community-based, networking-style activism. At the time, the continuities among these different types of women-led actions for the "environment" were not obvious (although Gibbs did present a speech to the Amherst conference), even though they showed evidence of the existence of different women struggling for many of the same socioenvironmental issues. In the 1990s, there is no visible, active, explicitly "ecofeminist" movement in the United States. Instead, ecofeminist efforts to theorize and strategize around a "women and environment" connection have remained almost exclusively within the realm of the production of theory. On the other hand, the movement for environmental justice, widely recognized as being driven and energized by women, yet not marked as a "women and environment" movement, continues to expand and develop its strategies, organization, and commitment to the grassroots on a national and international scale.[41]

I want to consider some implications for social movement theory and strategy that an understanding of the gendered nature of the EJM may suggest. To get at this question, it is important to look at the different struggles over environmental resources and meanings that these women undertake in this "unmarked"[42] women's movement. Women active in the EJM contest and redefine discourses and practices of not only environmentalism but also of gender, racial, and class stereotyping. They also question and reconstruct the concepts of "objectivity" and the "validity" of scientific expertise.

Many women in the movement evoke deep concerns about the health and future survival of their children and communities when explaining their initial or continued involvement in fighting for environmental justice. The identity and experience of being a "mother," and the outrage at watching local corporations and government officials exhibiting total disregard for the lives of their children, have significantly motivated many women to become politically active. Lois Gibbs explains her distress at the serious illness of her one-year-old son and the rare blood disease that her young daughter developed when her family moved to Love Canal:

> It was like what the hell's going on here, I did everything right, I prided myself in being a responsible mother, and then I found out about the dump. What really got me involved was the realization that my community and my family, specifically, could be sacrificed and that people who were in charge, whether they were health authorities or local government or state or federal authorities, all knew that my family was being poisoned and they still made a conscious decision that it was ok because of the cost involved in cleaning up. That just totally flipped me out, because I had always believed what was in your high school civics books. It was my children who motivated me clearly. I just got really outraged, nobody can say that there is a price on my children's heads, nobody can say that based on cost–benefit analysis, risk–benefit analysis, or whatever weird stuff they use, that they can justify the killing, the murdering of people.[43]

Penny Newman, in a moment of painful irony, has suggested that "one of the siting criteria for hazardous waste facilities is that they be within one mile of a school." She speaks of the "anger at having somebody else making decisions that affected my children, that the responsibility as a parent was taken away from me" as the factor that incited her antitoxics activism.[44]

Activists in the Mothers of East Los Angeles (MELA), a Mexican American community organization founded in 1985 to fight for a wide variety of community "quality of life" issues, assert that initially they did not consider themselves political but were "compelled to unite because the future quality of life for our children is being threatened. . . . If one of her children's safety is jeopardized, the mother turns into a lioness."[45] In MELA, however, the symbolic identity of "mother" is extended to women who are not biological mothers and to men in the community.[46] In other words, for the Mexicanas who make up the majority of MELA activists, the concept "mother" stands in for the protector of the community, the family, La Raza, no matter which gender. Founder Juana Gutierrez argues that love for her children, family, community, and La Raza, all social concepts that are indistinguishable in her discourse, motivate her to continue fighting against environmental injustice.[47]

Although most of the women in the EJM will, to some degree, assert that they are acting on behalf of the well-being of their children, their identity as simply "mothers" is by no means always the central focus of their activism. Cora Tucker, who works for environmental justice in her organization, Citizens for a Better America, explains that women's activism reflects the fact that women are the first people to act on "bread and butter issues."[48]

> People don't get all the connections. They say the environment is over here, the civil rights group is over there, the women's group is over there, and the other groups are here. . . . They say, "now Miss Tucker, what you really need to do is go back to food stamps and welfare, environmental issues are not your problem." And I said to him, "Toxic wastes, they don't know that I'm black." . . . [Most white people] say that black people are only interested in bread and butter issues. But nothing in the world is more bread and butter than clean air to breathe or having good water to drink.[49]

The question of community survival in the face of cultural imperialist attacks by the dominant, white, male, industrial complex figures conspicuously in the involvement of many women of color in environmental justice work. Women in many communities and cultures have customarily been seen to be the repositories of, or given the responsibilities for maintaining, local cultural traditions and histories. Janice Dickerson, a former resident of the now relocated African American community of Reveilletown, Louisiana, and current director of the Gulf Coast Tenants Association, speaks to this issue of cultural survival:

> I think that Reveilletown might be a focal point for a lot of [black people] to start looking and begin to realize what can happen to our history. I mean we lost our history once when we were shipped to America, and here we are, we've been settled here for over a hundred years and we are uprooted again, and we are uprooted by basically the same culprit, white American capitalists. So the cycle has to stop, somebody needs to develop an interest.[50]

Reveilletown, a post-Civil War community founded by ex-slaves, had, until recently, sustained its residents through an agricultural economy based on sugarcane production. The residents of the community were poisoned by vinyl chloride emissions produced by the Georgia Gulf petrochemical company, and were relocated in an out-of-court settlement in 1989.[51] For Janice Dickerson, therefore, the environmental consequences of Georgia Gulf's activities destroyed not only the health of many people of Reveilletown (as well as its land, air, and water) but also the integrity of a particular African American culture's way of life. In explaining her current commitment to fighting for the socioenvironmental rights of her community, Dickerson continued:

> The petrochemical industry in Louisiana is not only destroying the health of the people, it's destroying the environment, the air and the water, it's destroying the quality of life that people were accustomed to in the area, what we are talking about is just plain old survival for everyday people. From the perspective of the African American, it's a civil rights matter. Civil rights and the environmental movement are both interwoven because again we are the most victimized. There is no difference from a petrochemical industry located 2–3 hundred feet from my house and killing me off, than there is when the Klan was on the rampage just running into black neighborhoods hanging black people at will. I feel at this point in my life, there's so many basic issues that need to be confronted and dealt with from the basic survival of the African American along the petrochemical line here in Louisiana, that you can't really do it on a part-time basis. There are so many issues that you need to deal with, somebody needs to get out and do it full time, so I've decided to do this with my life.[52]

For many women-of-color activists, characterizing the EJM as a movement of women is not adequate, or, rather, it does not accurately describe the composition of the movement. The importance of the concept of "community," which includes men, women, children, friends, neighbors, and relatives, is often cited as the most appropriate locus of political concern. Identifying the leadership as "women" serves to isolate them and produces a rift in the social fabric of the community, which is the place where their loyalties lie. For whatever reason women in this movement make the decision to become politically active for environmental justice, they often find that their work makes them visible or identified as "women," and often *eccentric* women, whether they like it or not. Cora Tucker was accused of being a "hysterical housewife" as she lobbied the state legislature in North Carolina as an activist for voting rights in the early civil rights movement. Tucker and other women who are considered "oddities in their communities" have self-consciously appropriated the derisive and sexist slur of "hysterical housewife" as a political term and have redefined it as a powerful identity.

> I've learned that's a tactic men use to keep us in our place. So when I started this stuff on toxic waste and nuclear waste, I went back to the General Assembly . . . and I said, "You're exactly right. We're hysterical, and when it comes to matters of life and death, especially mine, I get hysterical." . . . If men don't get hysterical, there's something wrong with them.[53]

Newman contends that the women in her organization realized that "emotion was the strength" that they had on their side, and they translated offensive terms like "hysterical," "irrational," and "ignorant" to mean rationality and reasonableness in the face of genuine risks to the survival of their communities. This epistemological standpoint constructs knowledge through a translation practice. The women activists must translate, for themselves, the terms used to undermine them. What counts as rational knowledge, for these activists, relates to a strong sense of the need to consider the purposes for which the knowledge is enlisted.

Tucker, Newman, and other activists in the EJM recount that when they first enter the political arena they are confronted with the perception of themselves as being women outside of their "natural element" as mothers, housewives, ladies, and so on. Tucker expresses this experience clearly:

> If you're out of the norm they say, "what's wrong with this woman?" They think you're crazy. Most of your mommas would never have gotten up at a board meeting and say anything about toxic waste because they were trained that "ladies" didn't act that way. Ladies don't take on an issue. I don't know if "lady" is a compliment or not. I really don't like to be called a lady because my momma used to tell me that a lady was a woman who didn't know which way was up.[54]

Newman recounts the response to her as a woman acting outside what were considered acceptable roles for women:

> I had been involved in the community in civic organizations like the Junior League Women's Club and the P.T.A., but never considered myself an activist. The first time a newspaper referred to me as an activist people acted as if they had called me a communist, "she is not!" they yelled. Before Stringfellow, I had been active in acceptable kinds of activities, and had participated in the permissible ways that women were able to express their concerns, and to use their skills.[55]

Takvorian points to an interesting contradiction in the popular and academic representations of antitoxics and environmental justice women activists. Many of the women (who outnumber men eight to one) in her organization do not have children and can be considered "nontraditional." She argues that the standard depiction of the "naive, yet feisty, newly political housewife campaigning for her children" is a limited and often inaccurate representation.[56] Moreover, Takvorian contends that it is an essentializing and trivializing move to suggest that women come to environmental justice organizing because they are "naturally" caregivers:

> Women don't operate in very traditional ways when they operate as policy advocates and community activists, like Robin Cannon or Laura Hunter who runs our clean bay campaign, when they go out there to fight the power structure which is male. The all male port district authorities find Laura to be interfering, and they say that to her, "why are you here?" To say that somehow being in this role just comes

> naturally, I mean, nothing could be further from the truth. I mean, this is certainly not the role that my parents had in mind for me.[57]

Traditional notions of gender, race, and class are confronted and reexamined by environmental justice activists even though most do not explicitly take on a feminist agenda. By showing up in places considered by the dominant culture to be "unnatural" for women of color and low-income white women, they challenge the limitations of oppressive stereotypes. When these women assume leadership positions in the community and demand changes in family expectations and responsibilities as they "leave the house and enter the trenches," they break down traditional notions of gender, race, and class and construct new empowered identities. Newman recounts the story of a woman who joined "Concerned Neighbors" to fight for the cleanup of the Stringfellow toxic dump, yet ended up gaining much more in other aspects of her life:

> She is a young Hispanic woman who had never been involved in anything in the community. Getting involved in Stringfellow and having to speak in front of people, and to the authorities, and tell what she knew and how it impacted her kids etc., has given her a lot of confidence to challenge other things in her life. She's now gotten out of an abusive relationship, she's left her husband, and she never would have been able to stand up to him if it hadn't been for the experience. Those are what I see as the real successes of this. Yes, we've been able to stop the exposures and done a lot that way, but it's just as important that she was able to stand up to her husband who would have killed her if it weren't Stringfellow killing her. He was just as real a danger.[58]

A number of activists have identified a critique of gender and sexism as a central organizing skill and have produced a handbook, *Empowering Ourselves: Women and Toxics Organizing,* that addresses issues of community organization that specifically pertain to women's experiences of gender.[59] Newman has specifically formulated an analysis that integrates gender, race, and class issues in her longtime activism for environmental justice. Her work incorporates systematic documentation of the gendered and racialized nature of both the effects of environmental pollution and the community-based struggles against it.[60]

LOCAL KNOWLEDGE AND SCIENTIFIC EXPERTISE: POPULAR CRITIQUES OF ENVIRONMENTAL SCIENCE

Women activists in the movement for environmental justice not only produce new understandings of the relationships of gender and environment, they also, in the process, challenge government and corporate power structures and notions of "professionalism" and scientific expert knowledge. By confronting traditional boundaries between professional and scientific expertise and community-based experiential knowledge, they call into question standard research protocols and environmental policy analyses. In some instances, the arguments and the critiques of scientific research mobilized by women activists have worked strategically to raise suspicion in an "expert's" belief in his or her own knowledge claims. For example, Alston discusses the level of impact that she and others were able to have on the new head of the National Institutes of Health's (NIH's) environmental health division.

> The NIH tests one chemical at a time and determines whether it is carcinogenic, but our case to him was that we were contaminated by multisources, lots of different toxics and that this type of testing really doesn't speak to this. . . . At first he was very rigid like the rest of the scientists are, and he kept saying that there's no protocol, there's no model. We kept saying but that's the point, we need for you to go back to basics. . . . About two months later, he had gone down to Cancer Alley; he was sitting in this woman's kitchen and he looked out the front door and saw this one plant spewing out, then he looked out another window and he saw something else spewing out, and then he looked out the back window and there was something else from a different plant. So it was so clear to him that the one chemical exposure model was just ridiculous. Now he says that they have to go back to the very basic science that they were at 25 years ago when they were trying to figure out how to test for one chemical. There's been a lot of resistance within the agency.[61]

Women environmental justice activists, therefore, engage in serious critiques of science in their challenges of the value neutrality of scientific evidence. Newman has gotten firsthand knowledge about the supposedly "objective" technocratic rationality that governs much decision making related to assessing the risks of hazardous substances. She details her critique of the "environmental" discourse of "acceptable risk."

> When we allow discussions about an "acceptable risk" of 1 in 1,000 or 1 in 10,000 we are accepting that it is all right to kill 1 person in every 1,000 or 10,000. We have allowed the premise to be that it is all right for an additional person to die so that a facility can operate. These calculations are made for each individual chemical under perfect operating conditions. No calculations are made for the effects of people being exposed to two or more chemicals simultaneously, and of course the "kill rate" increases during accidents or "illegal" discharges. The law permits corporations to kill as long as they stay within set limits.[62]

In the process of her research on these "scientifically" devised industrial standards, Newman discovered that women and children are not considered in calculations of "acceptable risk." She learned that these calculations are determined by the CDC and other local health departments and "are based on occupational exposures for healthy males working an 8-hour day, 5 days a week, wearing safety equipment in controlled settings." In many low-income communities, specifically communities of color located near toxic sites, "children, pregnant women and the elderly are often exposed for 24 hours a day, 7 days a week with no protective clothing." Newman's political strategy, therefore, incorporates the effects of environmental racism in low-income communities of color simultaneously with an analysis of the differential impact of toxic exposure on women. She refers to this phenomenon as the "feminization of pollution."[63]

Community self-reliance has become one of the primary strategies in the grassroots movement for environmental justice because "you can't count on the agencies to 'prove' there's anything wrong." Communities have learned repeatedly that "the studies produce statistics to be analyzed away; that the tests produce numbers to be classified into safe levels or standards; and that 'experts' can find ways to explain away anything."[64] This practice was made clear to two sympathetic environmental health experts who worked on a collaborative health study with Newman's organization in Glen Avon, California. The two women researchers developed a protocol to accumulate scientific evidence that would allow the group to "get someplace in the future." Newman recalls that when these scientists were rebuffed by the Health Department "they were absolutely appalled, they couldn't believe that the knowledge they had generated wouldn't have been taken at face value."[65]

Women in the EJM also engage in critiques of science through their analyses of existing environmental impact and epidemiological research methods. This includes questioning standards of proof, modes of detection, research questions posed, and data interpretation based on standard conventions gauging statistical significance. Activists have argued that standard environmental health protocols ask the wrong questions or are not directed toward investigations that would provide relevant information to communities. Triana Silton, codirector of Center for Community Action and Environmental Justice, clarifies this point:

> Some community organizations are making demands for reforms in the Agency for Toxic Substances and Disease Registry (ATSDR), because they're pretty notorious for doing really crappy health assessments. They go into communities and use as a control group a community that sends their kids to school right by the contaminated site. They go into communities and don't actually talk to people but do the health assessment based on medical records. They don't look at routes of exposure like eating the fish, growing vegetables in the gardens, things like that. So communities have said, "this is what you need to be looking for, this is how you should be doing health assessments," and on the other hand we've said things like, "We think Lupus and breast cancer are environmental diseases, when are you going to start investigating that?"[66]

Robin Cannon, of Concerned Citizens of South Central Los Angeles, received assistance from Omawalle Fowles, an environmental scientist living in Los Angeles. Fowles helped Concerned Citizens to evaluate the effectiveness and legality of the environmental impact report (EIR) filed by the city of Los Angeles for a proposed 13-acre solid waste incinerator known as LANCER (Los Angeles City Energy Recovery Project). She discovered that the city had not developed a health risk assessment in its EIR document. Cannon explains:

> Omawalle said that we need to lobby for a health risk assessment and that was one of the first things we did at the City Council meetings. We asked for what she called a peer review committee, a panel of scientists who could review the work of the doctor that was doing the health risk assessment, so we could have a lot of different views. We also wanted to be able to select some of those scientists and that was one of our demands and that came as a result of Omowalle's experience and help. We didn't want a health risk assessment that was done on the average 21-year-old white male because that was not the makeup of our community. The community was largely African American and Latino who were suffering from respiratory ailments, hypertension, and heart disease. And our youth had a lower lung capacity than youth almost anywhere in the city or the nation.[67]

One of the most crucial deficiences that activists have noted in traditional epidemiological studies is the practice of studying single-source contamination. Most community activists are concerned with the synergistic effects of multiple-source contamination since that is the most likely scenario. Cannon continues:

> We wanted to know not only what the long term effects would be, we asked for what some of the short term impacts would be, this was something that no one else had ever gotten. We wanted to know what some of the synergistic impacts would be, so when you have mercury and lead combined, what kinds of things do you create and what can they do. We wanted to know about the compound impacts of chemicals, if we have this much lead in our bodies this year and this much next year, ten years down the line when I have so much lead in my body and I have a heart attack, could it have been a contributor? Omawalle knew quite a few minority scientists and we got these people to testify on our behalf, so by March when we had developed this coalition of people from all communities, we has a multiethnic panel of scientists of our own who were saying, this is what this community needs to know.[68]

Productive partnerships and new alliances between grassroots organizations and the scientific/medical establishment are able to emerge as activists shore themselves up with scientific, legal, corporate, economic, and legislative expertise. These diverse groups mobilize multiple means of gathering knowledge, including those of experiential "common sense" and experimental science. The "biases" and "partiality" of the different practices and actors become part of the epistemological framework—not that which must be expunged in order to get at the "truth."

Resource and training organizations such as Citizen's Clearinghouse for Hazardous Waste, the Southwest Network's newly constituted "grassroots training institute," and the CCAEJ offer community groups technical assistance, leadership training, and organizational development. Women in the movement who had little or no scientific education, or confidence in learning it, have begun to gain empowerment in developing a certain fluency in scientific discourse and in the environmental policy process. Cannon recalls her amusement and sense of satisfaction when she and another member of Concerned Citizens went to the public hearing of the peer review committee on LANCER's potential negative health effects:

> Charlotte and I went up to the mezzanine in City Hall with the council people and the peer review committee. We asked questions about the synergistic impacts and the compound effects and the short- and the long-term effects and all that, so when we were finished the guys thought we were doctors (*laughter*) to their amazement, and we said "no, we're just community members." A lot of people do gain a lot of valuable knowledge, and if we have the motivation nothing is that far from being comprehended.[69]

The level of organizational sophistication and strategizing emerging in this movement seriously counters the "expert" opinions of social research institutes such as Cerrell and Associates, who provide the state of California and private corporations with sociological criteria for targeting communities that are "least likely to resist" the incursion of potentially dangerous industries. Not surprisingly, the profile of the "least resistant" community goes something like this: small, rural town of less than 25,000 population; with many residents employed by the polluting facility, therefore deriving significant economic benefits from it; politically conservative, with a free market orientation; disproportionately above middle-aged, with a high school education or less; and with nature exploitative occupations, i.e., farming, mining, low-income.[70] Women activists, such as Cannon and Newman, repudiated this "expert" diagnosis. The stereotypes of "unaware," "unconcerned," and "compliant" poor communities were also exploded by Mary Lou Mares and Esperanza Maya, two founding members of El Pueblo Para el Aire y Agua Limpio, an environmental justice organization in Kettleman City, California. They succeeded, along with a coalition of lawyers, scientists, and environmentalists, in winning a historic suit against Chemical Waste Management, which had proposed to build the largest toxic waste dump in the nation in this 90 percent Latino, 40 percent monolinual Spanish-speaking, low-income community.[71]

Women activists in environmental justice organizations possess not only "experiential knowledge" of the effects of toxins in their environments and bodies but also extensive understanding and ability to apply scientifically generated knowledge of ecosystem dynamics, chemical production processes, and the specific uses of a variety of industrial technologies. For instance, Lee described the not inconsiderable library research that Cannon and her sister undertook in learning about toxic incinerators in South Central Los Angeles, concluding, "If you want to know about incinerators, just ask Robin." Similarly, in Tucson, Arizona's mostly Latino south side, Rose Marie Augustine, a 54-year-old "ignorant housewife," can discuss "everything there is to know about TCE (trichloroethylene)."[72] Augustine's research into TCE emissions produced by numerous defense contractors located in south Tucson, including a Hughes missile plant, has resulted in the corporations' forced installation of a $33-million air stripper and emission controller. She asserts, "Even though we don't have diplomas, we do study a lot."[73]

Silton describes the sense of empowerment that community activists gain when they increase their expertise in new technologies, such as learning the computer data base Toxics Risk Inventory (TRI). This data base, developed by the Right to Know Network, enables communities to educate themselves about the polluting facilities in the area, the toxicity of the chemicals the facility emits, and the risks of exposure. Silton insists that this development of technical expertise is "part of taking science into your own hands." Takvorian also argues that when community activists are able to use the discourses of science to inform their organizing, they become effective actors in both the scientific process and the policy process. She recalls moments when scientists and activists encountered one another and listened:

> I think a lot of scientists don't listen at all, but I've seen a lot of activists just hold their own with scientists to the point where the scientist has to take note and pay attention that this person knows something about what they're talking about. I have seen that we're able to collect enough information to stop policy makers from making decisions based on their expert's advice exclusively. They begin to understand that when you're putting comments in on an EIR that this could have some merit.[74]

These women activists produce a form of "popular science" or "popular epidemiology"[75] that offers a new path to scientific inquiry that bridges epistemological and methodological approaches and provides channels, albeit contentious ones, to bring scientists and environmental activists together in a more interactive knowledge production practice. This is actually a process, not only of collaboration, but of construction of a new scientific method—one that takes different perspectives, standpoints, and social justice arguments into account in efforts to understand and come to terms with complex socio-environmental problems.

CONCLUSION: TRANSFORMING ENVIRONMENTAL POLITICS

The mainstreaming of U.S. environmental politics since the first Earth Day demonstrations in the early 1970s has resulted in its steady deradicalization and capitalization. This has occurred largely at the expense of low-income communities of color who live, work, and play in marginal environments and have never been considered part of the "mainstream" of U.S. culture. Thousands of people living in these environments continue to be threatened by the dumping of industrial toxins in their communities and by the racist environmental policies, often "scientifically" justified, that support these hidden externalities of capitalist production. Moreover, numerous environmental indicators have demonstrated that the life-sustaining conditions of the global environment are deteriorating at an alarming rate. As Lee has argued, the mainstream environmental movement has not produced a "winning strategy" for social and environmental change. The new forms and strategies of grassroots politics being set into motion by the movement for environmental justice represent powerful interventions into the possibilities of creating such winning strategies.

Myriad voices of women activists from communities of color and low-income neighborhoods, only some of whose words I have included in this chapter, underscore the importance of thinking and working from the grassroots to effect social and environmental transformations. Mainstream environmentalism constructs highly globalized discourses and policies, but in the process it obscures the experiences and actions of the people struggling for local environmental conditions in marginalized communities. Activists from the EJM criticize this top-down, paternalistic approach that characterizes many prominent national and international environmental organizations and argue persuasively that it is not working. Women activists in the movement emphasize the significance of organizing from the grassroots, from the multiple, local, historically, and culturally specific contexts in which people are working to improve the social and environmental conditions of their lives. From these localized, community-driven efforts, a larger "movement" is being forged. As one activist has put it, the EJM's grassroots political culture is not an effect of the self-interested NIMBY phenomenon but rather the product of new forms of coalition politics.

The leadership of the mainstream environmental movement could benefit from listening to the multiple voices of the "hysterical housewives" who constitute the majority of people working for change in the expanding network of grassroots environmental justice organizations. These activists show that by single-mindedly focusing on slogans such as "save the whales" or "extinction is forever," the mainstream groups, perhaps inadvertently, but nonetheless, conceal or ignore their own accountability in perpetuating the discriminatory and even genocidal effects of environmental racism. The multiple struggles for material and cultural survival that these activists and their communities have been engaged in for years, in the face of massive social and environmental assaults, illustrate a commitment to addressing the fundamental problems underlying the current "environmental crisis." Women activists from diverse backgrounds are simultaneously challenging *and* redefining (and in the process, reinventing) discourses and practices of unjust environmental decision making, gender, racial, and class stereotyping, and dominant notions of scientific expertise. These women's voices from the EJM may provide valuable lessons and innovative political strategies for transforming U.S. environmentalism.

ACKNOWLEDGMENT

An earlier version of this chapter appeared in *Socialist Review, 22*(4), October–December 1992, pp. 93–130.

NOTES

1. The phenomenon of the predominance of women, specifically "marginalized" women, in environmental justice organizations has been documented by various sources. See, for example, Celene Krauss, "Women of Color on the Front Line," in Robert D. Bullard, ed., *Unequal Protection: Environmental Justice and Communities of Color* (San Francisco: Sierra Club Books, 1994), pp. 256–271; Lin Nelson, "The Place of Women in Polluted Places," in Irene Diamond and Gloria Orenstein, eds., *Reweaving the World: The Emergence of Ecofeminism* (San Francisco: Sierra Club Books, 1989), pp. 173–188; Jean Blocker and Douglas Lee Eckberg, "Environmental Issues as Women's Issues: General Concerns and Local Hazards," *Social Science Quarterly, 70,* 1989, pp. 586–593; Jane Kay, "Women in the Movement," *Race, Poverty and the Environment, 1*(4), Winter 1991; Barbara Ruben, "Leading Indicators: Women Speak Out on the Challenges of National Grassroots Leadership," *Environmental Action, 24*(2), Summer 1992, pp. 23–25; and Anne Witte Garland, *Women Activists: Challenging the Abuse of Power* (New York: The Feminist Press, 1988).

2. I am using the term "mainstream" in the sense of the commonly understood meanings and social organizations that constitute environmentalism in the United States. This would include ideas that embrace nature as a threatened wilderness separate from polluted, overpopulated cities and the preservation of wild animal species and the nonhuman world in general. "Mainstream" also refers to organizations that invoke a historical legacy that includes the writings and philosophies of figures such as John Muir, Aldo Leopold, and Gifford Pinchot. Such

organizations include the Sierra Club, the National Wildlife Federation, and the Nature Conservancy. When the term "mainstream" is used by EJM activists to refer to this set of priorities, it is meant to convey a distinctly pejorative tone. Although the influence of Rachel Carson's *Silent Spring* did spawn a branch of U.S. environmentalism concerned with industrial pollution, particularly in the area of pesticide abuse, her legacy is often unheralded in "mainstream" environmental discourse.

3. See Robert Gottlieb and Helen Ingram, "The New Environmentalists," *The Progressive,* August 1988, pp. 14–15.

4. Discourses of environmental preservation, protection, and the conservation of the "aesthetics of nature" dominate the environmentalism of mainstream groups, especially the "Big Ten" (Friends of the Earth, Wilderness Society, Sierra Club, National Audubon Society, Environmental Defense Fund, Natural Resources Defense Council, National Wildlife Federation, Izaak Walton League, National Parks and Conservation Association, Nature Conservancy).

5. In recent years and in response to the exhortations of many people-of-color organizations in the United States, the importance of addressing the complexities of "urban environments" and "urban ecologies" has appeared in some mainstream environmental discourse. Organizations such as Greenpeace, Sierra Club, and Earth Island Institute's Urban Habitat Program have begun to link inner-city needs with environmental concerns on many projects, thereby constructing an awareness of urban areas as "multicultural ecosystems" requiring specific environmental knowledge to ensure sustainable socially and ecologically sound development. See, for example, Rutherford Platt, Rowan Rowntree, and Pamela Muick, eds., *The Ecological City: Preserving and Restoring Urban Biodiversity* (Amherst, MA: University of Massachusetts Press, 1994); Richard Stren, Rodney White, and Joseph Whitney, eds., *Sustainable Cities: Urbanization and the Environment in International Perspective* (Boulder, CO: Westview Press, 1991). In addition, some environmental historians have expanded their objects of scholarly attention to include cities and metropolitan areas as rightfully "environmental." A good example is William Cronon's *Nature's Metropolis: Chicago and the Great West* (New York: Norton, 1991).

6. Author's interview with Dana Alston at the Public Welfare Foundation, Washington, DC, December 22, 1992.

7. Author's interview with Pam Tau Lee at the University of California's Labor and Occupational Health Program, Berkeley, CA, January 25, 1993.

8. Robert D. Bullard and Beverly Wright, "Environmentalism and the Politics of Equity," *Mid-America Review of Sociology, 12,* Winter 1987, pp. 21–37; Robert D. Bullard, *Dumping in Dixie: Race, Class, and Environmental Quality* (Boulder, CO: Westview Press, 1990); R. F. Anderson and M. R. Greening, "Hazardous Waste Facility Siting: A Role of Planners," *Journal of the American Planning Association, 48,* Spring 1982, pp. 204–218; U.S. General Accounting Office, *Siting of Hazardous Waste Landfills and Their Correlation with Racial and Economic Status of Surrounding Communities* (Washington, DC: U.S. General Accounting Office, 1983); and Sue Pollack and Joann Grozuczak, *Reagan, Toxics and Minorities* (Washington, DC: Urban Environment Conference, Inc., 1984).

9. Alston, interview, 1992.

10. For example, see Devon Peña, "The 'Brown' and the 'Green': Chicanos and Environmental Politics in the Upper Rio Grande," *Capitalism, Nature, Socialism, 3*(1), 1992, pp. 1–25; Bullard, *Dumping in Dixie*; Laura Pulido, *Latino*

Environmental Struggles in the Southwest (PhD dissertation, UCLA, 1991); Ward Churchill, *Struggle for the Land* (Monroe, ME: Common Courage Press, 1993); Marcy Darnovsky, "Stories Less Told: Histories of U.S. Environmentalism," *Socialist Review, 22*(4), 1992, pp. 11–54; and Robert Gottlieb, "Reconstructing Environmentalism: Complex Movements, Diverse Roots," *Environmental History Review, 17*(4), 1993, pp. 1–19.

11. Alston, interview, 1992.

12. Penny Newman described how her community organization "Concerned Neighbors," founded in the early 1980s to fight an unregulated toxic waste site in Glen Avon, California, chose to call themselves an antitoxics group or a community health and safety group rather than an environmental group to distance themselves from environmentalists. She recounted how many community organizations fighting toxics were often "exploited" by large lobbying groups such as the Environmental Defense Fund and the Natural Resource Defense Council to "fight for the environmentalists' favorite bills in Sacramento" but were ignored after the legislation was passed. Newman contended that her community was tired of being the "poster child" for the large environmental organizations who measured their effectiveness by their mastery of the policy process. Instead, Concerned Neighbors chose to network with other communities fighting similar battles. This networking process, in part, accounts for the emergence of the Citizen's Clearinghouse for Hazardous Waste (CCHW).

13. Lois Gibbs, *Love Canal: My Story* (Albany, NY: SUNY Press, 1982).

14. For histories of the antitoxics movement, see Robert Gottlieb, *Forcing the Spring* (Washington DC: Island Press, 1993), and Andrew Szasz, *Ecopopulism: Toxic Waste and the Movement for Environmental Justice* (Minneapolis: University of Minnesota Press, 1994).

15. Robert D. Bullard, "The Environmental Justice Movement Comes of Age," *The Amicus Journal, 16*(1), Spring 1994, pp. 32–37.

16. United Church of Christ, Commission for Racial Justice, *Toxic Wastes and Race in the United States: A National Report on the Racial and Socioeconomic Characteristics of Communities with Hazardous Waste Sites* (New York: Author, 1987).

17. Karl Grossman, "From Toxic Racism to Environmental Justice," *E Magazine,* May/June 1992, p. 31.

18. SNEEJ is made up of a coalition of environmental justice groups from eight southwestern states (California, Arizona, New Mexico, Nevada, Colorado, Utah, Oklahoma, and Texas) and numerous Indian nations. SNEEJ concentrates on demanding accountability from the EPA, fighting for the sovereignty and protection of Native Lands, developing sustainable communities, educating and developing youth leadership, and struggling for border justice for Mexico. SNEEJ members have structured the organization around gender and racial–ethnic justice, as well.

19. Richard Moore. "Confronting Environmental Racism," *Crossroads/Forward Motion, 11*(2), April 1992, p. 7.

20. Ibid., p. 8.

21. Pat Bryant, "Toxics and Racial Justice," *Social Policy, 19,* Summer 1989, p. 52.

22. Sharon Noguchi, "Birkenstockers Meet Ethnic Activists," *San Jose Mercury News,* February 5, 1993, p.9B.

23. Author's interview with Triana Silton, at the Center for Community Action and Environmental Justice, Los Angeles, CA, on June 20, 1994.

24. Alston, interview, 1992.

25. Gibbs, interview, 1992.

26. Author's interview with Diane Takvorian, director of the Environmental Health Coalition, in San Diego on March 30, 1993.

27. This was also the title of one of the early texts produced on the emergence of the EJM, by Dana Alston, *We Speak for Ourselves: Social Justice, Race and Environment* (Washington DC: Panos Institute, 1990).

28. Ibid.

29. Richard Moore, "Confronting Environmental Racism," p. 8.

30. Alston, interview, 1992.

31. Ibid.

32. Lee, interview, 1993.

33. Author's interview with Penny Newman, then West Coast field organizer for CCHW, in Glen Avon, California, on April 1, 1993.

34. Bullard, "The Environmental Justice Movement Comes of Age," p. 34.

35. See Larry Wilson, "Moving toward a Movement," *Social Policy, 19,* Summer 1989, pp. 53–57.

36. See Celene Krauss, "Women of Color on the Front Line," pp. 256–271; Karen Stultz, "Women Movers: Reflections on a Movement by Some of Its Leaders," *Social Policy, 20,* Winter 1990, pp. 36–37; Barbara Israel et al., "Environmental Activists Share Knowledge and Experiences: Description and Evaluation of STP Schools at the Highlander Research and Education Center," *PCMA Working Paper Series, 29* (Ann Arbor: University of Michigan, 1991); and Barbara Ruben, "Leading Indicators," pp. 23–25. According to many of the women I interviewed, the issues of gender and leadership were critically discussed at the Leadership Summit. The decision was made to appoint, on a rotating basis, one woman and one man who would act as cochairs for each of the regional network gatherings. In this way, many women would gain the opportunity to develop leadership skills in the process of their work. According to Israel et al., some women who attended the Highlander Center's STP (Stop the Pollution) schools (workshops that bring together environmental justice grassroots activists from around the country for leadership development and political strategizing) were not so convinced about the "gender sharing" approach to leadership training. Instead, they have requested the implementation of a *women's* STP school. For these women, confronting the specificities of gender and empowerment in their activism may best be accomplished in a "women-only" space.

37. The notion that women possess a unique consciousness of caring for the earth and its inhabitants creates a point of tension within ecofeminist movements—for example, some feminists ask, are women who protest "as women" against the bomb or against environmental destruction engaging in an effective use of society's own values against itself or are they accepting society's ideological definition of themselves as inherently more caring? Many feminists suggest that the adoption of the historical association of "women and nature" or "women and environment" by some ecofeminists is essentialist (that is, "naturalistic," positing a natural, universal woman's essence) in that it reproduces the age-old gendered relations of power where women and nature are once again relegated to secondary status. Many argue that this essentialist use of the women/nature connection maintains and perpetuates the gender dualism that also essentializes men as naturally warring and violent and destructive of the environment. Likewise, these critiques suggest that

essentialist positions ignore the ways that some women contribute to environmental problems and to the culture of militarism and the ways that some men fight against gender oppression and work for the environment. See Bina Agarwal, "The Gender and Environment Debate: Lessons from India," *Feminist Studies, 18*(1), 1992, pp. 119–158; Janet Biehl, *Rethinking Ecofeminist Politics* (Boston: South End Press, 1991); Elizabeth Carlassare, "Essentialism in Ecofeminist Discourse," in Carolyn Merchant, ed., *Ecology: Key Concepts in Critical Theory* (Atlantic Highlands, NJ: Humanities Press International, 1994); and Robyn Eckersley, *Environmentalism and Political Theory: Toward an Ecocentric Approach* (Albany: SUNY Press, 1992). In terms of political strategy, however, some feminists suggest that essentialism must be used "strategically" because the destructive discourses and institutions that associate women and nature still persist in many societies. See Val Plumwood, "Beyond the Dualistic Assumptions of Women, Men and Nature," *Ecologist, 22*(1), January/February 1992, pp. 8–13.

38. See Diamond and Orenstein, *Reweaving the World*; Elizabeth Dodson Grey, *Green Paradise Lost* (Wellesley, MA: Roundtable Press, 1979); Susan Griffin, *Women and Nature: The Roaring Inside Her* (New York: Harper Colophon, 1978); Carolyn Merchant, *The Death of Nature: Women, Ecology and the Scientific Revolution* (San Francisco: Harper & Row, 1980); Judith Plant, *Healing the Wounds: The Promise of Ecofeminism* (Philadelphia: New Society Publishers, 1989); Adrienne Harris and Ynestra King, eds., *Rocking the Ship of State: Toward a Feminist Peace Politics* (Boulder, CO: Westview Press, 1989); Karen Warren, "The Power and Promise of Ecological Feminism," *Environmental Ethics, 12,* Summer 1990; Rosemary Radford Reuther, *New Woman, New Earth* (New York: Seabury Press, 1975); Greta Gaard, ed., *Ecofeminism: Women, Animals, Nature* (Philadelphia: Temple University Press, 1993); Maria Mies and Vandana Shiva, *Ecofeminsm* (London: Zed Books, 1994); Mary Mellor, *Breaking the Boundaries: Towards a Feminist Green Socialism* (London: Virago Press, 1992); Joni Seager, *Earth Follies: Coming to Feminist Terms with the Global Environmental Crisis* (New York: Routledge, 1993); and Carolyn Merchant, *Earthcare: Women and Environment* (New York: Routledge, 1996).

39. For interesting analyses of the ideas of identity and difference as they are manifest in women's political activism, see Ann Snitow, "A Gender Diary," in Adrienne Harris and Ynestra King, eds., *Rocking the Ship of State,* pp. 35–73; and Johanna Brenner, "Beyond Essentialism: Feminist Theory and Strategy in the Peace Movement," in Mike Davis and Michael Sprinker, eds., *Reshaping the U.S. Left: Popular Struggles in the 1980s* (London: Verso, 1988), pp. 93–113.

40. Leonie Caldecott and Stephanie Leland, eds., *Reclaim the Earth: Women Speak Out for Life on Earth* (London: Women's Press, 1983), p. 6.

41. The awareness of women's connections to the environment and to the sustenance of natural resources and daily life, however, has flourished in various intellectual and activist domains. International nongovernmental organizations (NGOs) that concentrate on issues such as women and development, women and population, women and human rights, and women and health have emerged with a certain amount of legitimacy within the international relations/policy scene (e.g., Women's Environment and Development Organization [WEDO], USA; Development Alternatives with Women for a New Era [DAWN], Barbados; Women's Environmental Network, UK; International Policy Action Committee [IPAC], Brazil; ISIS International, Philippines). In the United States, women's particular interests in

the environment and health can be seen in the evolving coalition of environmental and women's health organizations that are developing projects examining the environmental causes of breast cancer in women (e.g., National Women's Health Network, Washington, DC; Women's Community Cancer Project, Boston; Long Island Breast Cancer Action Coalition; Breast Cancer Action, San Francisco). Courses in Women's Studies programs have arisen and focus on much more complex concerns than simply whether or not women are "closer to nature" or whether women are naturally more "peace-loving." For a supportive critique of ecofeminism, see Noël Sturgeon, *Ecofeminist Natures: Race, Gender, Feminist Theory and Political Action* (New York: Routledge, 1997).

42. I am using the expression "unmarked" to connote that these environmental justice activists do not necessarily identify themselves as engaging in a "women's movement." However, in another sense, clearly these women activists are *marked* in the dominant culture by their racial, class, and gender or ethnic backgrounds.

43. Gibbs, interview, 1992.

44. Newman, interview, 1993.

45. Cited in Gabriel Gutierrez, "Mothers of East Los Angeles Strike Back," in Bullard, *Unequal Protection,* p. 223.

46. Mary Pardo, "The Dialectic of Tradition: Latina Grassroots Activists and the Mothers of East Los Angeles," paper presented at the American Sociological Association, Los Angeles, CA, August 5–9, 1994.

47. Gabriel Gutierrez, "Mothers of East Los Angeles Strike Back," p.233.

48. Robbin Lee Zeff, Marsha Love, and Karen Stults, eds., *Empowering Ourselves: Women and Toxics Organizing* (Falls Church, VA: Citizen's Clearinghouse for Hazardous Waste, 1989), p. 5.

49. Ibid., pp. 5–6.

50. Interviews conducted by Greenpeace activists from the video, *First National People of Color Environmental Leadership Conference,* directed by Karen Hirsch and A. C. Warden, 1991.

51. Bryant, "Toxics and Racial Justice," p. 48.

52. Hirsch and Warden, *First National People of Color.*

53. Zeff, Love, and Stults, *Empowering Ourselves,* p. 4.

54. Ibid.

55. Newman, interview, 1993.

56. Elaine Wellin also has written on this point in her paper "Breaking the Silence: The Social Construction of Women's Activism in Grassroots Environmental Groups," American Sociological Association, Los Angeles, CA, August 5–9, 1994, and in her dissertation, Sociology Department, University of Michigan, 1995.

57. Takvorian, interview, 1993.

58. Newman, interview, 1993.

59. Ibid.

60. Penny Newman. "Killing Legally with Toxic Waste: Women and the Environment in the United States," in Vandana Shiva, ed., *Close to Home: Women Reconnect Ecology, Health and Development Worldwide* (Philadelphia: New Society Publishers, 1994); and Penny Newman, ed., *Communities at Risk: Contaminated Communities Speak Out on Superfund* (Riverside, CA: Center for Community Action and Environmental Justice, 1994).

61. Alston, interview, 1992.

62. Penny Newman, "Killing Legally with Toxic Waste: Women and the Environment in the USA," paper delivered at the conference on "Women and the Environment," India, July 17–22, 1992.

63. Ibid., p. 24.

64. Penny Newman, "Cancer Clusters among Children: The Implications of McFarland," *Journal of Pesticide Reform, 9*(3), Fall 1989, pp. 10–13.

65. Newman, interview, 1993.

66. Silton, interview, 1994.

67. Author's interview with Robin Cannon at Concerned Citizen's of South Central Los Angeles, in Los Angeles, on April 2, 1993.

68. Cannon, interview, 1993.

69. Ibid.

70. *Political Difficulties Facing Waste-to-Energy Conversion Plant Siting* (Los Angeles: Cerrell Associates, 1984).

71. Magdalena Avila, "David vs. Goliath," *Crossroads/Forward Motion*, April 1992.

72. Lee, interview, 1993.

73. Bob Ostertag, "Rose Marie Augustine: School of Hard Toxics," *Mother Jones,* January/February 1991, pp. 49–50.

74. Takvorian, interview, 1993.

75. For a discussion on "popular epidemiology," see Phil Brown, "When the Public Knows Better: Popular Epidemiology Challenges the System," *Environment, 35*(8), 1993, pp. 16–41. By popular epidemiology, Brown means "the process by which laypersons gather scientific data and other information, and also direct and marshal the knowledge and resources of experts" in order to investigate a case involving toxic contamination. Popular epidemiology, unlike traditional epidemiology, cites social structures of class, race, and gender as part of the "causal" chain of events in the destructive impacts of toxic contamination. Brown explains that, "many people who live at risk because of toxic hazards have access to data otherwise inaccessible to scientists. Their experiential knowledge usually precedes official and scientific awareness, largely because it is tied to the labor and domestic care of everyday life. Whether or not the health hazards in communities and workplaces are due to toxic substances, discovery most often stems from lay observation" (p. 19). Also see Patrick Novotny, "Popular Epidemiology and the Struggle for Community Health in the Environmental Justice Movement," Chapter 5, this volume, and other chapters of this volume.

Chapter 5

Popular Epidemiology and the Struggle for Community Health in the Environmental Justice Movement

Patrick Novotny

Among the most important social movements of the past several decades have been the various movements for environmental and occupational health. The occupational health and safety movement has engaged in both the critique of existing understandings of public health and the affirmation of alternative approaches to health that draw from the distinctive perspectives of working women and men. A growing number of women's health groups, such as the Boston Women's Health Collective and the National Black Women's Health Project in Atlanta, Georgia, have also forced a critical reexamination of women's health issues in their own respective communities. More recently, the emergence of a movement for environmental justice has drawn attention to the racial and social class disparities in environmental health.

In the past decade, growing numbers of low-income and working-class persons, particularly in African American, Latino, Native American, and Asian Pacific communities, have formed the environmental justice movement. The environmental justice movement is working with the problems of hazardous wastes, groundwater contamination, industrial pollution, and workplace safety by drawing together labor unions, tenants associations, and civil rights and community groups. This movement has formed a new understanding of environmental activism. In seeing environmental problems as connected to the concerns of social justice, the environmental justice movement has redefined "environmentalism," particularly in terms of cultivating an understanding of the environment as seen through the experiences of low-income and predominantly African American and Latino communities. An estimated 40 million persons, many of them low-income, working-class, and persons of color, live in close proximity to hazardous waste sites in the United States. With this renewed attention to disparities in environmental and occupational health, the field of popular epidemiology[1] is emerging in conjunction with the efforts of activists in the environmental justice movement to challenge traditional epidemiology and forge alternative empowering approaches to community health.

Through the movement's demands for environmental justice and in the face of environmental racism and classism, activists have focused their organizing efforts on problems that have for the most part not been understood as exclusively "environmental problems" in these communities. Environmental hazards, economic impoverishment, and racial discrimination are not considered separate in the environmental justice movement. The movement has instead fashioned an understanding of environmental justice that incorporates these social problems into its political concerns. The Southern Organizing Committee for Economic and Social Justice, a longtime civil rights and labor organization in the South that has worked on environmental issues in the past several years, is an example of an organization in the environmental justice movement that has redefined the meaning of environmentalism. This organization is typical of the manner in which "the environmental justice movement in the South has developed from the grassroots in myriad communities across the region and inherently incorporates all the other life-and-death issues its activists confront, including joblessness, abusive police practices, and the lack of health care, decent housing and equitable education. Thus, it bases itself on the new definition of the 'environment' which includes all the life conditions where we live, work and play."[2]

Much of the environmental justice movement is characterized by the leadership of tenants associations, welfare rights and housing groups, civil rights organizations, and labor unions with little prior involvement in environmental issues.[3] With many years of experience in political organizing and well-established histories of working with racial discrimination and economic impoverishment, these groups are redefining environmentalism to include the larger context of economic exploitation and political disempowerment. Lead poisoning, for instance, is not only understood as an environmental issue but as an educational and housing issue, as well, insofar as it is widely known to cause learning disabilities in young children, particularly those who live in low-income and public housing. Environmental issues are therefore understood as housing issues, educational issues, workplace issues, health issues, and poverty issues in these communities. By connecting these environmental concerns with the problems of racism, economic impoverishment, and political disenfranchisement, these communities have been more likely to organize through existing community groups that work on these problems rather than having to form new groups with a narrowly understood concern for the environment.

The movement for environmental justice is also drawing attention to the disproportionate incidence of environmental health disorders in low-income and working-class communities, particularly those of African Americans, Latinos, Asian Americans, and other persons of color. These environmental problems in poor and predominantly African American and Latino communities are seen as deeply embedded in the history of these communities, so that the movement for environmental justice is strongly linked with a much longer tradition of movements against racial discrimination and socioeconomic inequality. As Connie Tucker of the Southern Organizing Committee for Economic and Social Justice has reflected, "It's not a new struggle for us. It's a very old struggle against racism, against the targeting of people-of-color communities with the worst that society has from unemployment to drugs to toxic wastes."[4] Pat Bryant of the Gulf Coast Tenants Organization, a tenants group in southern Louisiana and a leading organization in the environmental justice movement, likewise reflects that "our problems are nothing new. It has been the environmentalism movement that has ignored our problems for so long."[5]

There is an overwhelming body of empirical evidence that people of color and lower incomes face much greater environmental problems. Study after study confirms that poor people and people of color bear the disproportionate burden not only of hazardous waste disposal and incineration but air pollution, groundwater and soil contamination, lead poisoning, pesticide poisoning, and garbage dumps. Neurological disease, reproductive disorders, and respiratory illnesses are frequently concentrated in communities with these environmentally hazardous facilities. Community activists, for instance, have documented a higher incidence of childhood leukemia, heart defects, and miscarriages in communities with a proximity to hazardous waste sites.[6] Environmental and occupational health hazards are also responsible for increased incidence of cancer, asthma, and leukemia in low-income and working class communities of color.[7] Many of the health effects of industrial pollutants have not been tested and evaluated thoroughly. Less than 10 percent of the chemicals in the workplace have been adequately tested for carcinogeneity.[8] Moreover, almost no research is conducted on the cumulative effects of chemical exposure in the workplace. Scientists and epidemiologists, according to Stephen Lester of the Citizen's Clearinghouse for Hazardous Waste (now the Center for Health, Environment, and Justice), "actually know very little about the health effects of exposure to combinations of chemicals at low levels."[9] The cumulative effect of these chemicals on public health is particularly acute due to the widespread application of pesticides, herbicides, and other compounds in agricultural production, which poses a substantial health risk not only to farm workers and their families but also to millions of consumers throughout the country.

The cumulative impact of hazardous wastes, industrial emissions, solid waste landfills, public transportation and other facilities in the context of community impoverishment, substandard housing, and inadequate medical facilities is to worsen identifiable health problems for communities that already lack even rudimentary health services. In some communities along the Mississippi River in Louisiana, for instance, the incidence of cancer mortality exceeds the national mortality by more than 200 percent.[10] And yet this poor and largely rural part of southern Louisiana has some of the most rudimentary social services and health care infrastructure imaginable, due in large part to industrial property tax abatements that have been used to lure the petrochemical industry to the region. An "insidious cycle" of economic impoverishment, industrial pollution, health problems, and environmental degradation is strikingly evident along the Mississippi River, with its dense concentration of more than 150 petrochemical plants. This is typical in many other industrial zones of the country.

Environmental hazards, groundwater contamination, and industrial pollutants also weaken the immune system and make individuals more vulnerable to health disorders, which is particularly disconcerting in light of the measles, tuberculosis, and other epidemics reminiscent of the nineteenth century that have devastated some of the nation's most impoverished communities in the past decade. More widespread immune system disorders such as asthma and diabetes have also increased in the past decade, particularly among young children. For this reason, the impact of these chemical hazards and environmental problems on the human immune system needs to be studied further, particularly for low-income and disadvantaged communities.

POPULAR EPIDEMIOLOGY AND "PEOPLE'S SCIENCE" IN THE ENVIRONMENTAL JUSTICE MOVEMENT

With this attention to disparities in environmental and occupational health, popular epidemiology is emerging through the efforts of activists in the environmental justice movement to challenge the limitations of classic epidemiology. Popular epidemiology is an empowering approach to health that places squarely in the foreground the knowledge that community residents, labor unionists and others have of the health and environmental problems resident to their communities and workplaces. The work of popular epidemiology is largely based on the assumption that broad-based political organizing is fundamental to solving the health problems of the workplace, community, and environment.

The inseparability of the physiological, psychological, and social effects of environmental hazards is foremost in the popular epidemiology of community activists in the environmental justice movement. Popular epidemiology is a critique of "classical" epidemiological research—a critique similar to those proposed by the progressive health movements, including the more recent interventions by the lesbian and gay health movement. Much as in the women's health movement, community activists in the environmental justice movement have worked to develop new ways of collecting health and epidemiological materials in order to gain greater community control over health information rather than having to rely on the work of outside researchers or health experts with little accountability or connection to the community. This "community-centered" approach is in contrast to what Linda King of the Environmental Health Network (EHN) and others refer to as classic epidemiology. Classic epidemiology is often understood as disconnected from local communities, as a scientific endeavor more concerned with "multiple regression" and "statistical significance" than the realities of ordinary people with these environmental problems. Through the work of residents and community organizers who are known to local residents, the work of popular epidemiology surmounts the mistrust that many people have of classic epidemiology. These communities have sought ways of collecting materials on the health of communities themselves and have a well-tempered distrust of health professionals and government agencies that work with environmental health.

Traditional epidemiology and conventional approaches to community health frequently obscure the connection of physiological and sociological factors to the incidence of health disorders. The work of traditional epidemiology and conventional conceptions of public health for the most part ignore what the Labor/Community Strategy Center refers to as the "unspoken categories" of class, ethnicity, gender, and race in environmental and occupational health, in particular the changing social demographic composition of the working class.[11] Epidemiological research often avoids disproportionate risks in workplace exposure by low-income and working-class persons of color, particularly for working women.[12] Much of the research in classic epidemiology, moreover, is limited to general or aggregate reviews of the incidence of health disorders and ignores the disparate concentration of health and physiological disorders in particular localities. Much of the existing health research in epidemiology, according to community activists, has been compromised since environmental hazards are studied under "normal" conditions that do not take into account the factors that affect these chemicals when they are released or disposed of. Where factories are densely clustered together in particular neighborhoods, local residents may have to contend with literally dozens of different chemicals. In older neighborhoods, groundwater and soil contamination from abandoned industries may also compound these chemicals. And factors such as transportation and nonindustrial facilities can also release chemicals and hazardous substances, thereby creating more dense concentrations of environmental hazards that are typically not included in aggregate reviews of the incidence of health disorders. In most areas, moreover, the victims of hazardous waste contamination are, not surprisingly, usually workers or residents of working-class neighborhoods. The limitations in conventional approaches to epidemiology and public health have set the stage for the emergence of the field of popular epidemiology in conjunction with local struggles for community health in the movement for environmental justice.

Much of the work of popular epidemiology is shaped by what community activists perceive to be limitations in classic epidemiology, particularly its tendency to reduce health problems to the incomprehensible statistics and abstractions of "risk assessments" and "environmental impact statements" that are all but meaningless to community residents. Under the rallying cry "our children are statistically significant," community activists have challenged the "acceptable risk" standards and "risk assessments" devised by industry and government. In their view, traditional epidemiology tends to present essentially unknowable standards of "acceptable risk" to people in communities living with a multitude of exposure to chemicals and other hazards. Focused on the pain of watching family, friends, neighbors, and community residents suffer from health problems, popular epidemiology, instead reflects upon the extent to which "risk based" approaches and perspectives tend to be much more likely to serve the economic interests of capital and the state rather than the communities whose health is most at risk.

By ignoring the "unspoken categories" of class, gender, and race (as well as ethnicity and age) in environmental health, traditional epidemiology fails to understand the extent to which particular groups are more affected by health threats than others. Where such sociodemographic factors are not mutually incorporated, findings can misleadingly overlook disparities in particular groups not simply contained by such obvious factors as race or income.

Traditional epidemiological research often understates the differences in workplace exposure by low-income and working-class persons of color, particularly for women in the workplace. Popular epidemiology is therefore critical of traditional epidemiology, particularly its tendency to leave unaddressed many of the social, cultural, and historical factors in a community that have an impact on health. For instance, such factors as family relationships and residential mobility are relevant to the health of a community, an aspect that classic epidemiology typically discounts. Traditional epidemiology is also seen as permeated by the limitations of health researchers themselves. Linda King of the EHN has emphasized the need to reevaluate the "attitudes and prejudices" of scientists involved in environmental health research.[13] Instead of focusing on acute health problems, activists such as King have sought to look at the underlying conditions that impact community health. In particular, community activists have been critical of the tendency of traditional epidemiology to account for patterns of health problems in terms of lifestyle rather than community exposure to environmental hazards.[14]

The inseparability of the physiological, psychological, and social consequences of environmental hazards is paramount in the efforts of community activists and health professionals to forge a popular epidemiology in the environmental justice movement. Popular epidemiology challenges the decontextualized individualism of classic epidemiology by focusing attention on the connections between workplaces and communities where the health and well-being of people are endangered. Popular epidemiology is therefore interested in the connections between environmental, occupational, and residential health disorders that might not otherwise be evident in traditional epidemiology. Too often, a tendency toward specialization results in researchers only studying one aspect of these health disorders in relative isolation from other factors. The "intimidating mathematics of multiple regression which dominate public health research"[15] can, for this reason, lead to ambiguous and misleading results without historical and sociological investigation of the incidence of health disorders. Much of the work of traditional epidemiology also tends to be overly cautious and tentative in its findings. Through a focus on methodological factors, this research often places many qualifications on its findings rather than coming to conclusions about the health effects. Further, in traditional epidemiology and approaches to public health, according to Nancy Krieger and Mary Bassett, "existing analytic techniques cannot address phenomena like class relations or racial oppression which cannot be expressed as numbers."[16]

In focusing between the connections of workplaces and communities where the health and well-being of people are endangered, popular epidemiology widens the purview of factors for people who are likely to have contact with hazardous chemicals in their workplace as well as have exposure to these chemicals in their communities. Health disorders, according to Richard Levins, "are structured differently in the various habitats of work, home and school, and vary across gender, class and race."[17] The interconnectedness of occupational and residential environmental hazards and the way these are entangled with social class, gender, race, and ethnicity are of crucial importance in popular epidemiology.

Popular epidemiology offers groups in the environmental justice movement an empowering approach to the health of their communities, an approach that typically combines sociodemographic and historical research with community health surveys to document the impact of industrial and environmental hazards in their communities. Activists draw from both their firsthand knowledge and experience with environmental health problems as well as their own self-education, often relying on the libraries of community colleges and nearby medical facilities for access to scientific and health research that they review. Popular epidemiology rests on the assumption that community residents are "experts" in understanding the environmental and health problems in their own communities. It validates their experiences and knowledge, thereby empowering them to take more of an active role in their own health. Popular epidemiology and such practices as health surveys bring together community residents to share their experiences with environmental problems and may be the basis for politicization of these problems. Through popular epidemiology, community activists have taught themselves about the intricacies of chemical production, industrial technologies, and the health effects of chemicals in the environment. Many activists in the environmental justice movement also have a startling knowledge of the health effects of environmental hazards such as hazardous waste incineration or solid waste landfills. It is not uncommon for government agencies in some instances to rely on the experience and knowledge that community residents have brought to bear on the discussion.

The health survey is a necessary part of popular epidemiology. Community health surveys are studies of the patterns of health disorders widely suspected to be linked with environmental and workplace hazards. Across the country, citizens are using health surveys to document for themselves the environmental problems in their neighborhoods. Community health surveys are citizen-led studies of the patterns of health disorders widely suspected to be linked with environmental and workplace hazards. An effective community health survey draws public attention to concentrations or "clusters" of health disorders and likewise is influential in pressuring government officials, public health professionals, and private industry to respond to the health concerns of community residents. Activist groups have become adept at conducting neighborhood health surveys, which are used to provide evidence of miscarriages, cancer, and other health problems in communities.

In many communities, popular epidemiology is the result of "kitchen table research" that can include health surveys, soil and water sampling, literature reviews of relevant medical and public health research, and the mapping of industrial pollutants and hazardous releases. The Citizen's Clearinghouse for Hazardous Waste has included in its publications on community health and "people's science" information that includes sampling devices to detect environmental contamination so simple in orientation that several have been entered as high school science projects. The environmental justice movement abounds with the stories of ordinary citizens who with their own water and soil samples along with videotape equipment have documented the release of hazardous substances from industry. Few community groups are without a bulging file cabinet filled with community surveys, health information, and related materials on the health of their communities.

The most unique aspect of such community health surveys is their ability to simplify the environmental and health problems facing communities in terms that are readily understandable by community residents themselves. Too often the studies by government agencies are either incomprehensible or inconclusive. Popular epidemiology enables ordinary people to have much more involvement in this process. The health survey is also crucial for the community-oriented work of popular epidemiology. It is often difficult to find evidence that links specific chemicals to particular diseases. Most of the diseases contracted by people exposed to hazardous wastes and toxic chemicals are relatively common. Leukemia, for instance, is linked to persons who work regularly with benzene and other hazardous chemicals. Leukemia, however, has many other potential causes. Many occupational and health problems, moreover, have long latency periods, sometimes as much as 30 years. For this reason, any detection of health problems with long latency periods such as cancer is extraordinarily complex. Hence, health surveys are crucial for bringing information together that can effectively document the exposure to environmental hazards and substances that can be potentially responsible for health disorders.

The residents of communities with environmental problems often have years of experience with the health problems and have worked to bring their perspectives into political organizing and community activism. At the same time, the definition of what constitutes an environmental health problem has considerably narrowed, to the point where interconnections among race, poverty, and environmental problems have largely been excised from our collective understanding of environmental health. The health survey and popular epidemiology can expand the range of factors considered in understanding environmental health. The limitations of such community health surveys, however, are that since they often have a limited time span, only acute health effects such as cancer or birth defects can be studied. And such problems have so many potential causes that it is difficult if not impossible to link health problems with environmental factors.

The health registry is yet another community-based effort similar to the health survey. Unlike the health survey, however, the focus is on the health problems of families and the collection of health information across a longer period of time. Health registries do not involve a control group like many epidemiological studies, but instead review the health of families and communities over a long period of time, documenting the deterioration of health problems as well as connections among families and neighbors in the kinds of health problems found.[18] In gathering a large amount of information from families in the same neighborhood, health registries can provide crucial data on patterns across the community. Health registries may also be used by community activists to demand a fuller study by government agencies of the problems in the community.

Health registries can also be designed for the environmental health problems in specific communities unlike the community surveys that require large numbers of people. Traditional epidemiology is more appropriate when the community being studied involves hundreds of thousands of people rather than a few dozen or a few hundred families, as is often typical with the communities living around hazardous waste sites. With the smaller number of people and local circumstances, health research has to be designed to be appropriate to these settings.

The community health survey is a relatively complicated undertaking, and leading figures in the environmental justice movement such as Lois Gibbs of the Citizen's Clearinghouse for Hazardous Waste encourage community activists to carefully weigh the costs involved in health surveys.[19] Health surveys typically take place over the course of several years or more, frequently in conjunction with other community organizing against environmentally hazardous facilities. Often, community residents must struggle with private corporations and with local, state, and federal government officials for accurate information on the health threats they face in their communities.[20] Corporations and government agencies responsible for environmental health hazards often try to deny problems and avoid accountability to community activists. In May 1992, an investigative study by the EHN and the National Toxics Campaign (NTC) was released entitled *Inconclusive by Design: Waste, Fraud and Abuse in Federal Environmental Health Research,* which extensively documents the way in which government health agencies and industry representatives have misled citizens with regard to the health effects of hazardous substances. Industries frequently provide workers and community residents with misleading information regarding the materials with which they are working and exposed to; thus, popular epidemiology is crucial for establishing patterns of health disorders for citizens who are understandably distrustful of private industry and government policymakers.

Popular epidemiology, like much of the activism in the environmental justice movement, more often than not has a transformative and empowering impact upon individuals. The loss of public accountability due to scientific and technical discourse is at least partially reversed through popular epidemiology. Likewise, citizens engaged in popular epidemiology and community health surveys undertaken in conjunction with community-based political organizing efforts "come to a larger understanding of the destructive roles that both corporations and government play in environmental degradation."[21] Many community groups in the environmental justice movement have served to shake up the common public understanding of "professional," "expertise," and "science" itself. Community groups have a well-tempered skepticism and implicit mistrust of trying to prove any conclusively about the health effects of environmental problems. Popular epidemiology or "people's science," according to Will Collette and Lois Gibbs of the Citizen's Clearinghouse for Hazardous Waste, "restores to people a sense of confidence in their own common sense."[22] Many activists in the environmental justice movement who pursue popular epidemiology are empowered and knowledgeable with regard to the complexities of scientific, epidemiological, and environmental health research.[23] Popular epidemiology is inextricably wedded to strategies for community health and political empowerment, in particular for low-income, working-class and persons of color. "The recognition of what is a health issue," according to Richard Levins, "is resolved not by some scientific method but in struggle."[24] The health survey and health registry can lead to larger concerns regarding the environment and community health, particularly campaigns that focus on particular industries and their impact on the local environment and community health.

The involvement of community activists in popular epidemiology has been extended to encompass a critique of public policy, scientific discourse, and even the limits of health and medical research itself. Much of the existing epidemiological research, for instance, makes little reference to the long-term health effects of hazardous chemicals not only on workers but on their families and communities as well. Also, many community groups have emphasized the fact that many cancers have extended latency periods. Finally, long-term exposure at low levels often in combination with other chemicals and environmental hazards can have implications for the health of workers and persons who reside in the vicinity of industry. As Linda King of the EHN reflects, "we have begun to change the science of epidemiology."[25] Through community health surveys and popular epidemiology, citizens are able to link their health disorders to environmental hazards. More importantly, popular epidemiology is a strategy for the politicization of community awareness of environmental hazards, as health surveys draw attention to diverse community grievances and direct corrective efforts toward immediate problems in surroundings familiar to workers, their families and friends.

In recognizing the detailed, intimate, and particular knowledge that people have or can have regarding their own communities, popular epidemiology is a community strategy for political empowerment. Through their activism in the movement for environmental justice, ordinary people in affected communities develop a "commonsense epidemiology." In their struggle for safe and healthy communities, according to Linda King, community activists recognize that "common sense is our best weapon."[26] Epidemiological studies, medical procedures for diagnosing and treating life-threatening diseases, and public health research may be unable to make connections with environmental problems that the public itself may know exist. At a demonstration of community activists at an incinerator in Gary, Indiana, a woman in the community spoke movingly of the health effects of emissions from an incinerator less than a block from her home. The woman lamented the fact that some breast cancers and other diseases may have a 30-year latency period and asked whether company officials would be in the community in 30 years when she and her neighbors began to suffer the health effects of exposure to incinerator wastes. Referring to the health studies conducted by the incinerator company, as inadequate, the woman charged that "we can't prove it, but our bodies know it and our children's bodies know it"[27] when disease is rampant. This willingness to stand up to contemporary health research and draw commonsensical conclusions from "commonsense" understandings of public health is an important part of community health struggles in the environmental justice movement and of the energies that have shaped much of the work of popular epidemiology. From the outset, community residents are often concerned that their health problems will be analyzed-away amidst a dizzying array of multiple regressions and statistical charts. Popular epidemiology is redefining the terms of epidemiological research away from a narrow emphasis on the physiological problems of the individual and toward the recognition of sociodemographic relations and environmentally hazardous facilities in determining patterns of health and illness in the community, thereby underscoring that the health of the individual is necessarily intertwined with the wider context of the social and economic life of the surrounding community.

The emergence of popular epidemiology and the organizing efforts for community health must reflect more than simply the accumulation of community health information and demographics. Understandable and well-documented information is a crucial part of environmental struggles in many communities, according to Bob Hall; however, "many groups fail to recognize that information alone will not win their fight."[28] Workers, their families, and communities are often exposed to unacceptably life-threatening environmental and occupational risks, and further studies may frustrate immediate demands for accountability on the part of authorities. For this reason, health studies are often seen as prolonging government inaction. Many community activists feel that such health studies are used to "quell public concern" and are more about public relations than community health and the environment. Thus, the work of popular epidemiology and health surveys must be incorporated into other political organizing strategies for community empowerment. The findings of popular epidemiology by environmental activists must be linked to effective strategies for dealing with environmental and community health problems. Illnesses, chemical sensitivity, and physiological disorders need to be effectively linked to workplace and environmental hazards for popular epidemiology to have a catalyzing effect on communities.

While many citizens have become remarkably informed through popular epidemiology and have challenged the findings of scientific experts, countless communities lack the capacity to do so. The rural poor, in particular African Americans and undocumented farm workers, suffer disproportionately from environmental hazards; yet, these groups face obstacles to effective mobilization against environmentally hazardous facilities and have historically proven to be difficult to mobilize politically. "The communities that face the most severe environmental health problems," according to the EHN in Virginia, "are mostly living in rural areas, are politically disadvantaged because of race, gender and economic status."[29] Moreover, community activists frequently endure harassment, intimidation, and threats as well as the enormous frustration and sense of isolation that is commonplace among activists. Nonetheless, activists seeking to confront environmental and occupational hazards have been aided in the past several years by the emergence of national clearinghouses and "larger than local" regional and statewide clearinghouses in the environmental justice movement that assist in popular epidemiology and the struggle for community health.

"LARGER THAN LOCALS" AND COMMUNITY HEALTH IN THE ENVIRONMENTAL JUSTICE MOVEMENT

The Citizen's Clearinghouse for Hazardous Waste (CCHW), the Environmental Research Foundation (ERF), and the EHN, along with a host of regional and statewide groups, offer scientific and technical advice for groups interested in conducting popular epidemiology and community health surveys. The clearinghouses, moreover, offer consulting to deal with the health research provided by government agencies and industry. Many clearinghouses have proved to be intermediaries between communities and health professionals. The staff of the CCHW, for instance, has conducted workshops and training programs in communities on environmental health. Its publications and newsletters are a vital link for communities throughout the nation. Peter Montague and the ERF have used computer technology to disseminate scientific and epidemiological work widely to community activists. Each month, the group reviews the major literature on the environment and public health research. These groups have provided assistance for community groups, including the analysis of thousands of pages of technical documents and medical research, and transformed it into understandable knowledge that can then be drawn on and used by hundreds of groups around the country.

The CCHW provides technical assistance and consultation, reviews detailed technical reports and risk assessments of hazardous waste sites, conducts educational workshops, and provides information on corporations and industrial technologies for more than 7,500 community groups across the country who "all face technical issues in their local environmental struggles." The CCHW publishes *Environmental Health Monthly* for community groups and health professionals, which includes detailed articles on the health effects caused by exposure to hazardous substances. The organization also produces a series of publications and resources for community health activists in addition to a monthly column by clearinghouse science director Stephen Lester in its publication *Everyone's Backyard*. Prior to its dissolution in April 1993, the National Toxics Campaign (NTC) maintained a Citizens' Environmental Laboratory, which provided low-cost, reliable testing for communities at risk from environmental contamination. The NTC's laboratory was the most extensive full-service laboratory available for communities with hazardous waste contamination, and activists in the environmental justice movement at the forefront of work in popular epidemiology have become dependent upon it (now housed with the Jobs and Environment Campaign centered in Boston). Regional and statewide "larger than local" environmental clearinghouses, such as People Against Hazardous Landfill Sites in Northwestern Indiana, the Southwest Research and Information Center in Albuquerque, and the Louisiana Environmental Action Network, maintain resources and files on community health surveys and epidemiological research for community activists. Among the most important activities of regional and national clearinghouses is that of educating local physicians about occupational and environmental health problems.

The EHN in Chesapeake, Virginia, is the most innovative clearinghouse at the forefront of organizing for environmental health. The EHN is dedicated to working with physicians and health care professionals in those communities that have environmental health problems. This clearinghouse is the only environmental health group in the environmental justice movement that focuses on empowering communities affected by exposure to hazardous wastes. The EHN provides technical, educational, and community outreach services to environmental and health groups, primary care physicians, individuals with chemically induced illnesses, and injured workers. "We bridge the gap," according to the EHN, "between patient and doctor, communities and government, grassroots groups and the science of environmental medicine, as well as injured workers and the compensation systems."[30]

The EHN helps in the prevention, diagnosis, and treatment of health problems for families. As part of its overall program, the EHN trains environmental activists to establish health registries, which are used to track disease patterns in communities where there are residential and occupational exposures to hazardous chemicals. The health registry is a link between occupational exposures and the community. This link is important, since many of the people exposed to hazardous wastes in the community are likely to come into contact with these same materials in their workplace. Health registries also collect data from family members, so that family histories of health problems can be documented.

The health registry is crucial for popular epidemiology, since often there can be long latency periods between exposure to a substance and the manifestation of health problems. This time lag makes it difficult to link a specific exposure with a specific disease or disorder. For this reason, organizers have to work extensively with community residents to carefully reconstruct their family history of health disorders. Community organizers may have to spend hours in the living rooms and kitchens of community residents, painstakingly collecting health histories from entire families. Such residential and occupational exposure over an extended period of time can have many adverse health effects on workers and residents; however, there are numerous uncertainties surrounding the documentation of such effects without a health registry. Thus, communities with exposure to hazardous substances are assisted by the EHN in establishing health registries to document any changes in health and the incidence of physiological disorders for exposed individuals and health problems in highly impacted communities.

The EHN has pioneered this innovative program for community health registries as an alternative to quantitative risk assessments and conventional epidemiology studies. Unlike quantitative risk assessment and epidemiology research, a health registry places detailed health information in the hands of the affected persons and their primary care physicians. Health registries are a useful component of popular epidemiology in the struggle for community health for those communities with environmental and occupational health problems. Through its health registry program, the EHN has also begun to focus on civil rights violations in the enforcement of federal, state, and local environmental regulatory statutes, as well as environmental health inequities in communities.

The EHN's most important project is its investigation of federal health agencies, including the Centers for Disease Control and Prevention and the Agency for Toxic Substances and Disease Registry, which has shown that these agencies routinely conduct epidemiological studies that are "inconclusive by design." Refer to the practice as "institutionalized malpractice," EHN/NTC research concludes that these agency studies "have been used by polluters and government officials to mislead local citizens into believing that further measures to prevent toxic exposures are unnecessary."[31] The EHN/NTC study concluded that federal health agencies typically have inadequate contact with the populations and communities studied, rely on environmental health testing techniques that are inappropriate to the hazardous exposure that communities face, rely upon quantitative and statistical methods of environmental health assessment "entirely unsuited" to hazardous waste problems, contract with scientific researchers who are widely regarded as biased against associating hazardous substances with disease, and have inadequately researched such important health problems as respiratory illnesses and reproductive disorders.

POPULAR EPIDEMIOLOGY AND THE "UNSPOKEN CATEGORIES": RACE, ETHNICITY, GENDER, AND CLASS

In conjunction with community health activism, popular epidemiology is drawing attention to the environmental and health hazards of the workplace and the social context of physiological disorders. From the work of ordinary people with little experience or education in environmental health, popular epidemiology is linking more directly the health effects of chemicals and industrial processes on workers with their effects on community residents. Popular epidemiology, in conjunction with organizing efforts such as health surveys and health registries, is providing a framework for understanding the broader causes of the environmental health hazards facing communities. Occupational and environmental hazards are seen in the context of a political and economic system that threatens the health and well-being of workers and community residents alike. The movement for environmental justice and its struggles on behalf of community health might therefore foster linkages with segments of the labor movement (such as the United Farm Workers, the Farm Labor Organizing Committee, and the Oil, Chemical and Atomic Workers). By extending the struggle for popular epidemiology and community health, such a movement might underscore the commonality of interests among workers and community residents in challenging industry disregard for public health and the environment.

With the deterioration of environmental conditions in many low-income communities, longtime activists and residents have also taken it upon themselves to organize projects in their own communities that address the health problems of residents and the underlying causes of these problems. Hundreds of health clinics, occupational health and safety committees, immunization programs, health education projects, and other community projects in the past decade have formed what is broadly understood here as a movement for people's health. The term "people's health" has come from the work of organizing for community health, particularly in low-income and African American communities in the South[32]; yet, this term can encompass a wide range of community-based political organizing that has placed the health of impoverished communities as uppermost in its concerns.

Too often, the health needs of the poorest communities in this country go almost entirely unaddressed while medical breakthroughs and technological innovations in the medical profession continue to widen the disparities in health care. Much like the work of popular epidemiology and the movement for environmental justice, this movement for people's health has focused on the environmental problems in communities throughout the country, recognizing that many of the communities that have the least access to health care also have the worst environmental problems. This movement emphasizes that efforts to resolve these problems has to include political organizing as well as projects such as walk-in health clinics for low-income communities, community surveys of health problems, and medical screenings to provide immediate care for those most impacted by environmental problems.

Much of the reason for the disparities in health mortality for low-income persons and persons of color has to do with the fact that there is little in the way of early detection or preventative screening for these problems. Simply put, the health problems of poor people are too often not detected until their later stages, at which time they have deteriorated to even more life-threatening conditions that require acute treatment and can often lead to complications more frequently than if caught in earlier stages. Many times, these groups also tend to rely on emergency medical services or inadequate health facilities and public hospitals. Instead of relying on the treatment of individual patients alone, many of these community projects have focused more on preventative health and the underlying social causes of health problems, encouraging individuals to take an active role in their own health and to place demands on health professionals. Much like popular epidemiology, this work is intended to help people understand the way in which their political organizing with their neighbors can have an impact on the health of their families and their community.

The work of popular epidemiology and the organizing effort for community health are significant aspects of the environmental justice movement and the activism of African Americans, Latinos, Asian Americans, and other persons of color in the movement, insofar as these groups are disproportionately exposed to environmental and health hazards. Lead poisoning endangers the health of nearly 8 million inner-city children, mostly African American and Latino children, who comprise 90 percent of lead poisoning victims in the United States.[33] African American women are disproportionately vulnerable to a multitude of physiological and reproductive disorders. African American men and women also have significantly higher blood levels of carbon monoxide and carcinogens.[34] Three of every five African Americans and Latinos live in communities with one or more uncontrolled hazardous waste sites.[35] The predominantly African American Southside of Chicago has the greatest concentration of hazardous waste sites in the nation, and persons of color in cities such as Detroit, Milwaukee, and East St. Louis have been affected by the health effects of hazardous wastes. Moreover, working class persons of color have historically been employed in those occupations and workplaces that have a higher incidence of vulnerability to hazardous exposure. Thus, for persons of color, the work of popular epidemiology and community health is inseparable from struggles for environmental justice as well as the struggles to reframe health concerns as political and community concerns.

The struggle of the environmental justice movement for community health and well-being is itself inseparable from the political struggle for affordable health care. The work of community activists to forge a popular epidemiology might be linked to the closure of medical facilities in urban communities that serve primarily low-income persons and persons of color. The wide-ranging scholarly research on race, ethnicity, poverty, and the incidence of environmental problems has been joined more recently by consideration of the ways in which these problems overlap with socioeconomic inequalities in the health care system. Hazardous waste and industrial pollution tend to be much greater in low-income communities that have limited health insurance or access to health care. Environmental problems are compounded in communities by the absence of accessible health care, adequately trained physicians, and other factors related to the health care infrastructure in low-income communities. The closure of inner-city hospitals and the decimation of the health care infrastructure in low-income communities have exacerbated the social inequalities already evident in the incidence of environmental problems.

The health care crisis and the deteriorating infrastructure of many of the nation's public hospitals have endangered the health of millions of inner-city residents across the country. The inner cities typically have some of the worst health problems, particularly with regard to environmental health, air pollution, and hazardous waste contamination. Many inner-city neighborhoods, for instance, have no primary care facilities. Most of the nation's primary health care facilities are located in suburban communities that present special transportation and economic difficulties for inner-city residents to use. Along with the movement for people's health in the South and elsewhere throughout the country, the environmental justice movement and its work for popular epidemiology may help to draw attention to the increased closure of medical facilities and their inadequate funding in communities with the worst environmental problems. More and more resort to popular epidemiology may similarly draw attention to the inadequacies of federal environmental regulatory practices and hazardous waste law enforcement, which historically have severely neglected low-income groups and communities of color.[36] With the Clinton administration's reassessment of environmental regulations and hazardous waste policy, racial disparities in the enforcement of federal hazardous waste laws will in all likelihood remain unless activism remains strong in the environmental justice movement. At this crucial juncture (with the collapse of the Clinton administration's proposals for health care reform), activists in the environmental justice movement should remain committed to a progressive agenda of political empowerment and social change that recognizes the necessity to transform conventional understandings of health. Health, as Richard Levins correctly observes, "is a bigger issue than medicine or health service."[37] The recent work of community activists and health professionals in popular epidemiology should herald the coming of an agenda and age in which the health and well-being of each community is regarded as the most highly prized commodity in the society.

The environmental justice movement and the work of related groups can play a significant role in reminding community activists throughout the country that hazardous chemicals and environmental degradation are central to many health problems, so that any meaningful health care reform has to take into account the environmental conditions of the country. The environmental justice movement has to look beyond environmental problems to address the underlying structures that make these problems even worse in many communities. Too often, medical and scientific approaches disempower low-income communities; hence, this organizing for community health has to define health as part of a wider social and political struggle. The health problems of communities may be more directly addressed through community organizing than working to improve health care per se. A program to improve housing conditions, for instance, may be the most important thing for community health. Domestic violence advocates might also improve the health of women in a community dramatically through educational programs and battered women's shelters. A campaign might focus on improving substandard housing conditions, in which both peeling paint and old plumbing contribute to lead poisoning. A nutrition or immunization campaign might also be effective in addressing some of the underlying health problems. With each of these examples, the central effect of this work might not be directly related to health care, yet collectively these types of interventions will have the greatest impact on community health over the long term.

The conditions of health are not merely medical or physiological but rather are closely linked to the social, economic, and political conditions of the community. Unemployment, workplace exposure to hazardous substances, inadequate housing, clothing, and nutrition, cultural denigration, despair, and resignation, discrimination, and inordinate levels of violence, crime, and incarceration are only a few of the many factors that impact the health of a community. "Health care professionals tend to consider the provision of medical services as being almost equivalent to a public health program. However, in many cases the most important contribution to the health of a community may be a tenants' movement or a job program."[38] So long as these are not defined as health, however, these struggles remain outside of the purview of health concerns. The crucial point is that it is not possible to focus on the health problems of these groups simply by focusing narrowly on these as health problems. Instead, it is crucial to organizing for the environment and health care in these communities to constantly make the connections between health and other forms of social inequality in these communities, much in the same way that the environmental justice movement has sought to expand our collective understanding of the "environment" to include many of the social problems and conditions prevailing in communities.

The environmental justice movement is in a unique position to intervene in the national debate on health care by using public education and awareness campaigns to establish the interconnections among environmental destruction, occupational hazards, racial and socioeconomic inequities, access to health care, and medical research appropriate to the diagnosis and treatment of health problems in communities impacted by environmentally hazardous facilities. As activists recognize that environmental, workplace, and health problems are overlapping and mutually reinforcing, it is likely that newer, more radical political coalitions will emerge. Through health registries, community health surveys, and other forms of popular epidemiology, community groups with little or no knowledge of medical science have been empowered to demand a valid accounting of the linkup between the health of their families and communities and the kinds of health hazards that are found in their workplace and environment.

The importance of the environmental justice movement is its refusal to isolate environmental problems from other problems. For many of the poorest communities in the nation, environmental racism manifests itself in more subtle ways than simply the siting of hazardous waste dumps. Housing, health care, education, nutrition, and lead paint in low-income and public housing are all considered environmental issues in the environmental justice movement. As Richard Moore of the Southwest Network for Environmental and Economic Justice reflects, "Rather than being a single issue, the environment has been treated as one of paramount importance alongside of issues such as affordable housing, youth crime, immigrant rights, employment rights, drug abuse and others."[39] Much of the work to address these concerns has been carried out as part of a "broader agenda" of social and economic justice. These environmental problems are seen as deeply embedded in the history of these communities, so that the movement for environmental justice is part of a much longer history of movements against socioeconomic inequality.

In remedying the deplorable condition of health care provision in the United States, activists in the movement for environmental justice should continue to work for a broader and more inclusive agenda of health care reform for the estimated 40 million persons who live in close proximity to hazardous waste sites. The EHN/NTC study of federal environmental health research concluded that a national health care program is "the only practical way to eliminate a portion of the injustices suffered in toxic exposed communities." Hazardous waste and industrial pollution tend to be disproportionately concentrated in low-income, working-class, and minority communities that have no health insurance or access to adequate and regular health care, thereby "exacerbating a cycle of ill health and financial problems that are created by the lack of a sensible and fair approach to health care in the United States."[40] Nothing less than far-ranging changes in national environmental policies on industrial pollution and hazardous waste prevention, the enforcement of federal environmental regulatory statutes, and comprehensive health care reform is needed to address the inequity that has been challenged by the environmental justice movement. The EHN/NTC study calls for a national pollution prevention system, reforms in worker compensation and disability law, and extended health treatment services through a national "environmental health service corporation" that would supplement health care provision in communities with severe environmental problems, any one of which are considerably more far-reaching than any of the health reform initiatives under consideration by the Clinton administration.

Regardless of what politicians do, the work of popular epidemiology and its use in local struggles for community health are crucial to environmental justice activists, whatever reforms are ultimately undertaken at the national level under the Clinton administration. Popular epidemiology and the struggle for community health in the environmental justice movement are crucial to the health and well-being of low-income, working class people and persons of color who may not otherwise be represented in reforms of the health care system over the next decade and who have shouldered the burden of health and environmental problems in this society for too long.

NOTES

1. Phil Brown, "Popular Epidemiology: Community Response to Toxic Waste–Induced Disease in Woburn, Massachusetts," *Science, Technology and Human Values, 12,* 1987, pp. 78–85; and Phil Brown and Edwin J. Mikkelsen, *No Safe Place: Toxic Waste, Leukemia and Community Action* (Berkeley: University of California Press, 1990).

2. Southern Organizing Committee for Economic and Social Justice, *In Communities Across the South* (Birmingham, AL: Southern Organizing Committee for Economic and Social Justice, 1993).

3. Environmental Careers Organization, *Beyond the Green: Redefining and Diversifying the Environmental Movement* (Boston: Environmental Careers Organization, 1992).

4. Connie Tucker, Executive Director, Southern Organizing Committee for Economic and Social Justice, interview with author, Atlanta, Georgia, September 8, 1993.

5. Pat Bryant, Executive Director, Gulf Coast Tenants Organization, interview with author, New Orleans, Louisiana, November 4, 1993.

6. Penny Newman, "Cancer Clusters among Children: The Implications of McFarland," *Journal of Pesticide Reform, 9,* 1989, pp. 10–13.

7. Beverly Hendrix Wright, "The Effects of Occupational Injury, Illness and Disease on the Health Status of Black Americans: A Review," in Bunyan Bryant and Paul Mohai, eds., *Race and the Incidence of Environmental Hazards: A Time for Discourse* (Boulder, CO: Westview Press, 1992), pp. 114–125.

8. Samuel S. Epstein, "Evaluation of the National Cancer Program and Proposed Reforms," *International Journal of Health Services, 23,* 1993, pp. 31–32.

9. Stephen Lester, "Lessons We've Learned," *Everyone's Backyard, 4,* 1986, p. 8.

10. Pat Costner and Joe Thornton, *We All Live Downstream: The Mississippi River and the National Toxics Crisis* (Washington, DC: Greenpeace, December 1989).

11. Eric Mann and the Labor/Community Watchdog Organizing Committee, "Class, Race and Gender: The Unspoken Categories of Public Health," in *L.A.'s Lethal Air: New Strategies for Policy, Organizing and Action* (Van Nuys, CA: Labor/Community Strategy Center, 1991); Nancy Krieger, Diane Rowley, Allen A. Herman, Byllye Avery, and Mona T. Phillips, "Racism, Sexism and Social Class: Implications for Studies of Health, Disease and Well-Being," in Diane Rowley and Heather Tosteson, eds., *Racial Differences in Preterm Delivery: Developing a New Research Paradigm* (New York: Oxford University Press, 1993), pp. 82–122.

12. Eric Mann and the Labor/Community Watchdog Organization Committe, "Class, Race, and Gender," p. 31; Wendy Chavkin, ed., *Double Exposure: Women's Health Hazards on the Job and at Home* (New York: Monthly Review Press, 1984).

13. Linda King, "Poverty and Race: Environmental Health Research and Health Care," *LEAN News, 5,* 1992.

14. See Eric J. Krieg, "A Socio-Historical Interpretation of Toxic Waste Sites: The Case of Greater Boston," *The American Journal of Economics and Sociology, 54*(1), January 1995, pp. 1–14.

15. Nancy Krieger and Mary Bassett, "The Health of Black Folk: Disease, Class and Ideology in Science," *Monthly Review,* April 1986, p. 84.

16. Ibid.

17. Richard Levins, "Toward the Renewal of Science," *Rethinking Marxism, 3,* 1990, p. 117.

18. Stephen Lester, "Assessing Health Problems in Local Communities, What You Can Do," *Everyone's Backyard, 12,* 1994, pp. 15–16.

19. Lois Gibbs, "Health Surveys: Think Before You Count," *Everybody's Backyard, 3,* 1985, pp. 2–3.

20. David Ozonoff and Leslie I. Boden, "Truth and Consequences: Health Agency Responses to Environmental Health Problems," *Science, Technology and Human Values, 12,* 1987, pp. 70–77.

21. Brown and Mikkelsen, *No Safe Place,* p. xvi.

22. Will Collette and Lois Gibbs, *Experts: A User's Guide* (Arlington, VA: Citizen's Clearinghouse for Hazardous Waste, 1985), p. 19.

23. Giovanna Di Chiro, "Defining Environmental Justice: Women's Voices and Grassroots Politics," *Socialist Review, 22*(4), October–December 1992, pp. 111–123.

24. Levins, "Toward the Renewal of Science," p. 116.

25. Linda King, Executive Director, Environmental Health Network, interview with author, Chesapeake, Virginia, May 19, 1993.

26. Ibid.

27. Participant comments, protests at toxic sites in East Chicago, Gary, and Hammond, Indiana, June 5, 1993.

28. Bob Hall, ed., *Environmental Politics: Lessons from the Grassroots* (Durham, NC: Institute for Southern Studies, 1988), p. 6.

29. King, "Poverty and Race."

30. Ibid.

31. Sanford Lewis, Brian Keating, and Dick Russell, *Inconclusive by Design: Waste, Fraud and Abuse in Federal Environmental Health Research* (Boston: Environmental Health Network and National Toxics Campaign, 1992), p. i.

32. Patrick Novotny, "Organizing for Community Health: New Work for Environmental Justice," *The Workbook, 19,* 1995, pp. 148–151.

33. Eric Mann and the Labor/Community Watchdog Organizing Committee, "Class, Race and Gender," pp. 33–34.

34. Wright, "The Effects of Occupational Injury . . . "

35. United Church of Christ, Commission for Racial Justice, *Toxic Wastes and Race in the United States: A National Report on the Racial and Socioeconomic Characteristics of Communities with Hazardous Waste Sites* (New York: Author, 1987).

36. Fred Strasser, Anthony Paonita, Joseph Phalon, and Mare Desmond, eds., "Unequal Protection: The Racial Divide in Environmental Law," special supplemental report, *The National Law Journal,* September 21, 1992.

37. Levins, "Toward the Renewal of Science," p. 115.

38. Richard Levins and Richard Lewontin, *The Dialectical Biologist* (Cambridge, MA: Harvard University Press, 1985), p. 251.

39. Richard Moore, *Toxics, Race and Class: The Poisoning of Communities* (Albuquerque, NM: SouthWest Organizing Project, 1991), p. 10.

40. Lewis, Keating, and Russell, *Inconclusive by Design,* p. 50.

Chapter 6

The Network for Environmental and Economic Justice in the Southwest

An Interview with Richard Moore

Paul Almeida

Richard Moore counts more than a quarter of a century of experience as a community activist and organizer. Today a national leader of what he calls the "environmental and economic justice movement"—and what some other leaders and organizations call the "environmental justice" or "environmental and social justice movement"— Moore was a central figure in the founding of the Southwest Network for Environmental and Economic Justice (SNEEJ), a coalition of dozens of community and other groups in the southwestern and western United States, and Mexico. Of Puerto Rican descent, Moore has resided in New Mexico since 1965. He has worked with many community-based organizations on welfare rights, police brutality, street gangs, drug abuse, low-cost health care, child nutrition, and the ongoing struggle against racism. Moore founded and served as Director of the Bobby Garcia Memorial Clinic in Albuquerque and helped to establish the Southwest Organizing Project (SWOP), a multiracial organization that seeks to empower the disenfranchised in the Southwest and to fight for racial and gender equality and social and economic justice. He is on the Board of Directors of the Environmental Support Center in Washington, DC, and a longtime member of the Eco-Justice Working Group of the National Council of Churches. Moore also played a key role in planning the historic First National People of Color Environmental Leadership Summit (in October 1991), the culmination of many years of scholarly research on environmental racism and practical work in communities and the workplace fighting for environmental justice.

On leave from his position as Co-Director of SWOP, Moore presently serves as Coordinator of SNEEJ. In the following interview, conducted in the summer and fall of 1993, Moore discusses his work with SWOP and SNEEJ, organizing strategies, lessons learned from 25 years of political work, his views on the interconnections between economic and environmental exploitation, and other subjects central to the theory and practice of "red green" politics in the United States.

Paul Almeida—How do you define environmental and economic justice?

Richard Moore—We [people of color/the environmental justice movement] define our environment as where we work, live, and play. We see the interconnectedness between environmental issues and economic justice issues. From our perspective you can't work on economic issues without working on environmental issues and you can't work on environmental issues, without working on economic issues. An example is high-tech industry and the use of glycol ether and other dangerous chemicals in the workplace. People are being poisoned in their community and workplace.

When Chicano workers went on strike at Medite Fiberboards in 1990 in Las Vegas, New Mexico, they wanted higher wages because they found out that white workers at a sister plant in Oregon were making five to eight dollars more an hour for the same work. Besides striking for a wage, these workers were striking for better health and safety conditions in the plant. Workers in this plant were working with sizable amounts of formaldehyde. When the Southwest Organizing Project [SWOP] did interviews with the workers and took tours of the plant in 1990, we found that the workers didn't use masks or protective gear. These same workers were dumping excess chemicals under the direction of their supervisor in ditches behind the plant. The workers were not only being exposed to chemicals in the workplace, but the chemicals they dumped also contaminated the local groundwater. Their families and the community were drinking the contaminated water that was coming from Medite. Examples like this show the interrelationship between environmental and economic issues that we are addressing in the movement. The workers themselves inside the plant have to decide how they want to tackle these problems.

If we want to eliminate the production of hazardous chemicals, the workers themselves must be involved in the process. A community–labor alliance makes the struggle even more powerful, in halting the production or use of hazardous chemicals. In the community, our children are impacted on a day-to-day basis by the air they breath and the contaminated soil they play in. We cannot separate these issues in our work. That is why SNEEJ is a broad-based network of youth, students, community-based organizations working on environmental and economic justice issues, and labor organizations. We need to have a clean, healthy workplace and community.

PA—Could you provide a brief history of the Southwest Network for Environmental and Economic Justice?

RM—The Southwest Network was founded in 1990. To backtrack a little, I am one of the founding members of the Southwest Organizing

Project, founded 12 years ago. We have been working for over half of this time on environmental and economic justice issues. My community here in Albuquerque [Mountainview] was contaminated with nitroglycerine. As we began to look at groundwater and workplace contamination, we asked whether these situations were isolated to Albuquerque or if there was a trend throughout the state. After several months of investigation, we concluded that this was a trend and not an isolated situation—that communities of color are primary targets for incinerators, landfills, dumps, and industrial facilities. In 1990, after several campaigns in New Mexico, we decided to call together a group of activists from eight states to discuss several questions, the most important being environmental racism and economic extortion.

Environmental racism means to us the disproportionate impact of polluting facilities in our communities. When we look at environmental issues we are looking at them in their broadest context. Many of our organizations do not call themselves environmental organizations; we consider ourselves community-based organizations working on environmental justice issues. Other groups refer to themselves as strictly environmental organizations. In 1990, we called together about 80 environmental activists from eight states: Oklahoma, Utah, Colorado, Nevada, New Mexico, Arizona, Texas, and California. Asian Americans, Latinos, Native Americans, and African Americans made up 99 percent of the participants in what we called at the time the "Southwest Regional Activist People of Color Dialogue on Environmental and Economic Justice Issues." About half of us at the Dialogue had worked together for many years on issues of police brutality, water rights, land grants, housing, health care, child care—all kinds of civil and human rights issues. The crucial point here is that, even though the Southwest Network was founded in 1990, many of us had been working and networking on social issues together for a long time prior to the formation of SNEEJ. Many of us come from different liberation struggles such as the Chicano movement, the Native American struggle and indigenous movement, the African American liberation movement, and the Asian American movement.

PA—Principle Six of the Principles of Environmental Justice, adopted at the multinational People of Color Environmental Leadership Summit in 1991, "demands the cessation of the production of all toxins, hazardous wastes, and radioactive materials, and that all past and current producers be held strictly accountable to the people for detoxification and the containment at the point of production."[1] This sounds similar to Barry Commoner's call for source reduction and the "social governance of production."[2] Is there any work that SNEEJ is doing to bring this about?

RM—We are in the process of negotiating our participation in a collaborative effort. This collaboration is a coalition between Silicon Valley Toxics Coalition, Campaign for Responsible Technology, SNEEJ, and a few of our affiliate organizations [e.g., PODER in Austin, the Southwest

Organizing Project in Albuquerque, and TONATIERRA in Phoenix]. We are taking on some of the largest multinational corporations in the world, as well as the U.S. military. We need to do research and identify the specific chemicals that are impacting our workers. The campaign is called "The Miscarriage of Justice" because of the effects of the chemical glycol ether on women workers. A study at the University of California [Davis] found that glycol ether was associated with miscarriages for women working in the high-tech industries.[3] It looks like we will adopt this campaign. We will then target high-tech industries that are using glycol ether and attempt to stop the use of it. We have to be cautious, though, about the international mobility of these industries. We need to have in place our contacts in other parts of the world so that there are other campaigns ready to address the possibility of relocation. Our strategy, then, is to concentrate on one chemical and work to have it banned in the United States and then throughout the globe. This campaign should not be led by environmental organizations; it should be led by environmental justice organizations that include worker and community input. Semiconductor Manufacturing Technology Consortium [SEMATECH; including IBM] would be one target of the campaign, as would smaller companies. We are still in the process of researching who is producing and using this chemical, as well as these companies' track records on worker health and safety.

PA—Why do you think it's imperative to bring people of color together first, before reaching out to others in social and environmental movements?

RM—It is important that SNEEJ document its own history. SNEEJ was officially founded in April of 1990, but many of the individuals and affiliates of SNEEJ have worked together over the last 20 to 25 years. SNEEJ has taken many forward steps since 1990, and our success is related to the fact we worked together in the past. We have chosen the people-of-color path because, if we want to build something that's going to be around for a while, we have to construct a foundation. This strategy is based on past struggles, when we didn't take the time to build a solid foundation because we had to act fast, such as during the Vietnam War.

We have to build the kind of foundation that creates a democratic situation—a situation that includes women in the leadership of SNEEJ. We don't want to make the same mistakes that we did in the past. We want to build a "net that works." We talk about weaving "nets that work" and how it takes time and caution to put the net together. We need to consistently be doing things that relate to our objectives. Part of this process is people having ownership of SNEEJ. Ownership is not going to happen overnight, but we are 5 years old now and people are beginning to feel that this organization belongs to them. Many of our members come out of organizations and relationships where they were treated as tokens. Their way was paid to conferences, by others, to represent an organization in some way. Some of our people have been spoiled in this way. SNEEJ is different; we are a poor people's movement. We are trying to build a movement of poor working class people in this country. SNEEJ is an instrument in the building of this movement. If this organization ever becomes an institution that sets back this movement, then it's time for SNEEJ to dissolve itself. That decision will be made by a body of SNEEJ. As we continue to build a movement, it is important for those involved to understand their relation to the movement. SNEEJ is helping to build a movement in this country and making it connect to an international movement.

We can act as though institutional racism doesn't exist in this country, but that doesn't do us any good. SNEEJ believes that we are building an inclusionary movement. The environmental movement historically has been an elitist, primarily white, movement. Through our building of an environmental and economic justice movement, as a continuation of the movement for social, racial, economic and gender justice, we need to bring people of color together. We have in some cases hundreds of years of conflict between Chicano and Native American people over land and water issues. We have a lot of conflict within the category of Latinos between Chicanos, Cubanos, Mexicanos, Puerto Ricans, and other Latin American people. We see the same problems within the Native American community with the Hopi, Navajo, and Shoshone. Many thought that SNEEJ wouldn't last for more than three weeks trying to bring the four primary ethnic groups together, because of these conflicts.

When I was speaking at Tulane University recently about inclusion, someone asked me how many whites SNEEJ had on its coordinating council. I told him we don't have any, but if that's a concern we will go back home and put one token white person on our council. We are now going through a process to decide when SNEEJ will open its structure to others. It won't be for tokenism or fundraising purposes. It will be because our people have studied the situation and decided that now is the time to open up. Sixty to seventy percent of the organizations in SNEEJ are multiracial organizations. Based on this, we have asked for the minimal space we need to come together and discuss our own conflicts as well as empower ourselves. We can pretend in building our foundation that racism doesn't exist in this country, but we would be lying to ourselves. Since racism does exist, and we are currently witnessing more of it, we know it is reflected in the environmental movement. What we have seen coming out of the 1960s is that, when we get involved in social movements, we are told everyone will be treated equally as men and women and ethnically. Unfortunately, that is not how things get carried out. Many of our people get discouraged and demoralized. What we are trying to do is strengthen ourselves. I think this is a major contribution in building a larger movement in this country. I don't want to be a token in anybody's organization, no matter what color they are. We are making a valuable contribution in the beginning stages of a movement by saying that, as people of color, we have the right to come together to debate, dialogue, disagree, and struggle to strengthen ourselves. We also have the responsibility as people of color to come together to build a structure that lasts by discussing everything across the board, including history, sexism, racism, and homophobia. We discuss these issues in our negotiations with high-tech industry. They ask us why we bring up all these issues besides high-tech. We tell them that these issues are critical in understanding the economic history of this country.

PA—SNEEJ's mission statement declares that "SNEEJ recognizes that the demand for a safe, clean environment and workplace can only be achieved by building a multiracial, multicultural and international movement that promotes environmental and economic justice." Do you think U.S. or global capitalism is capable of providing a safe, clean environment and workplace?

RM—We are working towards worker control, worker-owned companies and cooperatives in the long run. I think that if workers and communities participate in their own interests, there is the possibility of a safe, clean community and workplace. But obviously capitalism has not been built on the basis of a clean, safe, and healthy community or workplace. In this context, there will be a conflict with the institutions of capitalism to bring our agenda forward. I don't think the interests of capital and the interests of workers are consistent with each other. I think it's important that workers understand their class interests and where that consciousness lies in relation to the community, to the United States, and to the world. I think that certain changes can be made and fought for within this society and system. I don't think capitalism as a system is capable of moving forward a people's and workers' movement for long-range security.[4] We in SNEEJ have the responsibility to inform people and have them come to specific conclusions to decide for themselves if capitalism advances their agenda.

PA—What would an environmentally and economically just society look like to you?

RM—As I said, I don't think an environmentally and economically just society is possible under capitalism. I think we can set up isolated models or pilot projects under capitalism, but we need to be able to put all these pieces together and that is what we are attempting to do. For example, if you are building a community, you need to decide who you want to live in that community. Most planned residential communities today are for rich folks to live in. We should have energy-efficient communities. We should be developing solar instead of nuclear energy. We need to build ecologically sound communities in terms of the terrain and infrastructure. Children deserve good childcare facilities and schools. There should be economic development plans in our communities that would be sustainable in the long term. There is a whole set of alternatives that need to be looked at, but this is not easy in a capitalist society. The interests of workers and our communities are not the first things that this society considers when it makes planning decisions. We could have better regulations and policies and additional resources put into alternative models in our communities. I think it's very possible to do this.

In SNEEJ, it is important that our members get the opportunity to visit other regions and other parts of the world. There are many models and experiments that are ecologically sound going on around the world that our members can learn from and share. We have a priority in SNEEJ to try to make it possible for our people to travel to other countries. We have received an invitation, for example, to a social and economic justice conference in Cuba in January. We want our people to get beyond the misinformation that permeates this country. The Gulf War was an example of this, where the military took over the control of the television networks' coverage of the war. This is part of SNEEJ's global vision and strategy, to make it possible for members to travel.

PA—You mentioned that one of the unique characteristics of SNEEJ involves trying to merge the experiences and struggles of urban and rural communities together. Can you offer some specific examples?

RM—The task of bringing worker and community issues, as well as different ethnic groups, together has been difficult for us. Our youth campaigners have talked about creating summer camps to study history and culture in our rural communities. Farm workers are very important to us because they put the food on our tables, among other reasons. SNEEJ is looking very seriously at community-based economic development in Northern New Mexico, as a possible model for other communities.[5] We are also sending a delegation to the Havasupai community [an affiliate of SNEEJ] in the Grand Canyon, where they have been fighting against uranium mining. The Supai village was recently flooded. SNEEJ initiated a predominantly youth delegation to the village and helped in the cleanup efforts there. Many of the people going on this delegation were from cities and had a chance to see what rural village life was like. There are also plans in SNEEJ to focus on the Bureau of Land Management [BLM], to see how their policies have affected rural communities of color.

PA—What is the social class composition of the organizations in your Network?

RM—Most people in our Network are working class and unemployed. We assist people to advocate on their own behalf; that's a crucial distinction for us. Many of the people affiliated with our Network have been poisoned in their community. Many of them have illnesses and severe health-related problems. Many have been poisoned in the workplace and are not working in those facilities anymore.

PA—With the incorporation of Californian environmental justice organizations into SNEEJ do you think a more class-based politics, along with the struggle against racism, will emerge, in the sense that California has immense pockets in urban areas of workers who are unemployed, underemployed, or subemployed?

RM—Definitely. People in our affiliate in Richmond, California [West

County Toxics Coalition], an African American and Laotian community, have been poisoned by a Chevron oil refinery. We look at this situation as it relates to other cases. We ask how Richmond is related to the poisoning of farm workers in the Central Valley of California or southern New Mexico? Many of the pesticides and chemicals are being produced in urban communities of color that later contaminate the groundwater and workers in rural communities of color. This is another example of the necessity of merging urban and rural experiences together in the environmental justice movement. We need to bring community members from Richmond to the farm worker communities that are being affected by pesticides and herbicides. We have already brought farm workers from contaminated rural communities to Richmond to see how Chevron impinges upon urban communities. This will help people see the environmental and economic linkages that are affecting their lives.

PA—Isn't it difficult to focus on a poor people-of-color movement, given the possibility of being co-opted by political institutions and environmental organizations which decide to staff a few people of color because of your demands?

RM—Yes, it is very difficult. There are many difficulties we haven't discussed. One of them is the leadership structure of SNEEJ. In the 1960s and 1970s, we had Corky Gonzales, Cesar Chavez, José Gutiérrez, and Reies López Tijerina, all men, as leaders of the Chicano movement. Our movement today is decentralized; we have hundreds of leaders. When they ask for our leaders, it takes us a while to figure out which leaders we will send to the meeting. There was a call coming out of the People of Color Environmental Leadership Summit in 1991 for the formation of a national environmental justice organization. Many of the representatives of SNEEJ disagreed with this call. We feel that anything coming out of our movement should come from the bottom up. This is not mere rhetoric for us. If there is going to be a national organization, the call should be from those grassroots organizations that have come to the conclusion that this should happen. In the 1960s and 1970s, there were organizations that claimed to represent hundreds of thousands of people; these claims may or may not have been true. There was a real gap in those days between the constituency and the leadership. In SNEEJ's first year of existence, we used our minimal resources on leadership development. We knew we would be attacked because of the racial composition of SNEEJ, and there was an attempt to attack us on the race versus class question. It is unfortunate that people of color are targeted in this country by government and industry for polluting facilities. We know these are race *and* class issues. In Robert D. Bullard's study, he found that even middle-class black communities were being targeted for landfills and incinerators in Houston, Texas.[6] There will be a dialogue in SNEEJ within the next year-and-a-half that is going to be coordinated with many of our allies within the white community to discuss their role within the environmental and economic justice movement. We feel we have something in common to fight for with white folks, but we have to strengthen ourselves first.

PA—I agree with you that race and class cannot be divided within the environmental justice movement. The United States is an advanced capitalist society layered by social classes where institutionalized racism and "not in my backyard" [NIMBY] politics constitute the process in which environmental costs funnel disproportionately into minority communities.

RM—Very clearly, I agree. Many national environmental organizations like the National Wildlife Federation have the luxury of having a lot of money to travel around the world to Malaysia, Africa, and other countries. They go and talk in these countries in terms of their own interests. It is not a luxury for us—we have to make contacts in places like the Philippines or our work is not complete. Motorola has exposed workers here in Albuquerque; we know Motorola will be exposing workers in their new plant, one of their largest, in the Philippines. The Global Toxics Network [GTN] is an example of these kinds of contacts. This network is one way we are countering the international mobility of multinational corporations like Motorola. GTN was formed just recently to link grassroots organizations on a global scale to address the use of toxins in their respective communities, regions, and nations.

PA—Does SNEEJ have any plans to develop a campaign directed at the state and federal government for job creation in communities of color and low income communities?

RM—This would be part of SNEEJ's proactive strategy. A campaign for jobs would have to be called by the grassroots in SNEEJ before it would be implemented. In 1990, at the first Dialogue, we called for a training institute. Only in the last five or six months have our affiliates been talking seriously about forming a training institute. We could have started the institute three years ago, but it would have only belonged to the leadership of SNEEJ, not the individual member organizations. I give this example because many times our communities have to be reactive, putting fires out in their communities, and they don't have the opportunity to be proactive.

In my neighborhood of Mountainview in Albuquerque, we fought off a garbage transfer station that could have been disastrous to the health and safety of our community. One day we will have to decide for ourselves as a community what we want on this piece of property. This is the proactive side to our struggles. We look at job creation as a proactive strategy. It will also take an interregional effort to move the demand for jobs forward. There are several organizations in SNEEJ that have been working on this issue, such as the Labor Community Strategy Center in Los Angeles. Again, this type of campaign for jobs needs to be initiated by our affiliate organizations in order to come to fruition.

There is also going to be an increase in voter registration drives by affiliates of SNEEJ. Many of our organizations don't believe that the system totally works for them, but they want to make sure that politicians in their communities represent their constituency. In the 1960s and 1970s, when we marched into city halls, county buildings, and commission meetings, the politicians would ask how many of us were registered voters and only about five percent of us would raise our hands. We saw that politicians were not accountable to the people they perceived as not voting. So we will be pressing for more accountability by elected officials in the coming years as a focus of our movement.

PA—The Southwest Organizing Project [SWOP], which you helped found, originated as a community-based organization in New Mexico in the early 1980s, one linked to the Chicano liberation struggle. Why and when did SWOP begin to make a "green" turn and view itself as an environmental justice organization?

RM—SWOP is a community-based organization working on environmental and economic justice issues. Many of the organizations like SWOP that are members of SNEEJ generally do not define themselves as environmental organizations. Many of the founders of SWOP, including myself, were involved in social, environmental, and economic justice issues many years before SWOP formed. I was involved in fighting the siting of a sewage facility in the San Jose neighborhood of Albuquerque. The Brown Berets and the Black Berets marched in the 1960s against the smells coming out of this sewage facility. SWOP was formed because we saw a decline in community-based organizations from 1975 to 1980. We saw this as creating a gap where national advocacy organizations step in and make up the agendas and decisions affecting which issues will be worked on in our communities. National housing organizations would come from Washington to our communities with a preplanned agenda. Then they would develop a core group with a few locals, but the decisions came from above. In the late 1970s, we traveled around the state of New Mexico asking communities and activists how we could form a statewide organization and how it should conduct business. People said it should be controlled by New Mexicans.

SWOP has been very active with issues relating to high-tech industry and other issues affecting workers and communities here in New Mexico. I think the term environmental justice for SWOP is integrated into a larger process of struggling for racial, social, and economic justice.

PA—Can you describe some of the most important struggles waged and victories won by your affiliate organizations?

RM—There have been many successful struggles. Many of our people are living and working in a drastic situation. We have won more struggles than we have lost; we are an organization made up of winners. When we come together to share at our meetings and dialogues, we talk about some of these successes. We learn from other communities' strengths, weaknesses, obstacles, and victories at the local, state, regional, national, and international levels.

The Southwest Organizing Project in Albuquerque, for example, signed one of the only existing contracts in the United States between a community-based environmental and economic justice organization and a military installation [Kirtland Air Force Base in Albuquerque, New Mexico] to investigate the contamination of my own community. This was a long struggle with numerous meetings, petitions, voter registration drives, and door-to-door contact with the community. We used all the avenues a community organization has to create the power necessary to get that contract signed. There are other SNEEJ affiliates, for example, in Tucson, Arizona [Tucsonians for a Clean Environment], and Richmond, California [West County Toxics Coalition], which have gone up against the largest corporations in the world. One of our organizations in East Austin, Texas [PODER], had 80 to 100 gigantic gasoline tanks in their community. It's called a "tank farm" because there are so many of them. Exxon and other corporations that are responsible for those storage tanks lost in recent victories by PODER. The five corporations that own these tanks now have to relocate into an industrial zone. Exxon was the last corporation that refused to pull out, which caused a discussion about calling for a boycott against Exxon. The day before the boycott was going to be implemented, with a demonstration and press conference at their headquarters in Houston, Exxon announced they were going to remove their tanks. In East Los Angeles, the siting of an incinerator brought a community together [the Mothers of East L.A., another SNEEJ affiliate] to fight that incinerator off. Since then, they have fought plans to site a prison in their community.[7]

We have five campaigns and one project that SNEEJ is currently focusing on: EPA Accountability, High-Tech Industry/Workplace Hazards, Border Justice, Sovereignty/Dumping on Native Lands, Youth Leadership Campaigns, and the Farm Worker/Pesticide Project. Our project on pesticides and farm workers originated from a proposal made at the annual gathering of 1992 from the floor by an individual in the general body of SNEEJ. Almost all the states where SNEEJ has affiliates in the Southwest have a significant farm worker population. Although we had been working on pesticide issues with our EPA campaign, there was a call to concentrate on farm worker issues in a broader context, including pesticide and immigration issues. Right now this has project status, and later a request may be made to the coordinating council to decide if we should turn the farm worker project into a campaign.

The project has a chairperson and some initiatives are being made. Overall, a project is a research or trial phase to see if it will become a campaign. All the campaigns that we are currently pursuing came out of the initial Southwest Regional Activist People of Color Dialogue on Environmental and Economic Justice Issues in 1990, from the trends section. At this initial Dialogue, when SNEEJ was founded, we discussed how the EPA was unaccountable in all our communities, and so we formed our EPA accountability campaign. We also discovered at this gathering that high-tech workers were not being poisoned in just Albuquerque alone, but in many places. We then formed our high tech campaign to address these issues. Dumping hazardous waste and locating polluting facilities on Native American land was also brought up and discussed by the trends section of the Dialogue. We also discovered that youth were underrepresented in the movement for environmental justice, so we formed our youth leadership campaign. Also our fifth campaign, our border justice campaign, was formed out of the trends section of the first Dialogue. We found that children were being born without brains in Nogales, Juárez, Tijuana, Matamoros, and other locations up and down the United States–Mexico border. We also realized that the INS is not only beating, threatening, intimidating, and killing people in California, but also in Texas, Arizona, and New Mexico. The National Network for Immigrant and Refugee Rights has documented this. All our current campaigns come out of the trends section of the first gathering. This is a primary part of the democracy that exists within the structure of SNEEJ. The goals, objectives, strategies, and tactics of the campaigns come from the individual participants in the campaign. In some organizations, you have the crucial decisions being made from an office; within SNEEJ our affiliate organizations involved in specific campaign areas are discussing among themselves how to address environmental issues and then reporting back at the regional and national levels to develop our strategies. Each of our campaigns and issues overlap, so we can't talk about the EPA without talking about border justice, and we can't talk about border justice without talking about the high-tech industry. SNEEJ sees its work as very integrated and interrelated, which I think makes it a very unique organization.

Our position is that we don't speak for others, we create the climate for people to think and make decisions for themselves. We have integrated Mexican organizations into SNEEJ [e.g., the Tijuana-based Comité Pro-Restauración del Cañon del Padre y Servicios Comunitarios]. We just had a conference in Tucson to continue this process, as well as in our third annual gathering in August in San Diego and Tijuana.

PA—Many high-tech industries relocated to the Southwest in the last 15 years because of weak labor, tax, and environmental laws, and also because of worker and community struggles where these industries were initially operating. Can you inform us about some of the environmental and labor struggles your movement has encountered with these industries?

RM—I think that Silicon Valley is a good example of high-tech industry moving out here to the Southwest. Our understanding is that governor Bruce King had been paving the path for years for these companies to relocate in New Mexico. Governor King has played a part in watering down occupational health and safety standards in our state. Often the unions in these industries are company unions and do not represent workers in any legitimate way. These companies look for "right to work" states like Arizona, which demonstrates that they are only looking out for themselves. They target states where unions are projected to be weak, where administrations will weaken environmental standards for the high-tech companies, and where there is a surplus of workers. In high-tech industries inside and outside the United States, the primary targets are women, usually single and head of the household. There have been discussions in the high-tech industries about the educational standards of their employees. Women with less education, who can't read the labels or the warnings of the chemicals they are using, are recruited. The companies can threaten workers by telling them that there are lines of others waiting to be hired.

This is where economic extortion comes into play. When our Network begins to press for workers' rights and for environmental guidelines, companies respond by saying they will relocate to Costa Rica, the Caribbean, Africa, or some other place. It is not even a threat anymore, they were probably planning on moving there anyway. They may just use us as the reason to leave.

What we are trying to do in the Southwest Network is to create an exchange between workers. For example, we have brought workers together from the Motorola plant, here in Albuquerque with Motorola workers in Phoenix to talk about their mutual experiences, to learn from one another. In Guadalajara, Mexico, there is a large Motorola plant and we are going to bring together workers from Albuquerque and Phoenix with the workers in Mexico to exchange, strategize, and talk about each other's realities. This is part of our process, reacting to economic extortion. In Las Vegas, New Mexico, with Medite or in Albuquerque with Ponderosa Products Inc.[8] when we push forward with regulation they threaten to leave. They tell us that if we don't like our groundwater contaminated that they will go to another community and do it there.

We in the Southwest Network have developed a high-tech campaign. We have people in Phoenix, Austin, and Albuquerque working and communicating with workers and communities in Silicon Valley on a very organized structural level. We have targeted part of our campaign around SEMATECH,[9] which receives $100 million a year of our tax money. The consortium of partners, such as IBM, match that $100 million. Part of their mission is to develop a safe computer chip. We have grassroots organizations working on industrial policy from the bottom up, and we are confronting SEMATECH to make sure grassroots and worker participation moves forward in this process. We want the health of workers, the poisoning of workers and communities, and environmental and economic justice all to be discussed. We also have to look at the international aspect because of the nature of high-tech industry and its ability to relocate. These companies are not only on the United States–Mexico border, but also in the interior of Mexico, which makes it more difficult to monitor them.

PA––Can you give a specific example of a high-tech plant leaving Silicon Valley because of labor, environmental, and community struggles, or because of California state regulations, and relocating here to the Southwest?

RM—Intel [the world's largest producer of computer chips] is one example. Intel boasts about why it came to Rio Rancho, New Mexico, from Silicon Valley in terms of jobs and to help the New Mexico economy. The fact is that they are sitting on three Superfund sites that they are responsible for in Silicon Valley. When we look at these companies we have to look at the environmental and economic aspects. Not only are they running from their environmental problems, they are creating what we call a "whipsaw" relation between states. They have New Mexico competing with the city of San Jose, or Albuquerque competing with Austin, for how many tax incentives and breaks they can receive.[10] It is an environmental and economic reality, why they move out here to the Southwest. We don't believe that high-tech industry is very safe, environmentally or economically. The pattern we usually see is that these industries make a commitment to a very large number of jobs, then as time goes by they may decrease their commitment of two thousand jobs to four hundred. Many of these workers have no health benefits because they are hired on a contractual basis. There is also the problem of groundwater contamination.

Another example is GTE Lenkurt. If we had the relation 20 years ago with the Silicon Valley Toxics Coalition that we have now, we may have been able to prevent the tragic poisoning of workers that took place.[11] This company has also contaminated the groundwater here in Albuquerque. We can now ask the right kinds of questions when a plant moves out here from Silicon Valley because of our relations with organizations there. We sent a delegation recently of high-tech workers from Albuquerque to the Silicon Valley. One of the workers that went was Maria Chavez, who worked and was poisoned at the GTE Lenkurt plant.

Intel is also part of this process. Regulations from California played a part in their move out here; they also received tax breaks from the state of New Mexico. Our former governor, Bruce King, pushed new tax breaks through the state legislature to attract Intel to New Mexico. Senator Bingaman set up a meeting with Vice President Gore and Intel officials to make sure that air permitting requirements would not delay expansion of the Intel plant in Rio Rancho, New Mexico.[12] SNEEJ and our affiliate, the Southwest Organizing Project, then had a meeting with Senator Bingaman. We demanded that, since he arranged a meeting between Intel and the Vice President of the United States, he should then make it possible for high-tech workers to have a meeting with the Vice President on how they are affected by toxic poisoning in the industry. Bingaman sent a letter to Vice President Gore "suggesting" a meeting with representatives of SWOP to talk about the hazards of the high-tech industry.[13] One of SNEEJ's jobs is to bring organizations like SWOP together with organizations in Silicon Valley, so we can look at the high-tech industry in a much broader context.

PA—Can you provide a specific example where a polluting facility or high-tech industry moved to Mexico or the Third World because of a struggle for environmental justice here in the Southwest?

RM—GTE Lenkurt is an example. GTE Lenkurt spokespersons will tell you that they moved to Juárez, Mexico, for economic reasons. Motorola is another example—they have moved one of their facilities into Mexico. They are not just moving and operating on the border; many of these companies are moving into the interior of Mexico. We have to be careful as we develop our border justice campaign to keep in mind that relocation is going on to the interior of Mexico.

PA—One of the strategies for bringing new social movements together theorized by some thinkers associated with *CNS* is that they all have to confront the liberal democratic state and its agencies sooner or later. Examples include your battles with city governments, the military, and the EPA—the idea being that all new social movements share a common interest in democratizing the state, which means the objective conditions may exist for these movements to "sublate" some of their interests and form a larger united front.[14] Do you see SNEEJ networking with other new social movements and progressive movements to discuss certain common interests, such as democratizing the state?

RM—I think SNEEJ and other environmental justice organizations are involved in this process. We sent out letters to the peace movement, solidarity, and environmental movements to help us address the debt for nature exchanges that were being negotiated in the regions they are working in, such as Central America.[15] We also have reached out to organizations working on immigration issues. We don't want them to give up the issues they are working on, but we feel environmental justice issues have a lot in common with immigration issues. This idea of reaching out to other movements needs to be viewed in a broader context. We have tried to do this, to some extent, with our international contacts. Our strategy is to have an exchange of people. If workers and people of color travel and visit to learn about the impacts that are taking place in other workplaces and communities, we may be able to prevent unhappy recurrences in other communities. We want to share information with people; this makes the environmental justice movement unique, with its interrelated strategies. We learn a lot from others through this strategy.

PA—What do you think of the idea of the Southwest as Aztlán, a popular idea in the 1960s and 1970s in the nationalist Chicano cultural and political struggles?[16]

RM—I think it is still important. There are different sectors within the environmental justice movement. There are people who believe much more in the concept of nationalism than others. SNEEJ has brought together a wide range of people with different beliefs. I still think the concept of Aztlán plays a significant role. Many elements of the Chicano movement play a part in SNEEJ since we are based in the Southwest. Many of the leaders in SNEEJ come out of the Chicano movement. The concept of Aztlán serves as an important vision in that it helps ground people in an understanding of where they came from and where they are going. There is nothing wrong with having visions—that's why Martin Luther King, Jr., made his "I Have a Dream" speech. These ideas are important when we are envisioning a utopia, how we would like things to be. Many times our people are not permitted that opportunity. We should be encouraging these kinds of thoughts within our movement. If we want to build a new society, we need to challenge ourselves and our intellectual capacities in terms of what we think that society will be in two hundred years. I think this will help ground us in a more realistic situation and to explain to others what our thoughts are. I support many of the concepts that have developed out of the different movements that make up SNEEJ.

I think it's hard for people to understand the struggles of indigenous people without looking at the question of sovereignty. It's not just about white folks getting a handle on the concept of sovereignty, but everyone. This is a very important concept for Native American organizations. We don't have to agree on everything, but there are things we need to attempt to understand. Aztlán is a concept that fits into this, as do other concepts that came out of the 1960s and 1970s. It is important for SNEEJ to learn about the history and culture of people in the network. I think when you move into someone else's community you have the responsibility to learn about the history and culture of the region. It is very arrogant not to take the time to do this. In many cases, there is a very European approach to things in terms of how one history and culture overwhelms another history and culture.

PA—Activists and scholars in the Chicano and environmental justice movements have used the internal colonialism model to conceptualize the Chicano and African American populations in the United States.[17] Do you see Chicanos in the Southwest as a colonized people?

RM—Very definitely. New Mexico, for example, has been and continues to be treated as a military colony of the United States. There are many similarities between New Mexico and Puerto Rico. There are similarities between the Chicano struggle for land and the Palestinian struggle. There are clear linkages between struggles here in the Southwest to other struggles throughout the world. Chicanos have things in common with African Americans in the South over questions of land ownership and water rights. Through networking we find we have much in common with others.

PA—Besides the civil rights movement, then, there are other political shades and influences within SNEEJ, such as Chicano and black nationalism as well as Third World socialism. Have these different politics been difficult to reconcile within SNEEJ?

RM—Not so far. There was concern in the network over the letter SNEEJ sent to national environmental organizations.[18] Many in SNEEJ felt we were inviting white environmental organizations to work in communities of color. This was a very large debate within SNEEJ. Some honest political struggle took place around this issue. At the moment, we have positive experiences. People have learned a lot from one another, and that is the climate we are trying to create. No one political affiliation has overall leadership or decision-making power in the structure of SNEEJ. Neither does one political party or organization have control over SNEEJ.

PA—Changing focus now, is there anything SNEEJ is doing or can do in conjunction with the Native American organizations in SNEEJ and the Indigenous Environmental Network to address the monopolization of mineral resources and land use on reservations in the Southwest by what has been called the "triple alliance" [the state, outside capital, and tribal councils or local elites]?[19]

RM—SNEEJ has an educational campaign devoted to sovereignty/anti-dumping on Native lands issues. We understand that under sovereignty a Native Nation has the right to site a toxic facility, an incinerator, a landfill, or a nuclear waste dump on its own land. The United States government respects sovereignty in the context of siting incinerators and nuclear waste facilities on Native land. We have expressed our concerns about the siting of toxic facilities on Native lands and the impacts on surrounding communities. SNEEJ does not represent Native people; they represent themselves, so we try to play an assistance role in whatever capacity we are capable. At the same time, the exercise of sovereignty rights is also resulting in tribal government decisions that are highly beneficial. The Isleta Pueblo in New Mexico has fought successfully for state status, and their regulations were approved under the Clean Water Act. As Pueblos, there is the question of sovereignty of Native Nations. For any particular Pueblo to receive recognition as a state, like New Mexico or Arizona, they have to apply to the Environmental Protection Agency. Then, if approved, they can set up their own water-quality regulations. Isleta Pueblo has set a precedent for obtaining state status and developing regulations. Indigenous people that were here first, who were already practicing sustainable development, have in this case won back the right to environmentally regulate their own land. Anybody now who contaminates the Rio Grande before it reaches Isleta Pueblo will be held accountable and responsible to the Pueblo protection agency. What happens from here then if the San Juan Pueblo and Sandia Pueblo do the same? If, in fact, the San Ildefonso Pueblo by Los Alamos applies for State status, then Los Alamos is going to be in a different situation. The pollution coming out of the Los Alamos National Laboratories blows downwind to the San Ildefonso Pueblo. Los Alamos would then be accountable to the people of this Pueblo.

PA—Some on the environmental left see the need for an international radical green movement, even a "Fifth International." Do you see your international networking and organizing across the border as the beginning of such a process?

RM—It's possible. The whole thrust of the Southwest Network's view, whether it's policy or regulation, is that we see it in domestic and international terms, coming from the bottom up. We invest our time and modest resources into bringing people together for dialogue and exchange.

If out of this process the call for something much broader comes about, then we will head in that direction. There are now other networks like ours in existence, such as the Indigenous Environmental Network, the Southern Organizing Committee for Economic and Social Justice, and the Asian Pacific Islander Environmental Network. There has been a trend to consolidate regional networks, where the voice of local organizations comes from. If out of these networks there is a call to do something national, then the discussion will take place at the grassroots level; it will also continue to take place at the regional level, and out of this interaction would be the formation of a national organization. Then the next step would be to form an international organization. At this moment, we don't believe in having a national organization, but we do believe in having a national network of organizations. We also have the need to be networking at the global level. We are trying to keep ourselves accountable to the constituents that we represent. I see an international movement necessary, but I don't think we will see it today, tomorrow, or next week. We realize now that if we want to create a network that is going to last for many years, we must concentrate on building a strong foundation. We still have much work to do at the local level.

PA—Does SNEEJ have any working relationship with left-green organizations or movements such as the Green Party USA? The "green cities" projects, for example, in Detroit, seem as if they would have the support of the environmental justice movement.[20]

RM—SNEEJ has endorsed the Detroit Summer project. Many of our contacts in the environmental justice movement in Michigan are involved in Detroit Summer. We are very interested in the green cities projects as a possible model, or at least some aspects of this model, being carried over to other communities. We are in the early stages of having a dialogue with the Green Party USA. We have been invited to participate and speak at their annual national meeting. At this moment SNEEJ has no direct formal relations with the Green Party USA. The Green Party has been very supportive of many of SNEEJ's initiatives and I think will continue to be as we continue our dialogue.

PA—Many of the environmental justice struggles in the Southwest have been led by women, I understand. Can you give a few specific examples?

RM—Many of the leaders of SNEEJ are women. Struggles at the community level are often led by women.[21] For example, Rose Augustine, who is one of the leaders of Tucsonians for a Clean Environment, is fighting against groundwater contamination in her community, caused by Hughes Aircraft. When you look at McFarland, California, and the struggle to clean up a farm worker community that was built on a pesticide dump, you'll find Marta Salinas as a leading activist and founder of Help Save the Children of McFarland. Before her untimely death, Patsy Oliver was a leader in Texarkana, Texas, where a creosote facility that soaks railroad ties contaminated her African American community. There are many more examples of women leading the struggle for environmental justice, such as Robin Cannon in Concerned Citizens of South Central Los Angeles and the Mothers of East Los Angeles, which successfully fought plans for the siting of an incinerator in their communities.[22]

One of the things discussed at SNEEJ's last coordinating council meeting was to have a gathering of women in the Southwest to discuss their struggles in the environmental justice movement. Another idea SNEEJ has is to bring together women working in industries ranging from high-tech plants to Levi-Strauss and the garment industry. There is talk of trying to do this next year.

PA—Do any specifically feminist issues come up at the SNEEJ annual gatherings?

RM—The feminist movement plays a critical role in the education of our movement by identifying how women are impacted by the issues we are involved in. Beyond that, there really hasn't been much discussion of specifically feminist issues at the annual gatherings.

PA—It has been over three years now since the environmental justice movement sent letters to the "Group of Ten" demanding more representation of people of color and sharing of power.[23] Looking back now, how do you evaluate their response?

RM—We have learned a lot over the last three years. In both SNEEJ and SWOP, from the time we sent out the letters we discussed the race and class composition of the national environmental organizations. We knew that, because of their class composition, we would continue to be in conflict with one another. The call by national environmental and conservation organizations for clean air and water or for the protection of animals, forests, and mountains we know is in conflict with the protection of ourselves.

It has amazed me that nothing has changed in three years. I think they need to consistently reread the letter. The leadership of all the organizations we sent letters to were all white males. It doesn't surprise us that in three years they still don't want to share power since they have held power for 500 years. As we begin to talk about sharing power in its broadest context, we want to make sure we don't do the same things that have been done to us, such as tokenism. I don't think the Group of Ten has learned any of this. They still treat us in an extremely paternalistic fashion. In some ways, it's worse than it was three years ago. National environmental organizations send us memos that are so bad they should be used as examples in antiracism workshops.

We heard about a meeting in November 1992 where the national environmental organizations were going to get together and form their own environmental transition team for the incoming Clinton administration, without any environmental justice organizations being invited. We then sent them a memo asking if they had learned anything. We asked if there were any environmental justice representatives that were invited to this meeting.

I don't know if we can have any relations with these groups because of their deep lack of respect for our movement. We want to create a different society. The national environmental organizations are part of the institutions that need to change to build a new society.

The National Wildlife Federation [NWF] is a prime example of this. The NWF was the first environmental organization to put a waste trader [Waste Management Incorporated] on its Board of Directors. A few people of color have been assigned to boards of directors. This is nothing major because they have chosen people of color who represent the same class interests. We have seen some change in staffing with people of color and working class people being hired by these organizations. This is crucial because we are beginning to see struggle within these organizations over issues that affect our people, like debt-for-nature swaps and the regulating of Superfund sites. We are moving on, if they want to help us and share resources, then that's fine. We are building a movement that is much bigger than all of the national environmental organizations.

PA—Why do you think the EPA sees the environmental justice movement as a threat if you are just demanding that they do their job correctly?

RM—They see our movement as a threat for a few reasons. The EPA is an integral part of the U.S. government and one of its functions is to carry out U.S. domestic and foreign policies. The way that the environmental justice movement is headed poses a threat to these policies in several different ways. One is that we don't claim to speak for anyone but ourselves. Many of the national environmental organizations claim to speak for workers and communities in this country. At the national level, these environmental organizations are sitting down at the table with the CEOs of government agencies such as the EPA, and it is assumed that when they are negotiating an issue such as pesticide regulation they have had some contact with the workers affected, like farm worker organizations. If they did contact farm worker organizations, the national environmental organizations would not be permitted to make the compromises they are now negotiating with the EPA.

Another reason we are a threat to the EPA is that we have nothing to lose. We are not looking for positions in the Clinton administration or funds from the U.S. government, so we cannot be bribed. The EPA is finding it difficult to figure out our movement. Since the environmental justice movement comes out of the civil rights and human rights movement, we are not the normal advocacy organization that deals with government agencies such as the EPA, Department of Energy, and the Department of Defense. If we had the same class interests of "the Group of Ten," these government agencies could predict our political strategies. Our different class basis makes it difficult for government agencies like the EPA to understand and deal with our movement.

A third reason we are seen as a threat is that our movement merges environmental and economic issues. We are speaking for ourselves, not other people, which brings a different dynamic to the table when we negotiate with government agencies. We bring both an environmental and economic agenda to the table. This is problematic for them because they tell us that we have to break our issues down for each agency. The EPA tells us to go to OSHA for workplace issues and that they will try to help with community issues.

The EPA is accustomed to dealing with environmental organizations that take a "not in my backyard" approach to environmental problems. This has usually resulted in the export of the problem to communities of color. The position of the environmental justice movement has always been "not in anybody's backyard." In West Dallas, where an African American and Latino community had soil contaminated by a lead smelter, the community fought successfully to have it removed, but the soil was relocated to Monroe, Louisiana, where the community is 70 percent African American. We have to be careful when we are getting our communities cleaned up because we have the responsibility not to cause the poisoning of another community. At an international level, we know that the pollution will be moved to the Third World.[24] We have the responsibility to put in place contacts at the regional, national, and international levels. This is a necessary strategy for us if we want to live up to our principle of "not in anybody's backyard."

It is not easy for the EPA to lie to us, as they have done previously. Take the example of the North American Free Trade Agreement [NAFTA]. If the EPA has not protected the interests of working people in this country, then why under NAFTA would we expect the EPA to defend our interests in terms of protecting the air, water, and our communities on both sides of our border? The EPA also divides up our integrated, approach by separating issues of water, air, pesticides, worker health and safety, and communities. We want to talk about these issues as integrated and they try to box each issue into different categories. I honestly think we are a great threat to the EPA. We are talking about justice with the knowledge that communities of color are primarily targeted for polluting facilities. The EPA and their regulatory policies have played an instrumental role in allowing this to happen. We know they have some good policies in the books, but they are not implemented in our communities. We challenge their role in carrying out U.S. domestic and foreign policy.

PA—Last May, you along with other leading activists in the environmental justice movement had a meeting in Washington, DC, with Clinton's EPA chief, Carol Browner.[25] What did you discuss and how did she respond?

RM—On the one hand, it was a very positive meeting. For the first time, Browner had the opportunity to meet many of our leaders in the environmental justice movement. We told her what our concerns were and our criticisms of the EPA and of her leadership. We don't have the luxury of waiting and seeing how the EPA will act because of how drastic the situation is in many of our communities. We need to see immediate results and that is the "cop-out" of the Clinton administration as a whole, as well as of individual agency appointees. They tell us that it is going to take longer to resolve our problems than they anticipated. We have to react to the existing situations we have in our communities, as well as to be proactive in developing an agenda that will prevent many of the things that are taking place in our communities and workplaces.

There were several demands that were made to Browner. The first part of the meeting was to let her know that we didn't want to hear speeches, but to let her know the issues affecting our communities. Browner wanted us to stick to community-related issues because of the narrowness of EPA jurisdiction. We told her that we refuse to isolate and fragment our issues. One issue we discussed with Browner was the call by the three environmental justice networks [the Indigenous Environmental Network, the Southern Organizing Committee, and SNEEJ] to President Clinton and Vice-President Gore for a national environmental and economic justice summit. We asked Browner to endorse this call and she agreed to.

Another demand was that she visit communities of color in our regions, led by an organization in one of the three networks. Our recommendation was that her first visit should be in the South under the leadership of the Southern Organizing Committee. She has accepted this demand. Browner will also be visiting communities in the Southwest under the leadership of SNEEJ. In May, she visited the lead-poisoned community in West Dallas led by our affiliate, New Waverly Baptist Church. On July 28th, Browner visited Los Angeles. These are not the visits we demanded at our meeting. The visits we required were not public relations visits. There must be prenegotiations that take place with our organizations and the EPA about the issues we are addressing and what is going to happen to these problems once she leaves. Presently she has agreed to do this and we are in the process of negotiating how it will take place.

The third issue we discussed was that we want our environmental justice organizations to play a role in appointing regional EPA administrators. In the Southwest, we have regions Six, Eight, and Nine, which are headquartered in Dallas, Denver, and San Francisco. Recently, Browner called SNEEJ and shared with us a list of names the EPA is considering to head the Dallas regional office. We also told Browner we need more interagency communication and dialogue within the federal government. This was demanded to counter the way they split up our issues into different agencies, which keeps anything from happening. We want to see a cluster of government agencies, such as EPA, OSHA, HUD, air regulation, and civil rights enforcement come together to address our problems. We discussed previous problems with different federal agencies blaming one another for our problems, which diffuses responsibility. If you look at lead poisoning in communities of color, you'll see that HUD plays a role, as well as the EPA and ATSDR [Agency for Toxic Substances and Disease Registry].

We need to bring all these agencies to the table to talk about policies, regulations, and clean-up.

Al Gore recently announced at an NAACP conference in Indianapolis that he would meet with environmental justice activists. This is not what we demanded at our meeting with Browner. We specifically demanded a summit on environmental justice, not a meeting with the Vice-President. We want to bring doctors, lawyers, EPA officials, community representatives, and workers to testify about the toxic poisoning of communities of color and low-income communities. We are in the process of making clear to Gore that this is what we demanded. We want to be involved in the planning process of this summit and in setting the agenda. We also requested that we continue to have quarterly meetings with the EPA chief administrator. We demanded that women and people of color needed to be more represented in the staffing of the leadership positions in the EPA, and they have failed to do this so far.

PA—Were the people attending this meeting mainly grassroots environmental justice activists representing the three main networks?

RM—Yes, there were representatives from the three networks. This was the group that met three times with William Reilly, and this is a continuation of that process. This group primarily comes out of the People of Color Environmental Summit of 1991. Others included Charles Lee from the United Church of Christ, Bob Bullard from the University of California, Bunyan Bryant from the University of Michigan, Beverly Wright from Xavier University, and myself all attended this particular meeting. There are about 25 of us that attend these meetings.

PA—You have also recently launched a campaign against the EPA to recognize the dumping of pollutants and toxics in communities of color. How have they responded?

RM—We have developed action agendas, demonstrated, and negotiated. In the earlier part of the campaign, we called for accountability on the part of the EPA. In some cases, the EPA has good policies and regulations, but they are not being implemented within communities of color. We have a drastic situation with sick children, skin rashes, and cancer cases. We needed to initiate an aspect of the campaign to make the EPA administrators accountable. We sent out requests to the three regional administrators asking them to have meetings with representatives of our organizations to discuss the toxic poisoning of communities of color.

The Dallas EPA office is located in a bank building, with guards, that many of our community folk don't have access to. This is symbolic of the general character of the EPA and the U.S. government. The administrators refused to meet with us. We set up demonstrations in San Francisco, Denver, and Dallas. The momentum kept growing at each demonstration with more people and more ethnic diversity. At the same time, we sent a letter to William Reilly in Washington, DC, who at the time was the Chief Administrative Officer of the EPA.[26] We listed a series of statements in the letter on how the EPA played a major role in the environmental racism that existed in our communities. Many of the same people that had oversight over polluting corporations are now working for these corporations, the same corporations that have poisoned our communities. This was both a local and national campaign targeting the EPA to recognize environmental racism.

I would also like to share with you some information about a confidential internal EPA memo. In fact, when we finally met with Reilly, we told him he better fax any confidential internal EPA memos directly to our office because we will find out about them anyway. This memo discusses how dangerous they perceive our movement to be. The memo also lays out a co-optation strategy to reverse the trend of our movement. It mentioned goals, tactics, objectives, and strategies, who is responsible, and how to carry out this plan. We received this memo in November 1991, and released it to Congressman Waxman [D–CA]. At an EPA press conference in January 1992, while EPA officials presented their report on environmental equity, Congressman Waxman's staff person released this confidential EPA memo on co-optation. Several things began to happen after this incident. We were being contacted consistently to testify in Congress about environmental justice and environmental racism. We were also contacted to testify about problems we saw with the EPA. In addition, after this memo was released, we got a call from the Dallas regional EPA office a month later, saying that they were ready to meet with us. It took a year-long campaign to have one meeting with the regional EPA administrators. In March 1992, we met in San Francisco with the EPA regional administrator and twenty of his deputies, and we have been having some discussions with the Denver acting EPA administrator.

We took delegations to both the San Francisco and Dallas EPA offices. We told them about the case of McFarland, California, and how that Mexican American community was built on top of a pesticide dump and children were being born without limbs while women were having up to six miscarriages. We told the EPA about Alviso, California, and the flooding problems in this Chicano community. To prevent the flooding, they put in a levy ring that was contaminated with asbestos, which caused much illness. We told them about the African American and Laotian community in Richmond, California, where a Chevron oil refinery is located. There have been accidents and spills there, causing many health problems for the residents. We also told the EPA about west Dallas, where the African American and Latino community had one of the worst lead contamination situations in the United States. A battery recycling plant in the area had contracted out to use the casings of old batteries for fill in creating driveways and to cover holes where houses were built. If you go to west Dallas today, you will see ramp after ramp in the community. Many people's legs had been cut off, so there are many ramps for all the wheelchair users in the community. When we went into west Dallas, there was yellow tape all around the community, on the schools, playgrounds, and houses, like an accident scene. We asked a resident who was still living in this community what the EPA told him to do, since the whole community was contaminated. He said they told him to just run in and out of his house as quickly as he could.

The primary demand at the end of each of these meetings was that the regional administrator come and visit our communities and meet with the residents to see the kind of situation we are living under. If the regional administrator comes into Tucson and walks from one block to the next and finds 26 cases of cancer on one block, what are they going to do? Daniel McGovern, the previous regional EPA administrator, did come to visit communities in Tucson. One of the residents he visited died of cancer three weeks later. If these people who stay enclosed in their offices, the ones who are supposed to be directing the protection of our communities, come and look at the conditions we are living under, and see, taste, and smell for themselves, and it still doesn't do anything to their decision making, then we know we are not dealing with human beings. Then we have to develop different strategies and tactics to bring justice to an unjust situation.

The campaign was well organized and strategically carried out. The regional administrator in San Francisco visited nine different communities. We finally saw some movement. After up to ten years of struggle in some of these communities, people finally saw the EPA begin to move. Medite here in New Mexico, for example, was cited with fifteen or sixteen citations after the regional administrator visited. This continues to be a very successful campaign. We have to react and be proactive at the same time. In addition, we have to look at policy and regulations that would in a proactive way not allow this to continue.

We've seen things change already. The Earth Day statement that Clinton gave in Washington was almost all our language. Something from the bottom up is getting to the top. We sent Clinton and Gore a letter recently requesting a meeting with us. We called for a summit by Clinton and Gore on environmental and economic justice. They have now committed themselves to a meeting and negotiations. We are taking Clinton up on his offer of an "open door policy." Maybe he was just talking about corporations; if so, he should have said so at the beginning. One of the positive results to come out of our EPA campaign is our quarterly meeting with the EPA chief administrator.

PA—In conclusion, what role do you see for the journal *Capitalism, Nature, Socialism* [*CNS*] in the environmental justice movement?

RM—One of the major roles I see for *CNS* is to open up the debate. The whole movement for racial, social, and economic justice has been on an upswing in recent years. I have had the opportunity to do a lot of traveling lately, and I've seen a lot of energy for change consistently coming out of the communities I've visited. Much of this energy takes the form of single-issue organizations fighting, for example, for childcare or against the siting of an incinerator. There is an incredible amount of activity going on in this country right now. The journal can help people understand the dynamics of this, and open up the debate about what is currently happening in society. We don't have to agree with everything; that would be ridiculous. The journal allows a dialogue to take place. We also think there is a lot to be learned from community struggles that have been taking place in the United States for a long time. We need to build better communication networks, as well, to be able to exchange information in the movement. I think the journal assists in opening up the dialogue for people to exchange varying views and to open up this debate. From our perspective, we are always interested in continuing the dialogue. We look at the work aspect and the study aspect and we think they need to be balanced out.

ACKNOWLEDGMENTS

This is a revised version of an interview that originally appeared in *Capitalism, Nature, Socialism,* 5(1), Issue 17, March 1994. I express thanks to Richard Moore, Suk Rhee, Jennifer Fresquez, Bill Robinson, Felipe Gonzales, Devon Peña, Barbara Laurence, James O'Connor, and George Huaco for their unique contributions in constructing this interview.

NOTES

1. The multinational People of Color Environmental Leadership Summit was held in October 1991, with 650 organizations, leaders, and activists in the environmental justice movement in the United States in attendance. Seventeen principles of environmental justice were affirmed and adopted at the conference. A full listing of the principles can be obtained from the Southwest Network for Environmental and Economic Justice, P.O. Box 7399, Albuquerque, NM 87194.

2. See Barry Commoner, *Making Peace with the Planet* (New York: Pantheon, 1990).

3. There have now been three successive studies on diethylene glycol dimethl ether and ethylene glycol monethl ether acetate—both used widely in the computer industry as solvents to etch away material deposited on silicon wafers to make chips. Between 1986 and 1992, three studies were done to understand what health effects glycol ether had on workers in the computer chip industry. The first study, sponsored by Digital Equipment Corporation and performed by the University of Massachusetts, found evidence of significant health risks in chip making operations. The second study was sponsored by IBM at John Hopkins University. "The John Hopkins study looked at 30 female workers who handled the chemicals at IBM plants in East Fishkill, NY, and Burlington, VT, from 1980 to 1989. It found that the miscarriage rate among workers who did not use the solvents was 15.6 percent, compared to 33.3 percent among workers who did" (John Enders, "Chemicals Linked to Miscarriages," *Albuquerque Journal,* October 13, 1992).

The third study looked at the effects of glycol ethers on miscarriage rates of women computer chip workers. This latest study, which was more extensive than the first two, was sponsored by the Semiconductor Industry Association at the University of California (Davis). This 1992 study found 1.4 times as many miscarriages among women working inside "clean rooms" as compared to women working outside of manufacturing areas in microchip plants (Kirk Ladendorf, "Study Links Microchip Jobs to Miscarriage," *Austin American-Statesman,* December 4, 1992, p. A1 and sec. A16).

So far, no alternatives have been found by the computer chip industry to replace glycol ethers. Activist organizations in the United States (Silicon Valley Toxics Coalition, Campaign for Responsible Technology, and the Santa Clara Center for Occupational Safety and Health) have been warning against the use of glycol ethers in the computer industry for over six years.

4. See Daniel Faber and James O'Connor, "Capitalism and the Crisis of Environmentalism," in Richard Hofrichter, ed., *Toxic Struggles: The Theory and Practice of Environmental Justice* (Philadelphia: New Society Publishers, 1993), pp. 12–24.

5. See Devon Peña, "The 'Brown' and the 'Green': Chicanos and Environmental Politics in the Upper Rio Grande," *Capitalism, Nature, Socialism, 3*(1), Issue 9, March 1992, pp. 79–103, and Chapter 12 in this volume; and Laura Pulido, "Sustainable Development at Ganados del Valle," in Robert D. Bullard, ed., *Confronting Environmental Racism: Voices from the Grassroots* (Boston: South End Press, 1993).

6. See Robert D. Bullard, *Dumping in Dixie: Race, Class and Environmental Quality* (Boulder, CO: Westview Press, 1990).

7. For a detailed description of this case, see, Gabriel Gutiérrez, "Mothers of East Los Angeles Strike Back," in Robert D. Bullard, ed., *Unequal Protection* (San Francisco: Sierra Club Books, 1994), pp. 220–233.

8. Medite and Ponderosa are both fiberboard processing plants.

9. According to the newsletter *The Bargaining Chip* (June, 1992, pp. 1–2), SEMATECH is a "non-profit consortium of eleven semiconductor manufacturers, based in Austin, Texas. The membership companies are AMD [Advanced Micro Devices], AT&T, Digital Equipment, Harris, Hewlett-Packard, Intel, IBM, Motorola, National Semiconductor, Rockwell, and Texas Instruments. SEMATECH's mission is to build a smaller, faster microchip to challenge the growing success of the semiconductor industry's Japanese-owned competitors."

10. Paul Logan, "Intel Will Ask Sandoval for Record Bond Issue," *Albuquerque Journal,* June 5, 1993.

11. See Steve Fox, *Toxic Work: Women Workers at GTE Lenkurt* (Philadelphia: Temple University Press, 1991). GTE Lenkurt is a telephone electronic switchboard factory formerly located in Albuquerque, New Mexico. In 1987, a settlement was reached under which 115 workers (95 percent women, 75 percent Latino) received $2.5-million in compensation from GTE Lenkurt for health-related problems caused by toxins in the plant. A second products liability lawsuit has been filed by 226 former GTE Lenkurt workers against Dow, Dupont, and Shell, whose chemicals were used in the plant. The outcome of this suit should be known soon. GTE Lenkurt has since been bought out by Siemens-Stronberg-Carlson. Two-thirds of Albuquerque operations have been moved to Juarez, Mexico, the other third to Florida.

12. See Shonda Novak, "Analysis: Rio Rancho Too Good to Pass Up,"

Albuquerque Tribune, April 1, 1993, and "Intel Asked Gore for Reassurance," *Albuquerque Tribune,* April 5, 1993.

13. This letter was sent on May 12, 1993. There has been no response from the Vice-President.

14. See James O'Connor, "A Political Strategy for Ecology Movements," *Capitalism, Nature, Socialism,* 3(1), March 1992, and " 'External, Natural' Conditions of Production, the State and Political Strategy for Ecology Movements," *Conference Papers* (Santa Cruz, CA: Center for Political Ecology, Pamphlet 1, 1991), pp. 23–28.

15. See Daniel Faber, *Environment Under Fire: Imperialism and the Ecological Crisis in Central America* (New York: Monthly Review Press, 1993).

16. Ignacio Garcia writes: "Aztlán, according to Mexican mythology, was where the Aztecs had originated before they came to conquer the central valley of Mexico. All that was known about Aztlán was that it was north of Mexico. Some Chicano nationalists began claiming that the American Southwest was Aztlán and Chicanos were called by destiny to recover the homeland" (Ignacio M. Garcia, *United We Win: The Rise and Fall of La Raza Unida Party* [Tucson: University of Arizona Press, 1989], pp. 54–55).

17. See Mario Barrera, *Race and Class in the Southwest: A Theory of Racial Inequality* (Notre Dame, IN: University of Notre Dame Press, 1979), and Robert D. Bullard, "Anatomy of Environmental Racism and the Environmental Justice Movement," in Bullard, ed., *Confronting Environmental Racism,* pp. 15–39.

18. This letter was sent by SNEEJ on May 20, 1990, to the more progressive regional and national environmental organizations such as Greenpeace and the National Toxics Campaign. It was written in a friendly tone but made many of the same demands that the SWOP letter did to the Group of Ten, including diversifying staffs and not fundraising in communities of color.

19. See Kathy Hall, "Changing Woman, Tukunavi and Coal: Impacts of the Energy Industry on the Navajo and Hopi Reservations," *Capitalism, Nature, Socialism,* 3(1), Issue 9, March 1992, pp. 49–78; and Ward Churchill and Winona La Duke, "Native North America: The Political Economy of Radioactive Colonialism," in M. Annette Jaimes, ed., *The State of Native America: Genocide, Colonization, and Resistance* (Boston: South End Press, 1992).

20. For a brief overview of green cities projects, see Howard Hawkins, "Green Cities and Green Justice," *Z Magazine,* June 1993, pp. 18–19.

21. See Jeanne Gauna, "Unsafe for Women, Children and Other Living Things," *Response Magazine* (United Methodist Church), October 1991; and Giovanna Di Chiro, Chapter 4, this volume.

22. See Robert D. Bullard, ed., *Unequal Protection: Environmental Justice and Communities of Color* (San Francisco: Sierra Club Books, 1994), for descriptions of these struggles.

23. Between 1989 and 1991, three successive letters were sent to the ten largest national environmental organizations in the United States, or the "Group of Ten" (the Sierra Club, National Audubon Society, National Wildlife Fund, Environmental Defense Fund, Environmental Policy Institute, the Wilderness Society, Natural Resources Defense Council, Nature Conservancy, National Parks and Conservation Association, and Izaak Walton League), by the Coordinating Body of the Indigenous People's Organizations, Gulf Coast Tenant Leadership Development Project, and the Southwest Organizing Project (SWOP), respectively. These letters

outlined the historical role U.S. environmental organizations have played in taking land away from indigenous and traditional people in the western hemisphere. The letter sent by SWOP specifically called for the national environmental organizations in the United States to staff people of color in their offices, appoint people of color to their board of directors, and share power and resources with environmental justice organizations.

24. See John Bellamy Foster, "Let Them Eat Pollution: Capitalism and the World Environment," *Monthly Review,* January 1993, pp. 10–20.

25. This meeting took place on May 25, 1993.

26. This letter was sent on July 31, 1991.

Chapter 7

The Limits of Environmentalism without Class

Lessons from the Ancient Forest Struggle in the Pacific Northwest

John Bellamy Foster

Many prominent environmentalists today have adopted a political stance that sets them and the movement that they profess to represent above and beyond the class struggle. For example, Jonathon Porritt, the British Green leader, has declared that the rise of the German Greens marks the demise of "the redundant polemic of class warfare and the mythical immutability of a left/right divide."[1] According to this outlook, both the working class and capitalist class are to blame for the global environmental crisis (insofar as it can be traced to capitalist rather than socialist modes of production), while the greens represent a "new paradigm" derived from nature's own values, one that transcends the historic class problem. By removing themselves in this way from the classic social debate, these green thinkers implicitly embrace the dominant "we have seen the enemy, and it is us" view that traces most environmental problems to the buying habits of consumers, the number of babies born, and the characteristics of industrialization, as if there were no class or other divisions in society.

In contrast, it will be argued here—in the context of a discussion of the crisis of the old-growth forest and the timber industry in the U.S. Pacific Northwest—that rapid ecological degradation is an inherent part of the historically specific accumulation process that defines capitalist society and its class struggle.[2] An ecological movement that stands for the earth alone and ignores class and other social inequalities will succeed at best in displacing environmental problems, meanwhile reinforcing the dominant relations of power in global capitalism, with their bias toward the unlimited commodification of human productive energy, land, and the built environment, and the ecology of the planet itself. An earth movement of this kind will therefore contribute little to the overall green goal of forming a sustainable relationship between human beings and nature, and may even have the adverse effect—by splitting popular forces—of creating more opposition to the environmental cause.[3]

Nowhere is this overall dilemma of class versus ecology more evident today than in the Pacific Northwest, where the battle to save the last stands of ancient forest has left forest product workers and single-issue environmentalists at each other's throats. In timber-dependent communities, "preservationists" have been accused of being "enemies of the people," while single-issue environmentalists for their part have often characterized loggers and other forest product workers as "enemies of nature." "The northern spotted owl," Michael Renner observed in Worldwatch's *State of the World 1992,* "has become a symbol of the seemingly intractable conflict between jobs and environmental protection—and of the larger tensions between the health of the economy and that of the natural world on which it ultimately depends."[4]

The truth is that both a sustainable relation to the forest ecosystem and employment stability for workers in the industry are best achieved through the forging of an alliance between environmental activists and forest product workers around a common labor–environmentalist program aimed at the state. Yet, the narrow conservationist thrust of most environmentalism in the United States, the unimaginative business union response of organized labor, and the divide and conquer strategy employed by timber capital and its allies within the federal government against its two most powerful opponents—the working class and the environmentalists—have thus far combined to block the formation of any such coalition.

By 1992 environmentalists seemed to have won a resounding victory in their long struggle to protect the ancient forest of the Pacific Northwest. Attempts by the Bush administration to promote logging in areas that threatened the existence of the northern spotted owl had been overturned by the courts, while the election of the Clinton administration seemed to many to hold out the hope that the federal government would place much greater emphasis on environmental concerns. Six years later, however, the sense of victory on the side of environmentalists has been replaced by a growing sense of betrayal—if not yet outright defeat. The passage of the salvage logging rider of the Budget Rescissions Act, signed into law by President Clinton in July 1995, set aside the nation's environmental laws, opening up the ancient forest to rapacious logging once again, and threatening the existence of the marbled murrelet, the northern spotted owl, and numerous other species of the region. The Endangered Species Act itself is now under continual attack within Congress. Most threatening of all, the 1990s have seen the growth of a corporate-financed Wise Use coalition able to mobilize many thousands of workers, along with officers of resource-extracting corporations, landowners, cattle grazers, and realtors—providing a new political base and "populist" rationale for business-serving politicians seeking to undermine existing environmental laws.

Hence, more than ever before it is crucial to discover the means of forging a wider labor–environmentalist alliance, and to learn the lessons that the struggle for the ancient forest of the Pacific Northwest has to teach to environmentalists generally. To understand how a united front between forest product workers and forest ecosystem defenders might have been (and still could be) established in the Northwest it is necessary to explore this ecological crisis in its making, with particular attention to the role of capital and the state. Such an account would reveal the class origins of the ecological crisis, together with the general outlines of a progressive class-based response to the stranglehold that the jobs versus nature issue now maintains over the entire environmental movement.

ECOLOGICAL CATASTROPHE AND SOCIAL CRISIS

At the time of the Lewis and Clark expedition the ancient conifer forest, dominated by trees hundreds of feet in height and centuries—sometimes more than a millennium—old, covered some 20 million acres in western Oregon and western Washington alone. Today only around 12 percent or 2.4 million acres of fully intact "old-growth forest" remains—consisting of centuries-old trees, a multilayered canopy, numerous large dead standing trees, or "snags," and large downed trees on the ground and across streams—according to the most advanced old-growth inventory available from Peter Morrison of the Wilderness Society. Since private capital has cleared its land of nearly all of the original forest, the ancient forest that is left is to be found almost exclusively on public lands. Moreover, these last stands of late successional forest are largely confined to the higher elevations (above 2,500 feet) and are to be found in a crazy quilt of isolated patches—the result of patterns of land acquisition, logging, road building, and land clearances. According to data released in June 1992 by NASA scientist Dr. Compton J. Tucker, who has led a project comparing satellite photos of the Pacific Northwest and Amazon forests, the Northwest forest has been subject to "severe fragmentation" and "has been literally cut to pieces." "When you compare the situation in the Pacific Northwest to the Amazon of Brazil [in this respect], the Northwest is much worse." Biologists have drawn an analogy between the Northwest forests and a shirt perforated again and again, to the point that there are now more holes than cloth. In 1990 about 800,000 acres of the remaining intact old-growth forest, according to the Morrison estimates, were protected in parks and wilderness areas. The other 1,600,000 acres—more than half of which are already highly fragmented—were open to exploitation. In the 1980s, these stands of old-growth forest were disappearing at a rate of perhaps as much as 70,000 acres a year. If this rate of cutting had continued, the unprotected regions of the old-growth forest in Oregon and Washington would have been gone in less than 30 years.[5]

It was under these general conditions that two opposing forces converged in the 1980s to form a highly volatile situation with respect to the management of the old-growth forest. The first of these was evident in the implementation of a process of economic restructuring, arising out of the economic stagnation of the early 1980s, that required the ever more rapid liquidation of the old-growth forest, together with increased exploitation of forest products workers. Responding to a decline in the secular growth trend of the economy, capital in the Reagan period attempted to restructure the economy and state in ways that would remove any regulatory limits that had been placed on free-market exploitation of the natural and human "conditions of production."[6] As we will see, in the case of the Northwest national forests, this meant a subversion of the long-established principle of sustained yield, insofar as this could be interpreted as a "nondeclining *even flow*" of timber, and its replacement by a policy of increased cutting and rapid old-growth liquidation designed to maximize government revenues, bridge the gap in private timber supplies, and clear the ground for a "fully managed" system of plantation forestry in the national forests.

The second converging force took the form of a rapidly growing environmental movement determined to defend the ecological integrity of the Northwest forests. In the face of a stepped-up campaign of forest restructuring aimed at the liquidation of the remaining old growth, environmentalists in the 1980s struck back with every means at their disposal: blockading logging roads with their bodies, tree sitting, and filing a flood of legal proceedings designed to slow down and eventually halt the removal of ancient timber. A crucial turning point in the struggle came in 1988 when a federal court in Seattle upheld an environmentalist lawsuit claiming that the federal government had violated the requirements of the Endangered Species Act in failing to take steps to preserve the habitat of the northern spotted owl, a rodent-eating predator high up on the old-growth forest's food chain.

Environmentalists were aided not only by strong environmental law—the Endangered Species Act—but also by a series of scientific advances in the ecological understanding of the old-growth forest that strongly reinforced the case for preservation. With the release of the landmark 1981 study *Ecological Characteristics of the Old-Growth Douglas-Fir Forests,* authored by Forest Service ecologist Jerry Franklin and his associates, together with other related studies, it was demonstrated that the late successional or old-growth forest was by far the richest and most ecologically complex stage in the forest's existence, supporting a yet uncataloged diversity of life forms, many of which are now endangered as a result of forest fragmentation and destruction of critical habitat. Individual stands within the old-growth forest were discovered to be "unrivaled both in the size and longevity of individual trees and in the accumulation of biomass of individual stands." Among the coastal redwoods the old-growth coniferous forest was found to exceed that of any tropical rainforest thus far measured in total accumulated biomass per unit area by a ratio of seven to one, while forests throughout the old-growth coniferous region were found to support biomasses far beyond those of tropical forests (though the latter are unrivaled in the sheer diversity of life that they support). Moreover, it was revealed that the old-growth forest stored more carbon per unit area than any other terrestrial ecosystem thus far measured, making it a significant factor in the stabilization of the world's climate in the face of global warming. These and other new discoveries thus represented a scientific advance in forest ecology that seemed to point inexorably to the imperative of preservation.[7] Environmentalists became adept at disseminating this new ecological understanding—much of it the product of the work of government scientists, some of whom were drawn into the controversy as it unfolded—to an ever larger public through an impressive outpouring of critical articles, books, and videos. Biologists thus obtained the enmity of those determined to maintain high levels of cutting in the Northwest national forests. Yet, charged by the Endangered Species Act with evaluating the chances for preservation of the critical habitat necessary to maintain a threatened species, government scientists in study after study continued to confirm the dire threat to the northern spotted owl, and indeed to the entire Northwest forest, reinforcing the environmentalist argument.

The convergence of these opposing economic and ecological forces in the early 1980s therefore signaled the emergence of contradictory conditions of the kind that Carolyn Merchant has associated with "ecological revolutions." These are characterized by "widening tensions between the requirements of ecology and production in a given habitat and between production and reproduction."[8] As it became clear that the very existence of the ancient forest ecosystem was in danger, environmentalists, scientists caught up in the dispute, the judicial arm of the state (under the pressure of the Endangered Species Act and other environmental laws), and certain sections of the Forest Service, Bureau of Land Management, and Fish and Wildlife Service bureaucracies came to side with "the requirements of ecological reproduction," while the forces of capital and the command posts of the state (mainly within the topmost echelons of the federal executive) leaned toward the interests of production. The result was a widening ecological and class war as capital stepped up its efforts to exploit the old-growth forest, environmentalists responded on behalf of the forest, and the workers, caught in the middle, struggled to defend their economic livelihoods.

In April 1990, a scientific study carried out in conformity with the Endangered Species Act by an interagency panel of government biologists (known as the Jack Ward Thomas report, after the panel's chairperson) proposed setting aside more than 5 million acres of federal timberland in the form of "habitat conservation areas" to protect the northern spotted owl. If implemented this would have effectively doubled the amount of protected lands in the public forests of Washington, Oregon, and northern California, leading to an almost 50 percent drop in annual federal timber sales from the region. But even if this habitat conservation plan were fully adhered to, according to the biologists who prepared the report, the northern spotted owl's population could be expected to plummet by as much as one-half, from its estimated level of about 3,000 pairs, over the next several decades.

It is important to emphasize that, since the remaining old-growth acreage is not only limited but exists only in the form of scattered patches, the preservation of the owl habitat depends almost as much on the preservation of numerous "corridors" linking areas of widely dispersed old-growth forest (often occurring in a checkerboard pattern) as on the protection of the intact old-growth forest itself. Moreover, environmentalists have naturally struggled to preserve those acres of forest land that, while not conforming to the strictest definition of old growth—usually because the ecology had been damaged in some way—nevertheless embody a wealth of biological values, including the capacity to help support owl and other endangered species populations. Finally, in practice the issue has often boiled down to where to draw the lines on the map, raising practical, jurisdictional issues related to the extent and usage of various sections of the national forests. The battle to preserve the ancient forest in Washington, Oregon, and California therefore involved from the very start several times the area represented by the 2.4 million acres in Washington and Oregon that, according to the Morrison estimates, could be classified as fully intact old-growth forest. Environmentalists, in fact, tended to view the Jack Ward Thomas plan—despite its commitment to setting aside more than 5 million acres—as inadequate for the preservation of the old-growth forest ecosystem, since this plan had envisioned a further drastic decline in northern spotted owl populations over the ensuing decades.

The Jack Ward Thomas plan, the Forest Service estimated, would lead to the loss of 28,000 timber jobs over the next decade. At the same time, grossly inflated industry estimates placed the number of jobs to be lost due to the direct and indirect effects of the Thomas plan at more than 100,000. Soon the northern spotted owl was on the cover of *Time* magazine—under the sardonic heading "Who Gives a Hoot?"

Under pressure from the law, the environmentalists and the courts, the U.S. Fish and Wildlife Service, acting on the results of the Thomas report, officially listed the northern spotted owl as a threatened species under the Endangered Species Act in June 1990. From that point on, the crisis only seemed to intensify. In April 1991, the U.S. Fish and Wildlife Service announced that it would evaluate up to 11.6 million acres in Washington, Oregon, and northern California for possible protection to preserve the habitat of the northern spotted owl. Over the course of the following year, while court injunctions effectively banned most logging in the old-growth forests pending the adoption of plans in conformity with the Endangered Species Act, the number of acres under consideration for protection dropped from 11.6 to 8 to 7 million acres; and when the Fish and Wildlife Service unveiled its final recovery plan for the owl in May 1992, the amount of critical habitat to be protected had been reduced to 5.4 million acres—approximately equal to the Jack Ward Thomas plan—with projected job losses at 32,000. In contrast to the Thomas plan, however, the recovery plan estimated the loss of less than one-quarter of the total owl population, with the expectation that the remaining habitat would support 2,300 pairs of owls in comparison to the present 3,000. Moreover, the multidisciplinary scientific team responsible for the recovery plan presented a fairly optimistic scenario suggesting that the owl population would be sufficiently large and well dispersed for the owl to survive, replenish its numbers, and, at some point, be removed from the threatened species list.

Still, in the view of the Bush administration, the recovery plan provided by the Fish and Wildlife Service in conformity with the requirements of the Endangered Species Act—although a necessary step in getting the courts to allow a resumption of logging in the Northwest national forests—was not acceptable. The idea was to undermine it from the outset, as part of a larger campaign against the Endangered Species Act itself. Secretary of the Interior Lujan had publicly voiced the opinion that "maybe we should change the [Endangered Species] law. . . . The spotted owl business is probably the prime example." The first major thrust in the Bush administration counterattack, dubbed an "Act of God" by the Southern Forest Products Association, was to convene in 1992 (for only the third time in its history) the Endangered Species Committee, commonly known as the God Squad because of its power to override species preservation on the grounds of economic necessity. The second major thrust was to release a separate Interior Department plan at the same time as the Fish and Wildlife Service recovery plan—with the express purpose of undermining the latter.[9]

The God Squad's membership, as set out in the Endangered Species Act, includes the Secretaries of the Interior, Agriculture, and Army, the heads of the President's Council of Economic Advisers, the Environmental Protection Agency, and National Oceanic and Atmospheric Administration (all of whom are presidential appointees), and a representative from each affected state (in this case Oregon). On this occasion, the God Squad had been convened at a Bureau of Land Management (BLM) request to decide on whether to override the Endangered Species Act in the case of 44 sales of BLM timber. As reported in the *Portland Oregonian* (May 17, 1992), "The God Squad met . . . in the Interior Secretary's small, wood-paneled ceremonial conference room. Access was tightly restricted, but Lujan's staff reserved 10 seats for 'constituents.' All 10 were filled by representatives of the timber industry, labor unions and timber communities."

The result was as expected. In a largely symbolic attack, the main effect of which was to throw doubt on the Endangered Species Act, the God

Squad voted 5–2 (the head of the EPA, William Reilly, and the Oregon representative dissenting) in May 1992 to exempt 13 BLM timber sales from the requirements of the Act.

Immediately following the God Squad vote, Lujan released both the recovery plan mandated by the Endangered Species Act and the rival Interior Department plan promoted by Lujan himself. In the Lujan plan, prepared by a small team of Interior Department officials that included no biologists, the area to be protected would be slashed by nearly one-half (to only 2.8 million acres), reducing the number of habitat conservation areas from 196 to 75, while the surviving owl population, as estimated by the plan, would decrease to a maximum of 1,300 breeding pairs out of the 3,000 pairs now existing. According to Lujan, this Interior Department plan would result in the loss of only 15,000 jobs. However, since the Lujan plan would fail to protect the threatened species throughout its range, it represented a clear break with the provisions of the Endangered Species Act and would require special congressional legislation to be put into effect. Environmentalists immediately labeled the Lujan proposal an "extinction plan." Those who saw the Interior Department plan in these terms included scientists responsible for the preparation of the Thomas and recovery plans. In the cautious estimation of Jonathan Bart, who headed the government's multidisciplinary recovery plan team, the Lujan plan by providing insufficient habitat would "eventually result in extinction" over many decades of the northern spotted owl.

Confident that the wind was changing in their direction, supporters of the timber industry greeted the Lujan plan with only a lukewarm response. Although some timber industry representatives declared that the Interior Department plan was a "step in the right direction," other defenders of the Northwest industry, such as Republican Senator Bob Packwood from Oregon, refused to support even the Lujan plan on the grounds that it would eliminate too many jobs, claiming, "It comes down to this: are you for people or for the bird?" Still others declared that even the 75 conservation areas to be set up in the Lujan plan were unnecessary, on the spurious grounds that the owl could survive in second-growth forest.[10]

For many, however, the virulence of the Bush administration's assault on environmental legislation, the northern spotted owl, and the old-growth forest no doubt came as a considerable surprise. Indeed, what has made the nature of the ancient-forest crisis so mysterious from the beginning has been the tendency for most establishment discussions to focus in fetishized fashion on timber, owls, loggers, and environmentalists while ignoring the major historical agent of change: capital itself including the capital–state partnership ("a *partnership* between two different separate forces, linked to each other by many threads, yet each having its own separate sphere of concern").[11]

The existence of a capital–state partnership of this kind explains the continuity between the Bush and Clinton administrations in their management of the Pacific Northwest forests. After presiding over a Forest Summit in Portland, Oregon, and considering eight options to deal with the crisis of the ancient forest and the threat to the existence of such endangered species as the spotted owl and the Northwest salmon, Clinton insisted on the development of an "Option Nine." Ostensibly a proposal for protecting the old-growth forest—one that would reduce historic logging levels—Option Nine was nonetheless so full of holes that it left more than 40 percent of the remaining old growth completely unprotected, with much of the rest also opened up to aggressive thinning and salvage logging operations. As many as 200 different species of plants and animals remain in danger of extinction. Meanwhile, the administration refused to move against log exports—the one measure that would have done the most to expand employment opportunities in the industry. In addressing the ecological crisis, the Clinton administration thus stayed well within the bounds dictated by the corporations. Indeed, by June 1995 Clinton was to declare that "I've done more for logging than any single person in the country." A month later he signed the Budget Rescissions Act, which included the notorious salvage logging rider, setting aside the nation's environmental laws under the cover of a "forest health crisis," and thus opening up vast tracts of the national forest to logging, further endangering such species as the marbled murrelet, the northern spotted owl, and the Northwest salmon.[12]

From the beginning of this struggle, the giant forest products firms deliberately stayed behind the scenes, leaving the defense of their interests to their major political lobbying organizations, the American Forest Resource Alliance and the National Forest Products Association. Meanwhile, few mainstream commentators have thought it worth their while to explore the historical dimensions of this ecological catastrophe brought on by the accumulation of timber capital. The public is thus left with the distinct impression that the whole problem can be reduced to an irreconcilable conflict between workers and environmentalists, between owls and jobs—a conflict in which the state is presumably neutral and capital is notable mainly by its absence. It is this great silence with respect to timber capital's historic role, including its partnership with what might be termed the "natural resource state," that must be penetrated if a realistic understanding of the fate of the forest is to emerge.[13]

MONOPOLY CAPITAL AND ENVIRONMENTAL DEGRADATION: THE CASE OF THE FOREST

Most forest land in the United States is privately owned. The largest part belongs to farmers, ranchers, and small owners, while a handful of giant timber corporations, owning only a small portion of the whole, but in control of vast tree plantations in the most productive tree-growing regions in the Southeast and the Northwest, dominate timber production nationwide. These "even-aged industrial plantations" with their monocultures of pine and fir have been dubbed "forestry's equivalent to the urban tower block."[14]

Such concentrated control of the conditions governing the production and marketing of timber by a relatively small number of firms at the apex of the industry, Veblen argued early in the twentieth century, emerged in accordance with "the characteristic traits of the American plan [of natural resource exploitation]—initial waste and eventual absentee ownership on a large scale and quasi-monopolistic footing." In the Northwest, the giant private forest holdings were formed during the monopolistic drive at the turn of the century, with the largest tracts emerging from railroad property. In 1900, the Weyerhaeuser Timber Company came into being when the Northern Pacific Railroad sold 900,000 acres of virgin timberland to a group of Midwestern logging entrepreneurs headed by Frederick Weyerhaeuser. Today six companies, led by Weyerhaeuser, own more than 7 million acres of forest in the Northwest. As a result, these firms are able to dominate the entire forest product industry in the region—from the growing and harvesting of trees, to the operation of lumber and plywood mills and pulp and paper factories, to the marketing of the final products. Smaller forest product companies, lacking significant private forest lands of their own, must rely almost entirely on access to public timber to feed their mills.[15]

From the beginning, the power of the large timber firms depended on their ability to limit competition and prevent prices from falling by keeping an oversupply of timber from reaching the market. By the late 1920s, however, the Northwest timber industry was experiencing a serious glut of supply, followed by a depression in 1929. Timber capital therefore encouraged the federal government to add tens of billions of additional board feet of "standing timber" to the national forests (150 billion board feet [bbf] were added in 1933 alone) to be harvested, in contrast to the more rapid rate of cutting on private lands, only on a sustained-yield basis. In this way, the major corporations were able to achieve the following three objectives: (1) limiting the supply of timber on the market; (2) maintaining higher prices for their own timber; and (3) establishing timber community stability (hence the existence of a readily exploitable labor force) in what were essentially company towns.

With the coming of World War II, market conditions changed and total national timber production ratcheted upward, from a low of 17 bbf in 1933 to 36 bbf in 1941. Timber production continued to climb after World War II as a result of pent-up demand for housing and programs such as Veterans Administration mortgages. It was the Korean War boom, however, that produced the peak in private timber harvests in the Northwest. In 1952, corporations removed enough board feet from private lands in Oregon alone to house both Oregon's two million people and San Francisco's

700,000. From this point on, private timber harvests declined sharply. Yet, corporations continued to cut trees at a frenetic pace and were slow to replant prior to the 1960s. As a result, timber companies and homebuilders began to demand more intensive harvesting of high-value old growth on public timberlands to compensate for the shortages in private supplies. Annual removals of national forest timber rose from 3 bbf in 1945 to 13 bbf in 1970. Yet, this was not enough for the corporations. In 1970, a Nixon administration task force, bowing to pressures from industry, wrote that "a goal of about 7 billion board foot annual increase in timber harvest from the national forests by 1978 is believed to be attainable and consistent with other objectives of forest management."[16] By the 1980s, this "mining" of ancient timber had produced a sharpened contradiction between ecological and economic requirements. On the one hand, an environmental movement grew by leaps and bounds as a result of growing concern over the vanishing forest, reinforced by a more sophisticated scientific understanding of the late successional forest ecosystem. On the other hand, conditions of economic stagnation in the late 1970s and 1980s—reflected in a drop in housing starts from two million in 1976 to one million in 1982—put renewed pressure on capital to restructure its relation to both labor and the environment, speeding up its exploitation of both.

In this developing contradiction, it was the immediate economic imperative that initially had the upper hand. During the Reagan years, increased sales of national forest timber were seen as a means of lowering lumber prices and overcoming a severe slump in housing. At the same time, the pull of the world market was exerting increased pressure on U.S. timber supplies. More and more timber was finding its way abroad in the form of unprocessed logs destined mainly for Asia, where the selling price for logs was up to 50 percent higher than in the United States. In 1987, 3 bbf of logs were exported from U.S. Pacific ports to Pacific Rim countries—almost 70 percent to Japan alone. By 1988 this amount had reached 4 bbf (equivalent to about 60 percent of the total harvest from federal lands in Oregon and Washington). Meanwhile, U.S. imports of Canadian lumber between 1975 and 1985 rose from less than one-fifth to one-third of U.S. softwood lumber consumption.[17] Although the government prohibits the export of logs from federal forests, the fact that logs from private lands are being shipped in large quantities abroad means that the overall demand for timber is increased and local sawmills are forced to rely more and more on public timber from the old-growth forests for their supplies.

Eager to exploit growing world demand for logs and at the same time force down U.S. lumber prices, the Reagan administration pursued every means at its disposal to accelerate federal timber harvests. The man appointed to accomplish this as Assistant Secretary of Agriculture for Natural Resources and the Environment (hence the boss of the Forest Service), was John Crowell, Jr., formerly general counsel for Louisiana-

Pacific, the largest purchaser of federal timber. No sooner was his appointment confirmed than Crowell proposed a doubling of the rate of harvest from federal forest lands in Oregon and Washington, from an annual rate of 5 bbf to 10 bbf by the 1990s. Since this rate of cutting was far beyond what could be regarded as sustained yield, it was immediately apparent that Crowell stood for the quickest possible liquidation of the remaining stands of ancient forest. The chief barrier to "more efficient National Forest management," Crowell asserted, "has been the timber policy of 'non-declining even flow'. . . . The volume of wood present in these old-growth forests far exceeds what would be present as growing stock inventory once the forest is in a fully managed condition."[18] Or, as he stated more succinctly elsewhere, "If you cut the old growth you're liquidating the existing inventory and getting the forests into a fully managed condition."[19]

The entire Reagan strategy of increased exploitation of the U.S. national forests—it is crucial to understand—depended on a vastly accelerated rate of cutting in the Northwest in particular, since it was from these national forests that the great bulk of the net proceeds from Federal timber sales were obtained, although most Federal timber placed on the market came from forests in other parts of the country. In 1987, 90 percent of the net receipts from Forest Service timber sales came from the 12 Northwest forests, which nevertheless accounted for only one-third of the timber harvested from U.S. national forests that year.[20] Costs associated with timber sales (road building, etc.) depend on the area sold, but revenues depend on the volume of timber sold as well as wood quality. Both volume/area and quality are very high in the Northwest old-growth forests, which make them by far the most profitable area of U.S. Forest Service operations. Profit criteria therefore demanded higher rates of cutting in these forests. And since almost everywhere else in the United States the Forest Service was, in fact, selling timber at a complete loss, continued sales of high-value old-growth timber in the Northwest were essential to keep the overall timber sales budget in the black and thus to prevent enormous losses elsewhere—and hence the full extent of the federal timber subsidy to capital—from becoming visible.

But in order to carry out its plan of increasing sales and harvest levels in the Northwest national forests, dictated by all of these factors, the Reagan administration found it first necessary to deal with the crisis in the timber industry brought on by the depression in the national homebuilding market, which had been badly hit by the effects of skyrocketing interest rates in the early Reagan era. And this meant lowering the price charged to timber companies for federal timber from the Northwest still further. Contract arrangements for federal timber have traditionally allowed firms to purchase cutting rights for standing timber and to delay harvesting for two to five years until market conditions were favorable—a policy that has encouraged widespread speculation. The housing crash of 1982 thus left timber firms sitting on vast inventories of federal timber that were now overpriced in relation to depressed domestic lumber prices. Through the timber contract bailout of 1984, signed into law by President Reagan, the federal government made it possible for firms to profit from this situation by releasing them from contracts for billions of board feet of uncut timber and then reselling the same trees back again to the companies at bargain-basement, recession-level prices. Profits soared as corporations and Northwest members of Congress forced the sale of high volumes of low-cost federal timber (with both sales and harvests reaching near record levels) throughout the remainder of the 1980s and in the first year of the following decade.[21]

Meanwhile, internal BLM plans in 1983 to trim cutting and introduce longer rotation times in the forests in western Oregon under its jurisdiction, in the face of dwindling agency timber supplies, were suddenly scotched late that same year (some of those involved believe that the BLM's parent agency, the Interior Department, then headed by James Watt, was responsible) and timber harvests were instead increased. Thus, it comes as no surprise that internal BLM memos made public in 1990 warned that the agency had been harvesting at unsustainable levels and was running out of trees to cut. "In some cases there is no place to go after 1991," one internal memo noted.[22]

Equally disastrous from the standpoint of sustainability was the granting, beginning in 1984, of federal subsidies for private log exports—under rules pertaining to a wide variety of export commodities—which allowed timber firms with foreign-based sales operations (i.e., multinationals) to obtain tax exemptions of 15–30 percent of their export income. By 1992, this was costing the U.S. Treasury $100 million a year in lost revenue. According to U.S. Representative Les AuCoin (D–OR), Plum Creek Timber (formerly Burlington Northern) used these subsidies for log exports to export in effect over 5,000 U.S. forest product jobs in the 1980s while pocketing $33 million in tax savings.[23]

ECOLOGICAL CONFLICT AND THE CLASS STRUGGLE

The first real sign that the traditional, rather peaceful give-and-take over U.S. forest lands between accumulation and conservation had been radically transformed by the 1980s in ways that suggested the emergence of an era of confrontation occurred in April, 1983, when four Earth First!ers appeared out of nowhere in the Siskiyou National Forest in Oregon and took their stand between a running bulldozer and a tree. Before long, radical environmentalists were sitting on company dynamite to prevent blasting, tree spiking (driving large nails into trees in order to hinder the cutting and processing of timber), tree sitting, chaining themselves to timber equipment, and forming human barricades on logging roads by setting their feet in cement-filled ditches or inserting themselves in rock piles.[24]

While Earth First!ers chose a path of direct confrontation, other environmental groups relied on legal action. Soon federal agencies found themselves immersed in a flood of lawsuits and administrative appeals. In 1987, 25 environmental groups filed the first of three spotted owl lawsuits through the Sierra Club Legal Defense Fund, thus setting in motion the chain of events leading to the release of the Jack Ward Thomas report and the listing of the owl as threatened in the spring and summer of 1990.

For Northwest forest products workers these actions by environmentalists were naturally viewed with growing anguish. There can be no doubt that the impending "locking away" of millions of acres of public timberland increased the economic insecurity of thousands of workers. Soon, frustration with what they saw as an extreme preservationist ethic was inducing many workers to display angry bumper stickers such as "I Love Spotted Owls—Fried"—often seen in timber areas of the Northwest. On a number of occasions, owls (not northern spotted owls, because they are hard to find) have been found killed and nailed to trees or road signs. One was discovered with its head placed in a noose.

Timber firms have generally sought to reinforce this rage of the workers against environmentalists, adding fuel to the fire at every possible opportunity, with sawmill owners actually sponsoring antipreservationist lectures during working hours at the mills. More tragic, however, is the fact that environmentalists have sometimes fed this rage through the insensitivity with which they have occasionally greeted the plight of the workers. For example, the Native Forest Council—well known throughout the Northwest for its radical environmentalist publication *Forest Voice*—has argued that the problem of workers threatened by displacement can be left to the condign sanctions of the market:

> A market economy does not maintain an industry simply for the sake of employing workers. When a product becomes obsolete or a resource runs dry, the economy adapts. Companies and industries have been changing or shutting down for 200 years, and workers always find new jobs—the nation is not lacking in jobs; it's a natural, necessary component of capitalism. Chopping down forests for the sake of jobs is nothing more than social welfare—not something our nation prides itself on.[25]

Such unsympathetic attitudes toward workers are not uncommon among those who see themselves as representatives of "deep ecology." This can be seen in a position taken by Dave Foreman, cofounder of Earth First! (but now departed from the organization):

> One of my biggest complaints about the workers up in the Pacific Northwest is that most of them aren't "class conscious." That's a big problem. . . . The loggers are victims of an unjust economic system, yes, but that should not absolve them for everything they do. . . . Indeed, sometimes it is the hardy swain, the sturdy yeoman from the bumpkin

> proletariat so celebrated in Wobbly lore who holds the most violent and destructive attitudes towards the natural world (and towards those who would defend it).[26]

Despite the radical rhetoric, there can be no doubt that Foreman exhibits an extremely condescending attitude here toward workers (the so-called bumpkin proletariat) and their efforts to maintain their economic livelihood. It is surely inadequate to say that environmentalists are not contributing to the economic insecurity of workers when jobs are being threatened as a result of environmentalist actions, with environmentalists as a whole doing very little directly to aid the workers caught in this situation. It is equally objectionable to complain about lack of "class consciousness" and an absence of resistance among workers while turning a blind eye to the concrete struggles actually taking place. Nor should one be overly hasty to condemn forest product workers, the majority of whom believe in promoting a sustainable relation to the forest at some level, for adhering to destructive attitudes toward the natural world.

Not just deep ecologists but also mainstream environmental groups commonly distance themselves from workers. Less inclined to adopt the language of class, the latter seldom express their disdain for workers as openly, but their "that's not our problem" attitude—not to mention their interlocking directorates with major corporations and their white, upper-middle-class membership base—suggest many of the same biases.

As the late Judi Bari, the leading figure within Earth First! in the 1990s wrote, one of the major obstacles to labor–environmentalism is "the utter lack of class consciousness by virtually all of the environmental groups":

> I have even had an international Earth First! spokesman tell me that there is no difference between the loggers and the logging companies! . . . I have heard various environmentalists say that working in the woods and mills is not an "honorable" profession, as if the workers have any more control over the corporations' policies (or are gaining any more from them) than we do. As long as people on our side hold these views, it will be easy pickins for the bosses to turn their employees against us.[27]

The problem goes deeper than a mere failure of most environmental groups to align themselves with labor, however. When environmental organizations have sought to develop alliances with workers, they have often been targets of repression—emanating both from vested economic interests and the state. In May 1990 Judi Bari and Dale Cherney, both of whom had been involved in organizing the major environmental protest known as Redwood Summer, were injured and Bari nearly killed when a bomb that antienvironmental terrorists had placed under the driver's seat of Bari's car exploded. In a classic case of state repression, the FBI arrested Bari and Cherney for the bombing. They were later released.

Further, in those cases where forest ecosystem defenders and forest product workers have gotten together the mass media has provided very little coverage. In the words of longtime environmental activist Hazel Wolf, "As Secretary of the Seattle Audubon Society, I have attended joint environmental–labor press conferences that were poorly attended by the press and consequently ignored by the media. This is a story that is all too familiar and accounts for the success of the timber industry in driving a wedge between the environmentalists and loggers using the myth of jobs vs. owls."[28]

The failure of most environmentalist groups to link their demands with those of workers in the industry is most evident in the reluctance to take sides in the fierce battle being waged between employers and employees in and around the Northwest forests. In the 1980s, forest product workers in the Northwest were hit by a process of industrial restructuring that seriously undermined their economic positions and their capacity to engage in effective class struggles. These included: (1) a drastic drop in housing starts; (2) increased exports of unprocessed logs coupled with rising excess capacity in Northwest mills; (3) a vastly stepped-up rate of imports of lumber from Canada (which had the effect of creating deep fissures between Canadian and U.S. workers within the International Woodworkers of America); (4) rapid declines in employment due to mechanization; (5) wage competition from Southern wood workers (who earned almost $3 an hour less on average in 1986 than their Northwest counterparts); and (6) a general shift of the industry from the Northwest to the Southeast, where faster growing pine plantations and right to work laws provide a greater "comparative advantage" in timber production.

Of all of these factors affecting Northwest timber employment, automation has been the most important. In 1987, it took only eight workers to process one million board feet of timber, compared to ten workers a decade earlier. In 1976, a total of 15 bbf of timber was harvested from all sources in Oregon and Washington, giving employment to 150,900 workers in the lumber and wood products and paper and allied products industries. In 1989, the same total harvest level employed 135,700, or about 10 percent fewer workers. In Oregon, the state with the largest old-growth forests, employment in the lumber and wood products industries declined by 21.9 percent between 1978 and 1990, with 71 percent of this decline occurring between 1978 and 1988, before the northern spotted owl became a major issue.

Not surprisingly, capital chose this period to launch a wider class offensive. In 1983 Louisiana-Pacific demanded 8–10 percent rollbacks and the creation of a two-tier wage structure at its 15 Northwest mills, forcing the unions to strike. With no agreement after a year, the union locals at Louisiana-Pacific's mills were decertified. In 1985, Weyerhaeuser demanded wage and benefit cuts of about $4 an hour at a number of mills. When the unions resisted, the mills were closed. Having demonstrated its clout,

Weyerhaeuser in 1986 was able to force an agreement with the unions that involved wage and benefit cuts of $4 an hour plus the implementation of a complex "profit-sharing" scheme. Although strikes continued to break out at Northwest mills in the late 1980s, it was clear that the unions had suffered a great reversal in their class war with capital.[29]

During these fierce battles between forest products firms and their workers, environmentalists were generally nowhere to be seen (with the exception of certain Earth First! activists like Judi Bari) and scarcely seemed to notice. Few in the green movement saw this as an occasion to demonstrate their solidarity with workers. Indeed, middle-class environmentalists sometimes seem to go out of their way to separate themselves from workers.[30]

The political and organizational consequences of this environmentalism without class, separating environmentalists from workers, is particularly evident at the grassroots in the Northwest. Today the conflict at the popular level in Oregon is visible more and more in terms of the opposition between two large coalitions: on the one side, the Oregon Natural Resources Council (ONRC), the most powerful regional environmental organization in the country, embracing some 6,000 members and representing some 80 different conservation groups; on the other side, the Oregon Lands Coalition (OLC), a predominantly conservative, proindustry coalition, embracing over 72,000 members and encompassing 47 different organizations. While the ONRC is closely tied through local chapters to such national conservation organizations as the Sierra Club, the Audubon Society, the Wilderness Society, and the National Wildlife Federation, the antienvironmentalist OLC has forged close links to Republican figures in the Northwest congressional delegation and to the American Forest Resource Alliance—as well as more tenuous relations with AFL-CIO locals (tenuous because of the reputed antiunion orientation of some OLC member organizations, such as the procapital Yellow Ribbon Coalition).

The deep divisions that have emerged in this way between the labor and ecology movements explain much of the success of the state—under both the Bush and Clinton administrations—in containing the environmentalist assault on the timber industry. Exploiting to the full the divisions among popular forces, the Bush administration early on adopted a strategy of staving off the separate threats represented by environmentalists and workers to the interests of capital through a policy of divide and conquer. Thus, on the one hand, George Bush announced that he was concerned above all with the jobs of workers threatened by efforts to protect the endangered spotted owl: "We want to save the little furry-feathery guy and all that but I don't want to see 40,000 loggers thrown out of work."[31] On the other hand, the administration had repeatedly let it be known that the President was opposed to special legislation designed to assist displaced workers.[32]

Despite the release of the Thomas report in April 1990, and the listing of the northern spotted owl as a threatened species in June 1990, the executive branch ordered the Forest Service in the spring and summer of 1990 to stop working on an owl protection plan. At the same time, the White House suppressed a Forest Service/BLM report on the northern spotted owl that had come up with numerous ways to offset the job losses experienced by workers. According to what Democratic Congressional Representative Peter DeFazio has called a "reliable source" in the Forest Service, the Forest Service/BLM study was killed by the administration. A May 1990 draft of the suppressed report contained over 52 pages of concrete recommendations, including an $86-million public works program modeled after the Civilian Conservation Corps of the 1930s, bans on log exports "for all ownerships," increases in the share of revenues from timber sales to be returned to timber-dependent communities, extensive retraining programs, and money for road reclamation projects. Since this report pointed to the fact that a political solution to the crisis that would meet the needs of both environmentalists and workers was perfectly feasible, "someone in the White House, . . . I'd lay even money on John Sununu," DeFazio said, simply killed the report.[33]

Overall, the Bush administration was clearly on the defensive on the owl question from the spring of 1990 to the spring of 1992, when the Lujan plan was released. Its initial strategy was to encourage the main federal agencies involved—the Forest Service, the Bureau of Land Management and the Fish and Wildlife Service—to delay the adoption of plans to safeguard the northern spotted owl. Such delays would allow the timber companies to extract huge quantities of additional timber from the Northwest forest. For a year this delaying tactic seemed to work. But, beginning in April 1991, the administration strategy collapsed as federal courts (laying the blame on the Bush administration) ruled that the federal agencies were not in compliance with the law, with the result that bans were placed on Forest Service (and later Bureau of Land Management) sales until agency plans to protect the owl were formulated and put into effect. From that point on, the Bush administration focused its efforts on the more aggressive strategy of undermining the long-awaited Fish and Wildlife Service recovery plan and the Endangered Species Act itself through the convening of the God Squad, and the launching of its own plan for slow extinction of the owl in the interest of greater timber extraction.

The demise of the Bush administration seemed to many to mark a turning point in the struggle to save the ancient forest. Not only had a new president been elected who promised to end the conflict, but the entire struggle, which had long centered on the preservation of the northern spotted owl, had been successfully transformed by environmentalists into the wider issue of the preservation of an entire ecosystem, with the listing of other Pacific Northwest species as threatened or endangered by the fall of 1992. In October 1992 the marbled murrelet, a tiny seabird that nests in the coastal old-growth forests of the Pacific Northwest, was listed as a threatened species. Although two-thirds of the murrelet's range overlapped with that of the northern spotted owl, the effect of this new listing meant that certain coastal forest areas not included in the set-asides for the owl were affected. Also in 1992 three wild salmon stocks (the spring, summer, and fall Snake River Chinook) were listed under the Endangered Species Act—with immediate significance for logging operations along the streams of Northwest national forests, where the salmon returned each year to spawn. This served to dramatize the fact that more than 100 generally recognized wild salmon stocks in the region were extinct, while most of the rest were in danger of extinction—with logging constituting one of the most important contributing factors, after dams. As a result of these developments the issue of the ancient forest came to be seen as related to the management of entire watersheds, making it clear that it was no longer a question of the northern spotted owl alone, but rather a matter of life and death for numerous species within the Pacific coast bioregion.[34]

If these developments generated a period of great expectations on the part of environmentalists engaged in the struggle for the ancient forest, this was soon to change to one of deep disappointment, even anguish. After presiding over a forest summit in Portland, Oregon, and considering eight options to deal with the crisis of the ancient forest and the threat to the existence of such endangered species as the spotted owl and Pacific coast salmonids, President Clinton decided that none of these permitted enough logging to take place, which led to the development of the Clinton forest plan, based on the notorious Option Nine. As Alexander Cockburn and Ken Silverstein have written:

> While Option Nine reduced the amount of logging allowed on national forests, it failed to set aside any permanently protected old-growth forest preserves, and permitted clearcutting in the most ancient groves and in the most vital spotted owl and salmon habitat. In fact the environmental analysis accompanying Option Nine admits that this strategy places hundreds of species at increased risk of extinction, including the spotted owl, marbled murrelet, and dozens of stocks of Pacific salmon and steelhead.

The little that the administration did for workers, such as its much promoted "Jobs in the Woods" program, proved to be woefully inadequate. According to the Pacific Rivers Council, an environmental organization involved in the program, Jobs in the Woods, despite very limited funding and other constraints, "got characterized [misleadingly] as an employment program, bringing first false hopes and later frustration and anger to dislocated workers and community leaders who thought they would see immediate benefits."[35]

The biggest disappointment for environmentalists came with President Clinton's signing of the notorious salvage logging rider to the Budget Rescissions Bill in July 1995. Only two months earlier Clinton had condemned the salvage rider for provisions that "would basically direct us to make timber sales to large companies, subsidized by taxpayers, . . . [and] that will essentially throw out all of our environmental laws and the protections we have that surround such timber sales." Rather than veto the Budget Rescissions Bill (a second time), the president, however, signed it into law, and the salvage rider, now in effect, required the release of nearly a billion board feet of timber sales in the ancient forest of Oregon and Washington that had been held back because of environmental laws, including those sales that threatened such endangered species as the northern spotted owl, the marbled murrelet, and Pacific salmon stocks. In this way around 180 timber sales in Oregon and Washington with high environmental impact were mandated by Congress and the president. Some proponents of this legislation, such as U.S. Senator Slade Gorton (R–WA), did not hesitate to explain that economic motives dominated over ecological ones. "There comes a time," he has long contended, "when you just have to say enough is enough, and let a species go extinct." In a major giveaway to business at the expense of taxpayers in general, the rider required the Forest Service to price the sales at 1990 price levels.

In the case of the marbled murrelet the sales demanded by the salvage rider encompassed 15 percent of the murrelet's remaining habitat, with murrelet populations expected to drop by a similar fraction as a result. The danger to the murrelet is particularly great because 59 of the sales mandated by the rider are within the Siuslaw National Forest where the best unfragmented old-growth forest, making up the murrelet's habitat, is to be found, and where most of the murrelet population resides. Consequently, the marbled murrelet, as Scott Greacen of *Wild Forest Review* has written, seems "doomed by greed."[36]

With a new assault on the forest taking place the front line of the environmental struggle shifted to the militant environmental activists of the direct action groups like Earth First! In the sugarloaf area of the Siskiyou National Forest environmental activists seeking to defend 669 acres of old growth confronted the U.S. Forest Service, police, and Boise Cascade, resulting in numerous arrests beginning in October 1995 and the largest instance of environmental disobedience in Oregon history. Similar confrontations occurred in the Umpqua National Forest, where, day after day, protesters from Earth First!, Cascadia Forest Defenders, Siskiyou Forest Defenders and other groups attempted to halt logging with car blockades and other work interferences in the spring of 1996. Meanwhile, in an attempt to prevent the logging of the Warner Creek area in the Willamette National Forest Earth First!ers built road blockades, complete with a stockade and moat, and dug in on a long-term basis. The standoff ended

11 months into the blockade, after the Clinton administration announced in late July 1996 that Warner Creek would not be logged under the salvage rider and that an attempt was being made to swap the Warner Creek timber sale (made to the Thomas Creek Lumber and Log Co.) for another timber sale. Nevertheless, environmentalists—still on their guard—continued to occupy the site. In mid-August 1996 federal officials moved in, closing off the entire area to the public and denying access to the media. They then proceeded to bulldoze "Fort Warner," arresting five protesters along with two reporters from the major regional paper, the *Eugene Register-Guard*, for trespass. Strong-arm tactics by law enforcement authorities included keeping the two journalists handcuffed for three hours (detaining them for six) while seizing their film and notes, which were then developed and copied—all clearly in violation of federal and state constitutional guarantees protecting freedom of the press.[37]

Despite such militancy, environmentalists seeking to preserve the old-growth forest are clearly on the defensive at present, and the reasons why are not difficult to discern. A fairly organized antienvironmental backlash has developed over the past few years, rooted in the Wise Use movement based in the West. Wise Use backers, as earlier noted, consist mainly of officers from resource-extracting corporations, small landowners, sawmill owners, ranchers, farmers, and mining interests, with some support among workers in the timber and mining industries. Among their goals are clear-cutting wilderness areas, protection of grazing rights, deregulation of mining, exemption of private lands from environmental regulations (and compensation for any "takings" resulting from such regulations), and the selling off of large tracts of public land to the highest bidders. It is no secret that the Wise Use movement is bankrolled by corporations; thus, People for the West! relied on corporations for 95 percent of its 1990 budget. Nevertheless, the Wise Use movement has a grassroots basis. As Mark Dowie explains in his book, *Losing Ground,* Wise Use activists

> are organizing in the mostly rural communities while the mainstream environmental movement continues to rely for its support on the mailing lists of liberal magazines, suburban charities, and other environmental organizations. . . . However deceptive it may be, the appeal of the Wise Use message is very real to hard-working farmers, loggers, and miners—the very people environmentalism needs to reach if it is to survive as a relevant and effective movement, particularly in the West.[38]

Put simply, the success of the Wise Use movement can be attributed in part to the failure of big environmentalism (and big labor) to forge a strong labor–environmentalist alliance. All of this suggests that the time when environmentalism could prosper as a single-issue movement ignoring issues of class (and issues of race, gender, and international inequality) is clearly over.

TOWARD A STRATEGY OF ECOLOGICAL CONVERSION

If the foregoing analysis is correct and the environmentalist cause has been impeded by the executive arm of the state acting in tandem with the large corporations, while workers and endangered species are being forced to bear the main costs of the crisis, it is essential for environmentalists and workers to join forces around a common platform. A progressive class-oriented response to the old growth crisis would have to focus on an ecological conversion program that could be enacted at the level of the state. As Victor Wallis has argued, the term "*conversion* . . . has traditionally referred to the switch from a military to a civilian economy." But the concept can be applied more broadly to the socially planned redirection of the economy necessary to create a sustainable society.[39]

There is no doubt that an ecological conversion strategy of this sort could be adopted in relation to the old-growth forest crisis. Moreover, there are progressive, ecologically concerned voices within the workers' movement who would back such a strategy. This is illustrated by the position taken by William Street, a policy analyst for the International Woodworkers of America (IWA). Writing in May 1990 in his column in the IWA's paper, *The Woodworker,* Street explains:

> We . . . know . . . a worker's forest policy . . . starts by recognizing the need for a sustainable and renewable forestry. It recognizes that each portion of the planet must produce its proportional share of the resources it uses. The proportion should be produced as environmentally sound as possible. . . . A worker's forest policy would harvest at a sustainable rate and insure that those mature trees that are harvested are used for those socially desired products for which there are no substitutes. By thus restricting the use of older trees, harvest pressure would be diminished without contributing to unemployment.[40]

This position taken by a progressive figure within a forest product union does not represent a solution to the tragedy of the ancient forest, since it does not fully take into account the fragility of the remaining old-growth forest. Yet, it represents a view that includes ecological and social components that are crucial to any attempt both to save the forest and the safeguard the livelihood of workers. It constitutes a viewpoint, moreover, that is a far cry from those that single-issue environmentalists often attribute to workers.

One thing that Street's "worker's forest policy" makes clear is that, once the narrow profit-making goals of corporations are no longer seen as the primary constraint in working out solutions to problems of the environment and employment, all sorts of new rational possibilities open up, allowing for the development of common ground between workers and environmentalists. Clearcutting could conceivably be replaced by the "new forestry" techniques promoted by ecologist Jerry Franklin, in which the aim is to mimic natural processes by leaving behind large standing trees, snags, and fallen trees. Restrictions could be placed on the uses to which mature timber could be put—so that old growth could not be logged and then pulped to be converted into products like disposable diapers. Highgrading, or the selective cutting of the oldest and most valuable timber alone, could be prohibited. The use of herbicides and the burning of slash could be eliminated. Current bans on federal log exports could be followed up by bans (or export duties) on private log exports. Early-retirement programs could be designed for older workers in the industry, coupled with guaranteed annual employment programs for those in the smaller work force that remains. Larger shares of forest revenues could be returned to local communities. A Civilian Conservation Corps could be established to construct recreation facilities and carry out ecological restoration projects in the forests. Roads could be reclaimed in habitat conservation areas. Conversion funds could be provided to convert old-growth sawmills into more modern plants equipped to process second growth. A windfall profits tax could be placed on timber corporations that see the value of the timber on their tree plantations rise as a result of curtailments of public timber supplies. Extensive education and retraining programs (a workers' GI Bill) could be established for displaced forest product workers. Economic development grants and loans could be made available to distressed communities. Federal programs could be developed to help manage timber more effectively on nonindustrial private forest lands. Current federal timber contract practices could be altered to ensure that timber would be sold at its full value and to decrease speculative purchases. Federal subsidies to timber capital through road-building budgets could be sharply curtailed and the freed-up funds redirected to social services in timber communities. Funds could be allocated for the expansion of national forest lands to be managed on a nonprofit, ecologically sustainable basis, with revenues from the land base being used to support working communities. Finally, international agreements could be promoted to establish uniform practices of sustainable forestry and to reduce global competitive pressures that encourage deforestation and forest fragmentation.

What is important to recognize is that only a few relatively minor steps in this general direction would go a long way toward solving the employment problem and community instability caused by the "set-asides" for the protection of the northern spotted owl. A unified labor–environmentalist strategy that would meet the needs of both the forest ecosystem and forest communities is therefore perfectly feasible. What is necessary to make this possible is for society to invest some of its economic surplus in assisting workers whose jobs and communities are being undermined by new ecological requirements.

Unfortunately, people such as William Street are somewhat isolated within union circles, and organized labor in the Northwest has been reluctant to put its full weight behind ecological conversion (or industrial transition) programs when limited efforts have been made in this direction, since this is seen as an unnecessary concession to preservationists who wish to reduce logging levels. A sign of the times is that Irv Fletcher, President of the Oregon AFL-CIO, threatened U.S. Representative Elizabeth Furse (D-OR) with withdrawal of support for her reelection if she continued to oppose the salvage logging rider.[41] Matters are made still worse by the fact that the major environmental organizations have shown little direct concern for the plight of the workers and have only recently begun to think in a rather modest fashion about industrial transition. Under these circumstances, it is perhaps not surprising that the labor unions themselves have been overshadowed in this area by conservative coalitions that are unabashedly antipreservationist and procapital. Thus, when the God Squad announced in May 1992 that it would exempt 13 BLM sales from the Endangered Species Act, a representative of the Oregon Lands Coalition was quoted as saying, "This decision is a victory for the workers of Oregon, however small it may be." What is noteworthy about this statement is not so much the position taken as the fact that a conservative, probusiness, and antienvironmentalist citizen's alliance such as the Oregon Lands Coalition (which includes groups such as cattle grazers and realtors, as well as the antiunion, grassroots timber industry organization, the Yellow Ribbon Coalition, in its membership) should become—in the absence of a progressive trade union response to the crisis—the main voice for the "workers of Oregon" on the old-growth question.

This failure of the regional unions to push hard for an ecological conversion program is partly explained by the fact that such a program is an extremely difficult strategy for unions in a natural resource industry in an out-of-the-way area of the country to pursue on their own—particularly under circumstances of a declining natural resource base, economic crisis, capital relocation, union decline, and growing environmental controls. Ultimately, the pursuit of an ecological conversion strategy requires not imaginative initiatives in a depressed community so much as coordinated action on a national scale, and this involves finding the means to force the channeling of economic surplus into ecological conversion programs throughout the country. That sufficient surplus for this purpose exists can scarcely be doubted.[42] Recognizing this, the Oil, Chemical and Atomic Workers Union has proposed the creation of a "Superfund for Workers" that would offer up to four years of support to people displaced by environmentally destructive industries in order to enable them to pursue vocational retraining, or even an entire career shift by means of extended education. Other possible variations on this Workers' Superfund program include assistance to help form small businesses and income supplements for individuals who decide to pursue less well paid work. The annual cost for a million workers might be $40 billion.[43]

The actual trend in the United States in recent years, however, has been in the opposite direction—toward less and less support for displaced workers. Federal outlays for worker retraining are far below what they were when Reagan was first elected. Under these circumstances, workers end up carrying a larger and larger share of the total cost to society of industrial transition. In 1987, public spending on employment and retraining as a percentage of Gross Domestic Product (GDP) was 1.7 percent in Sweden, 1 percent in West Germany, 0.7 percent in France, Spain, and Britain, and a miniscule 0.3 percent in the United States.[44]

This situation is a problem not simply for workers and trade unions but for any environmental movement worthy of its name. Capitalism as a system devoted to accumulation without end is inseparable from a capital-intensive, energy-intensive economy—and thus necessitates growing throughputs of raw materials and energy, along with the creation of excess capacity, surplus labor, and economic and ecological waste. This should be differentiated from the basic needs of the broad majority of people, which have to do with the availability of steady and worthwhile employment and an improving quality of life, therefore having no inherent link to an intensive process of ecological degradation. Northwest timber workers, for their part, want above all to protect their livelihood and communities. In this respect the export of unprocessed logs, the relentless drive for ever higher levels of automation, the emphasis on clearcutting as opposed to "new forestry," the use of chemical weed killers, the burning of slash, and so on make no sense, from a worker's standpoint.

The "job blackmail" that often seems to compel workers to adopt an antienvironmental stance can therefore be seen to be tied to a system that promotes profits by means of the exploitation of both human beings and nature.[45] The direct route to the creation of a mass environmental movement is one that seeks to break the seemingly intractable conflict between jobs and environmental protection (a conflict symbolized nowadays by owls versus jobs) by placing ecological conversion—the planning of new ways of working with nature while fulfilling social needs—at the very core of each and every ecological struggle. This necessarily means moving away from the attitude that environmentalism can somehow stand above and beyond the class struggle.

A shift toward a broad movement for ecological conversion and the creation of a sustainable society also means that the partnership between the state and the capitalist class, which has always formed the most important linchpin of the capitalist system, must be loosened by degrees, as part of an overall social and environmental revolution. This partnership must be replaced, in the process of a radical transformation of the society, by a new partnership between democratized state power and popular power.[46] Such a shift requires revolutionary change that must be more than simply a rejection of capitalist methods of accumulation and their effects on people and the environment. Socialism—as a positive, not just a negative, alternative to capitalism—remains essential to any conversion process, because its broad commitment to worldwide egalitarian change reflects an understanding of "how the needs of the various communities can fit together in a way that leaves nobody out, but that also satisfies the environmental requirements that are global. Within a socialist framework, the sources of the largest-scale and most severe environmental destruction could be dealt with head-on, in a way that has already shown itself to be beyond the capacity—not to say against the interests—of capital."[47]

From an eco-socialist perspective there is no difficulty in seeing that the rapid destruction of the old-growth forest is not about owls versus jobs but ecosystems versus profits. Ecology tells us that the destruction of a complex ecosystem rooted in a climax forest that took centuries and even a millennium or more to develop involves thresholds beyond which ecological restoration is impossible. We must therefore find our way to a more rational economic and social formation, one that is not based on the amassing of wealth at the expense of humanity and nature, but on justice and sustainability. Whether the issue is species extinction, death on the job, women's control of their own bodies, the dumping of toxic wastes in minority communities, urban decay, Third World poverty, the destruction of the ozone layer, global warming, nuclear contamination, desertification, soil erosion, or the pollution of water resources, the broad questions and answers remain the same. As the authors of *Europe's Green Alternative* have written, we must choose between two logics: "on the one side, economics divorced from all other considerations, and on the other, life and society."[48]

ACKNOWLEDGMENTS

This is a revised, expanded, and updated version of John Bellamy Foster, "The Limits of Environmentalism without Class: Lessons from the Ancient Forest Struggle of the Pacific Northwest," *Capitalism, Nature, Socialism, 4*(1), 1993, pp. 11–41. I would like to thank Michael Dawson, Chuck Noble, Doug Boucher, Alessandro Bonanno, and Judi Bari for their comments at critical stages in the preparation of this chapter.

NOTES

1. Quoted in Martin Ryle, *Ecology and Socialism* (London: Century Hutchinson Ltd. 1988), p. 13. See also John Bellamy Foster, "Britain's Green Budget," *Capitalism, Nature, Socialism, 3*(2), June 1992.

2. See John Bellamy Foster, "The Absolute General Law of Environmental Degradation under Capitalism," *Capitalism, Nature, Socialism, 3*(3), 1992.

3. For an earlier article on which portions of the following argument are

based, see John Bellamy Foster, "Capitalism and the Ancient Forest," *Monthly Review, 43*(5), October 1991.

4. Michael Renner, "Creating Sustainable Jobs in Industrial Countries," in Lester R. Brown et al., *The State of the World 1992* (London: Earthscan, 1992), p. 138.

5. Peter Morrison, in Joint Hearings, Subcommittee on Forests, Family Farms, and Energy of the Committee on Agriculture, and the Subcommittee on National Parks and Public Lands of the Committee on Interior and Insular Affairs, U.S. House of Representatives, U.S. Congress, 101st Congress, First Session, *Management of Old-Growth Forests of the Pacific Northwest,* June 20 and 22, 1989 (Washington, DC: U.S. Government Printing Office), pp. 270–278; *Portland Oregonian,* October 15, 1990; Tucker, quoted in *New York Times,* June 11, 1992. The Peter Morrison/Wilderness Society estimate, based on the analysis of satellite pictures, takes account of the extreme fragmentation of these forests. Other estimates, including some by the Wilderness Society, that are broader and less rigorous in their methodology have placed the remaining old-growth forest acreage as high as 4.7 million acres or more (World Resources Institute, *The 1992 Information Please Environmental Almanac* [Boston: Houghton Mifflin, 1992], pp. 143–145). The crazy quilt pattern of the forest was partly the result of railroad land grants, which had given every other square mile of land along the path of the railroads to private railroad interests, resulting eventually in a checkerboard pattern of public and private forest lands. See Derrick Jensen and George Draffen, *Railroads and Clearcuts* (Spokane, WA: Inland Empire Lands Council, 1995).

6. For the larger theoretical significance of this see James O'Connor, "Capitalism, Nature, Socialism: A Theoretical Introduction," *Capitalism, Nature, Socialism, 1,* Fall, 1988.

7. R. H. Waring and J. F. Franklin, "Evergreen Coniferous Forests of the Pacific Northwest," *Science, 204,* 1979; Elliot A. Norse, *Ancient Forests of the Pacific Northwest* (Washington, DC: Island Press, 1990), pp. 20–24, 27–32, and 141–144; Catherine Caufield, "The Ancient Forest," *New Yorker,* May 14, 1990, pp. 46–49; David Kelly and Gary Braasch, *Secrets of the Old-Growth Forest* (Salt Lake City, UT: Peregrine Smith Books, 1988), pp. 21, 36–37, and 63; and Chris Maser, *The Redesigned Forest* (San Pedro, CA: R & E Miles, 1988), p. 53.

8. Carolyn Merchant, *Ecological Revolutions* (Chapel Hill: University of North Carolina Press, 1979), p. 5.

9. Lujan quoted in Jonathon King, *Northwest Greenbook* (Seattle, WA: Sasquatch Books, 1991), p. 53; and Southern Forest Products Association, *Newsletter,* October 7, 1991.

10. *Eugene Register-Guard,* May 15 and 24, 1992; *Portland Oregonian,* May 15 and 17, 1992.

11. Ralph Miliband, *Divided Societies* (New York: Oxford University Press, 1991), pp. 30–34.

12. Jeffrey St. Clair, "Any Which Way You Cut It," *Wild Forest Review, 1*(4), March 1994, p. 14; Jeffrey St. Clair, "Clinton and the Ancient Forests," *Lies of Our Times, 5*(3), March 1994, p. 14; and Jeffrey St. Clair, "Salvage Dreams," *Wild Forest Review, 3*(1), January/February 1996, p. 13.

13. For the notion of the "natural resource state" organized in the United States around the Department of Interior in particular, see Christopher Manes, *Green Rage* (Boston: Little, Brown, 1990).

14. Edward Goldsmith et al., *The Imperiled Planet* (Cambridge, MA: MIT

Press, 1990), p. 94; and Richard E. Rice, "Old-Growth Logging Myths," *The Ecologist, 20*(4), July–August 1990, pp. 143–145.

15. Spider Burbank et al., *A Study of the Weyerhaeuser Company as a Multinational Corporation* (Olympia, WA: The Evergreen State College, June 1975), p. 1; and Thorstein Veblen, *Absentee Ownership and Business Enterprise in Recent Times* (New York: A. M. Kelley, 1923), p. 194.

16. David A. Clary, *Timber and the Forest Service* (Lawrence: University Press of Kansas, 1986), pp. 80–93 and 110–111; William B. Greeley, *Forests and Men* (Garden City, NY: Doubleday, 1951), p. 206; Con H. Schallau, "Sustained Yield versus Community Stability," *Journal of Forestry, 87*(9), September 1989, p. 18; Keith Ervin, *Fragile Majesty* (Seattle, WA: The Mountaineers, 1989), p. 123; and Daniel R. Barney, *The Last Stand* (New York: Grossman, 1974), pp. 88–89.

17. U.S. Forest Service, Pacific Northwest Research Station, *Production, Prices, Employment and Trade in Northwest Forest Industries,* various issues; Marcus Widenor, "Pattern Bargaining in the Pacific Northwest Sawmill Industry: 1980–1989," in Steven Hecker and Margaret Hallock, eds., *Labor in a Global Economy* (Eugene: University of Oregon, Labor Education and Research Center, 1991); and Widman Management Limited, *Markets 89–93: The Outlook for North American Forest Products* (San Francisco: Widman Management, 1989), pp. 79, 107.

18. John Crowell, "Excerpts from a Speech by John B. Crowell, Jr.," in Bureau of Governmental Research and Service, University of Oregon, *Old Growth Forests: A Balanced Perspective* (Eugene: University of Oregon, 1982), pp. 133–136.

19. Quoted in *Portland Oregonian,* October 15, 1990. Crowell's position, while extreme, reflects the dominant Forest Service/timber industry view that a steady or even accelerated cutting of old growth in the national forests is necessary to close the "window" of a temporary shortage of timber, brought on in the past few decades by past failures of forest management and the slow growth of trees in the Northwest. This window, it is believed, will be closed in the first quarter of the twenty-first century when enough second growth-timber will be available to sustain production indefinitely—a point that will be reached, however, only when all of the old growth in the national forests has been removed and the entire timber economy had been put on a fully commodified, tree-plantation basis. In federal forest management this approach is justified as broadly consistent with "sustained yield" forestry. In practice, however, it has little to with sustainability in either ecological or economic terms, and has become little more than an additional rationale for pursuing timber capital's age-old policy of cutting as much timber as the market will bear.

20. Caufield, *The Ancient Forest,* pp. 69–70.

21. Joe P. Mattey, *The Timber Bubble That Burst* (New York: Oxford University Press, 1990), pp. 3–9; and *Portland Oregonian,* October 15, 1990.

22. *Eugene Register-Guard,* May 24, 1992.

23. *Eugene Register-Guard,* October 26, 1992.

24. Manes, *Green Rage,* pp. 10–15, 86–88, 99–102, and 210–211. Tree spiking is extremely controversial since it is life threatening to workers, who can be injured or killed when saw blades come into contact with the spike. Since March 1990, when Judi Bari publicly renounced tree spiking at an Oregon conference at the urging of timber worker Gene Lawhorn, Earth First!ers in the Northwest have repeatedly repudiated this tactic. See Rik Scarce, *Eco-Warriors* (Chicago: Noble Press, 1990), pp. 74–78 and 83; and Clay Dumont, *Loggers and Radical Environ-*

mentalists: Cultural Struggles in Timber Country (PhD dissertation, University of Oregon, 1991), pp. 79–84 and 135–137.

25. Native Forest Council, *Forest Voice,* 2(2), 1990, p. 5. See also Dumont, *Loggers and Radical Environmentalists,* pp. 85–91.

26. Dave Foreman in Steve Chase, ed., *Defending the Earth: A Dialogue between Murray Bookchin and Dave Foreman* (Boston: South End Press, 1991), pp. 51–52. Not all within Earth First! would agree with Foreman's point of view. Judi Bari is the most famous of those Earth First!ers who have adopted a labor–environmentalist stance. On Bari's ideas and position within the movement, see Scarce, *Eco-Warriers,* pp. 80–85, and Timothy Leigh Ingalsbee, *Earth First!: Consciousness in Action in the Unfolding of a New Social Movement* (PhD dissertation, University of Oregon, June 1995).

27. Judi Bari, *Timber Wars* (Monroe, ME: Common Courage Press, 1994), p. 14.

28. Ibid., pp. 286–328; David Helvarg, *War against the Greens: The "Wise-Use" Movement, the New Right and Anti-Environmental Violence* (San Francisco: Sierra Club Books, 1994), pp. 331–339; and Hazel Wolf, "Relief for Unemployed Timber Workers," *Monthly Review, 43*(9), February 1992, pp. 46–47. On the repressive role of the state, see Michael Parenti, "Popular Sovereignty vs. the State," *Monthly Review, 46*(10), March 1995, pp. 1–16. Not just environmentalists but workers too have been subject to repression when they have attempted to build a common cause with environmentalists. On this the case of Bari's ally, former mill worker Gene Lawhorn, is instructive; see Helvarg, *War against the Greens,* pp. 112–117.

29. Renner, "Creating Sustainable Jobs . . . ," pp. 150–151; U.S. Forest Service, Pacific Northwest Research Center, *Production, Prices, Employment and Trends*; *Eugene Register-Guard,* May 3, 1991; Widenor, *Pattern Bargaining,* pp. 252–261. For examples of certain tentative attempts by Earth First!ers to take action in support of workers, see Ingalsbee, pp. 361–366.

30. For example, in their popular ecological photo essay, David Kelly and Gary Braasch falsely glorify environmentalists as a nature-loving "bourgeoisie," which "donned backpacks *en masse* and headed for the hills, where they discovered the environment was in trouble" (Kelly and Braasch, *Secrets of the Old Growth Forest,* pp. 57–59).

31. Quoted in the *Times* (London), May 28, 1992.

32. On administration opposition to special assistance for the displaced workers, see the *Christian Science Monitor,* June 6, 1991.

33. U.S. Forest Service and Bureau of Land Management, *Actions the Administration May Wish to Consider in Implementing a Conservation Strategy for the Northern Spotted Owl,* unpublished, May 1, 1990; and *Eugene Register-Guard,* May 4, 1991.

34. Andy Kerr, "Saving Pacific Salmon," *Wild Fish,* May/June 1993, p. 3.

35. Alexander Cockburn and Ken Silverstein, *Washington Babylon* (London: Verso, 1996), p. 221; and Pacific Rivers Council, *Analysis and Recommendations for the Federal Land Jobs in the Woods Program* (Eugene, OR: Author, March 1995), p. 15.

36. President Clinton quoted in Stephen T. Taylor, "Blame the Democrats," *Washington Post,* October 29, 1995; Jeffrey St. Clair, "Salvage Dreams," *Wild Forest Review, 3*(1), February 15, 1996, pp. 8–21; Jeffrey St. Clair and Scott

Greacen, "(K) Is for Killer," *Wild Forest Review,* February 15, 1996, pp. 22–25; and Scott Greacen, "Bye Bye Birdie," *Wild Forest Review,* February 15, 1996, pp. 26–28.

37. Cimmaron and Raven, "Umpqua Ancient Forest Spared," *Earth First!, 16*(5), May 1996, p. 11; John Green, "Warner Creek," *Earth First!, 16*(6), June 1996, p. 9; and *Eugene Register-Guard,* August 18, 1996.

38. Mark Dowie, *Losing Ground* (Cambridge, MA: MIT Press, 1995), pp. 93–98 and 101–103. For a fuller development of the argument on the Wise Use movement presented in this chapter, see John Bellamy Foster, "Wise Use and Workers," *Dollars & Sense,* March/April 1996, p. 7.

39. Victor Wallis, "Socialism, Ecology and Democracy," *Monthly Review, 44*(2), June 1992, pp. 15–18. See also Raymond Williams, "Socialism and Ecology," in *Resources of Hope* (London: Verso, 1989).

40. Raymond William Street, "Ecology Is Not a Four Letter Word," *The Woodworker,* May 20, 1990. Since the original version of this essay was written the IWA, once one of the great radical unions of the West Coast, has closed down due to declining membership—a symptom of the larger crisis of the industry.

41. Fletcher quoted in St. Clair, "Salvage Dream," pp. 19–20.

42. See Michael Dawson and John Bellamy Foster, "The Tendency of the Surplus to Rise, 1963–1988," in John B. Davis, ed., *The Economic Surplus in the Advanced Economies* (Brookfield, VT: Edward Elgar, 1992), pp. 42–70. An abbreviated version of this research appeared under the same title in *Monthly Review, 43*(4), September 1991, pp. 37–50.

43. Renner, "Creating Sustainable Jobs . . . ," pp. 153–154.

44. Ibid.

45. Renner, "Creating Sustainable Jobs . . . ," p. 139.

46. Miliband, *Divided Societies,* pp. 228–229 and 233.

47. Wallis, "Socialism, Ecology and Democracy," pp. 16–17.

48. Penny Kemp et al., *Europe's Green Alternative: A Manifesto for a New World* (London: Merlin Press, 1992), p. 16.

Chapter 8

Remapping North American Environmentalism

Contending Visions and Divergent Practices in the Fight over NAFTA

Michael Dreiling

The implementation of the North American Free Trade Agreement (NAFTA) on January 1, 1994, firmly established a major policy victory for the architects of neoliberal markets.[1] For over two years (1991–1993), the U.S. Trade Representative's office and prominent corporate supporters faced a barrage of legal and political assaults on their efforts to pass NAFTA. Recognizing these challenges as a threat to their planned trade zone, a powerful corporate mobilization swiftly disabled their opponents and established support in Congress. Following congressional approval of the agreement in November 1993, the Heritage Foundation declared "Ronald Reagan's vision realized."[2] With similar enthusiasm, the U.S. Council for International Business served up international leadership awards to corporate executives for their role in "the business community's successful effort to pass NAFTA."[3] Indeed, while leading segments of capital celebrated, an elaborate public relations campaign disassociated big-business interests from the agreement and concealed the disastrous socio-ecological consequences of NAFTA and the larger economic strategy on which it rests.

Beneath these shrouds of neoliberal *celèbre,* the march of unhindered capitalist markets, and in particular NAFTA, continues to generate unjust and ecologically destructive economic practices. Contrary to claims that environmentally destructive growth will be curbed by market incentives in NAFTA and its side accords (supposedly the "greenest" trade agreement ever), the concentration of toxic industries along the border, rural and industrial dislocation, and the conversion of indigenous lands to cultivation of export crops have all accelerated under NAFTA.[4] The reduction of nature and community to commodities and cost factors—exacerbated by the promotion of unfettered trade and greater capital mobility—override any "green" incentives intended to deter capitalist zeal from boosting profits by displacing costs onto society and nature. The ease and flexibility with which capital may invest, relocate, and subsume nonmarket goals through NAFTA merely facilitates and encourages such cost externalization.

The sustained struggle against this neoliberal project, however, ignited internal transformations in key social movements in North America and pushed important movement organizations toward more democratic, tactically offensive, and internationalist practices. As early as October 1990, a broad range of social movements coalesced to assert a claim for "fair trade" and submit the imperatives of profit and accumulation to a "Continental Social Pact." The ability to construct and deploy "fair trade" within the social movement sectors of Mexico and the United States was conditioned during the early 1990s as two national historic conjunctures unfolded. The left populism of *neocardenismo*[5] in Mexico met a U.S. global environmentalism—recently challenged by a movement for environmental justice—and integrated popular opposition across political boundaries. They "met" at the border—the free trade zone—where political and economic crises in both countries collided with nation and ecology.

Indeed, by 1992 hundreds of cross-border links among *grassroots* groups constituted one important base for the national-level coalitions in opposition to NAFTA. Across North America, mounting opposition in the polls, on the streets, and in the workplace made legislative passage of NAFTA an uncertain affair, particularly in the United States and Canada. In the United States, the American Federation of Labor–Congress of Industrial Organizations (AFL-CIO) placed the trade agreement on their *legislative* agenda immediately upon news of the agreement, but environmentalists were at the forefront of mobilizing a critique of free trade, particularly in response to the extensive environmental degradation along the United States–Mexico border.

While this chapter focuses on the importance of environmentalism in politicizing global development and trade policy in the United States, the role of organized labor in harnessing resources and mobilizing opposition to NAFTA cannot be underestimated. Moreover, during the period between 1989 and 1992, while many environmental organizations were forging grassroots ties across borders, the AFL-CIO made an important shift in policy toward the labor movement in Mexico. The AFL-CIO broke a strong relation with the ruling party's (Partido Revolucionario Institucional—PRI) official union federation in Mexico, the Confederación de Trabajadores Mexicanos (CTM). At the same time, activists proposed that the AFL-CIO get involved in coalition efforts to expose the exploitation and abuse of Mexican communities and workers by transnational corporations along the border. This coalition effort led to one of the most important binational groups along the United States–Mexico border, the Coalition for Justice in the Maquiladoras (CJM). Consequently, the AFL-CIO's financial support and participation in the CJM forged a series of "weak ties" with other social movement sectors and activists, ties that would become highly salient in the development of national anti-NAFTA coalitions and continental networks. Thus, the labor movement was crucial to the strength of the anti-NAFTA challenge, but remained slow to cultivate a public discourse in opposition to NAFTA. In some ways this was a strategic error, which became apparent by early 1992 when Ross Perot began his drive for the presidency and monopolized the rhetoric on jobs and NAFTA, thus effectively excluding the potency of an autonomous challenge by labor.[6]

To summarize our main questions, what were the conditions that prompted the broad, cross-border environmental challenge to NAFTA? In what ways did environmentalist participation in the anti-NAFTA coalitions, both nationally and internationally, inform the political strategies of the fair trade movement? How was environmental mobilization subsequently disabled as a formidable opponent?

In answering these questions, this chapter focuses on environmental mobilization and countermobilization within the United States. In addition, systematic contrasts with NAFTA opposition in Mexico demonstrate the underlying tensions between movements that focus on the practice of social justice versus those defined by conventional tactics.[7] For example, recent grassroots movements for environmental justice have altered the terrain of environmentalism by offering an alternative to the conciliatory, market-oriented environmentalism being defined by several dominant mainstream groups. Furthermore, not only has the challenge of environmental justice introduced a political ideology that is critical of the environmental mainstream, but also it has taken a form that aims resources away from conventional channels of power toward the grassroots. Emulating some of these environmental justice tendencies during the NAFTA fight, groups such as Greenpeace and Friends of the Earth (FOE) pursued a strategy of international solidarity and cross-movement alliance building. Together, a movement intersection between environmental justice and the global focus of groups such as Greenpeace and Friends of the Earth shaped a strategy emphasizing grassroots mobilization and a global containment of neoliberalism. The struggle over NAFTA facilitated this position.

Yet, these practices and orientations were not the only forces defining opposition to NAFTA. The uneven response by mainstream environmental organizations to the challenge of fair trade and, by implication environmental justice, shaped their stance in the NAFTA fight. Those groups disposed to compromise with policy elites and the advocacy of market incentives found little difficulty in rejecting the multi-issue concerns being expressed from the broad coalitions against NAFTA. On the other hand, those groups facing internal ideological and grassroots leadership schisms, such as the Sierra Club, were pulled further away from the co-optive strategies of capital and the state policy elite. Similarly, in Mexico, social movement efforts to dissociate from clientelistic forms of state control found increased autonomy and leverage following the *neocardenista* mobilization since 1988. This increased autonomy and demand for democratization signified a break from the solid postrevolutionary power of Mexico's ruling party. Amidst heightened state repression in Mexico, the struggle against NAFTA reinforced earlier efforts to split from entrenched forms of political co-optation. Coupled with a search for autonomous politics and recent grassroots mobilizations, political and cultural concerns for environmental justice in the United States and a broad, left populism in Mexico were harnessed as "fair trade" strategies aimed at forging cross-border alliances and circumscribing neoliberal markets.

Three interrelated points define my aims in this chapter. First, social and ecological dislocations associated with widespread implementation of neoliberal markets, particularly along the United States–Mexico border, have generated local political responses with repercussions affecting both the language and political practice of leading North American environmental groups. Second, this confluence of movements, neoliberal markets, and environmental degradation entered the NAFTA debate and stimulated the construction of an internationalist vision of "fair trade," emphasizing the subordination of unregulated trade and capital accumulation to principles of social and ecological democracy. Finally, by making apparent the distinction between free trade and fair trade, the cultural rift between neoliberal environmentalism and environmental justice was further exposed, revealing important political splits and contradictions within North American environmentalism.

Indeed, underlining the political threat that "free market environmentalism" represents to struggles for environmental justice, the *passage* of NAFTA was facilitated by the capacity of neoliberal capital to deflect environmental opposition by harnessing the political resources of several large environmental groups in both the United States and Mexico. By exposing the loyalty of these leading pro-NAFTA environmental groups to "market incentives" and political "compromise," however, the struggle ignited effects that rippled throughout key sectors of North American environmentalism. During the NAFTA fight, the progressive organizations—Greenpeace, Friends of the Earth, and most surprisingly the Sierra Club—expressed the ability and desire to adopt principles of environmental justice and forge meaningful cross-border and cross-movement alliances. While many important changes resulting from the sustained mobilization remain evident in movements throughout North America, there is strong evidence that those seemingly secure changes are insufficient. In order to combat co-optation by neoliberal strategists and avert an ever more encompassing enclosure of society and ecology from the snares of neoliberal capitalism, the more progressive environmental organizations must return to the grassroots and adopt both the culture and tactics of ecological democracy.

ENVIRONMENTALISM, *NEOCARDENISMO*, AND NEOLIBERALISM

The fractures and realignments among major environmental groups that occurred during the fight against NAFTA emerged from developments in the environmental movement that began to take form during the 1980s. Despite a relatively stable growth in membership throughout the 1980s, the environmental movement, according to many authors, was "losing ground."[8] A growing rift between the grassroots and the nation's largest groups exposed very different momentums, strategies, and resources dispersed throughout the movement. Mark Dowie refers to this rupture in terms of two contending approaches—"third-" and "fourth-wave" environmentalism—aimed at capturing the nation's environmental imagination.[9] Third-wave environmentalism, according to Dowie, "represents nothing so much as the institutionalization of compromise" during the 1980s and may well be remembered as "a brief attempt by corporate America to capture the nation's environmental imagination."[10] Third-wave groups, such as the Environmental Defense Fund, National Wildlife Federation, and the Natural Resources Defense Council, have proved quite effective in designing "pollution prevention" strategies that rely on "market-based incentives" such as pollution permits and allowances—in short, the right to pollute, then sell the pollution.[11] A fourth wave, Dowie argues, is emerging and has shuffled beneath the "Big Green" groups with an increasing sense of urgency and breadth of social constituents. This fourth wave is emerging with an uncompromisingly anticorporate ideology, "democratic in origin, populist in style, untrammeled by bureaucracy, and inspired by a host of new ideologies."[12] For Dowie and others, recent mobilizations for environmental justice embody this emerging "wave" of the movement. In order to understand the ways in which third-wave groups were challenged and destabilized in the NAFTA fight, it is worth examining these broad divisions and related developments in the movement, particularly in relation to neoliberalism.

First, during the 1980s several mainstream environmental organizations targeted large banks and, more generally, international finance capital in an effort to mediate the consequences of "ecological imperialism," most typically as manifested in the politics of biodiversity and tropical deforestation, trade in toxics, global warming, and a whole host of debt-related assaults on the biosphere.[13] The tactics employed in these global causes varied, largely reflecting the organizations' specific commitments to the grassroots. Groups such as Greenpeace and Friends of the Earth (and its global affiliates, e.g., Rainforest Action Network) typify a form of global environmentalism that is not confined to "polite" politics and in fact encourages grassroots mobilization. On the other hand, the National Wildlife Federation, World Wildlife Fund, and several others pursued a politics of compromise and negotiation with representatives of Northern global finance capital. Alongside both tendencies was an expansion of relations with a rapidly expanding environmental sector among social movements in the global South. Moreover, a menu of global environmental concerns in the monthly newsletters of these large U.S. organizations formed an effective marketing strategy, helping stimulate high membership growth rates among those organizations focused on international issues.[14]

Second, this two-pronged globalization of environmentalism was coupled with a "fourth-wave" grassroots current—most recently articulated as the environmental justice movement—linking local sites of conflict through regional coalitions "typically led by women, working-class people, and people of color."[15] Unlike conservationist efforts to maintain *distance* between social issues and species (or habitat) protection, the environmental justice movement "does not treat the problem of oppression and social exploitation as separable from the rape and exploitation of the natural world."[16] By 1989, the antitoxics and environmental justice leadership initiated a challenge to "third-wave" leaders, demanding recognition of the social injustices associated with unequal toxic exposures induced by poverty and environmental racism.[17] The mainstream response was uneven, at best.

Lastly, neoliberal intellectuals and corporate foundations began cultivating and refining the ideological tools of the third wave and forged a "free-market environmentalism;" an environmentalist discourse subsumed under neoliberal ideology.[18] Standing on the defensive in 1981, the CEOs of the largest conservationist organizations formed the "Group of Ten" (G-10) with the explicit aim of creating a niche in the neoliberal policy agenda under President Ronald Reagan. Succumbing to the political dominance of neoliberal capital, the "Group of Ten" increasingly leaned to market solutions for environmental regulation, turning nature into "a commodity to be bought and sold" and defining objectives "in terms of their economic value."[19] Under this logic, the "property rights paradigm" provides the foundation for exalting market *authority* through "market incentives," which form the "key" to both protection and cleanup. Jay Hair—then president of the National Wildlife Federation—urged that "our arguments must translate into profits, earnings, productivity, and economic incentives for industry."[20] Similarly, numerous right-wing intellectuals, sponsored by the Heritage Foundation (and affiliates), the Center for Strategic and International Studies as well as newly directed federal funds, began a politico-ideological offensive on the politically threatening sectors of the environmental movement.[21] As the Group of Ten's defensiveness subsided, overtures to industry became routinized through the boards and funding sources of these organizations.[22]

Neoliberalism was deployed quite effectively in international institutions, as well. This was particularly apparent with the construction of a new orthodoxy at the International Monetary Fund (IMF)—an orthodoxy that gave full authority to the "free market" to relieve the "distortions" of state-directed national development. Following the debt crises of the early

1980s, debtor governments had to "present a plan for economic stabilization conforming to conditionality and structural adjustment criteria demanded by the IMF and World Bank as the basis for certification of creditworthiness."[23] These criteria were consistent with the restructuring of Mexico's debt in 1983 and 1989, which the U.S. government and the IMF administered.[24]

In 1983, the Mexican government formally adopted a structural adjustment package labeled the Plan Integral de Recuperación Economica, with policy-implementing instruments corresponding to the orthodoxy of the IMF.[25] The program entailed a dramatic reduction in public expenditures, privatization, a reduction of the fiscal deficit, increases in real interest rates, cuts in real wages, and a shift of macroeconomic policy in favor of the export sector. Under increasing pressure from growing debt and economic crisis, the PRI had effectively come under the heavy influence of global capital, a condition widely perceived as an intrusion on national sovereignty.[26]

The passivity of the PRI to the requisites of international powers is not enough to understand the weakening of state corporatism and the crisis in the PRI's legitimacy. The ongoing struggles of popular movements to shape their linkage with the political system, and the corporatist strategies to manage these struggles, have contributed to changes in the institutional configuration of both civil society and the state. Foweracker suggests that popular struggles in the past decade have largely strengthened civil society, as groups have shifted from petitioning the *cacicazgos,* or the clientelistic systems of power, to demanding democracy and the normative application of law, or an "*estado de derecho.*"[27] This shift in institutional terrain toward increasing the "civilizing capacity" of revolutionary constitutional law has indeed created new spaces for civil society.[28] At the same time, as Laborde argues,[29] this new space in society has allowed for an extensive political project on the part of new business groups demanding a more restricted neoliberal state.[30] On the other hand, unlike some countries where

> a similar neoliberal . . . project to restrict the political arena has coincided with the "new social movements" aspirations to autonomy and noninstitutional politics, in Mexico the popular movements have taken the field to combat this project and defend the . . . social pact inscribed within [the corporatist state]. And in this regard, it appears possible that in Mexico the project is leading not only to less economic protection but also to less political "protection" and, in particular, to a far less effective corporative control of the electoral arena. Thus, the contingent outcome of this project of political exclusion is to open up new political spaces for the strategic initiatives of popular organizations.[31]

Thus, the emergence of a "neoliberal authoritarian state"[32] in Mexico—in the context of a complex of national and international forces—established an arena whereby a national coalition of social movements would press the political system into electoral crisis in 1988.

The election year of 1988 witnessed large popular-sector mobilizations, such as the first appearance of the National Front for Mass Organizations in Mexico City.[33] These mass movements coalesced under a left-populist leadership that emerged from an intraelite dispute within the ruling party and formed the National Democratic Front (FDN)—the movement precursor to the Party of the Democratic Revolution (PRD). The mass organizations that joined the FDN campaign formed a national level network between the administrative center of Mexico City and the country's northern regions. The PRD's presidential campaign–movement for Cardenas, ultimately becoming *neocardenismo,* thus "reversed the tendency of social movements to sectoralize their demands by permitting their insertion into a national project that aggregated all criticisms of the economic and social policy of the regime *and* arguments for the defense of national sovereignty and the rejection of the state party and corporatism."[34]

Following the defeat of popular forces and Cardenas's presidential campaign through widespread electoral fraud and repression, it was clear that the PRI's successor, President Carlos Salinas de Gortari, would push the neoliberal project to new limits. His administration oversaw the privatization of hundreds of state enterprises and the liberalization of national economic frontiers (ultimately by urging a NAFTA), which further intensified capital investment in the northern states and border regions.[35] This created a disparate but important socio-spatial link to the North from the political center, and ultimately the socioeconomic space for a deeper construction of "bridges" across the border. Salinas's "reform" of Article 27 of the Mexican Constitution pushed the neoliberal program into the most precarious of contexts—the legal foundation for the *ejido,* the communally regulated landholdings of indigenous *campesinos.*[36] This "reform," considered a necessity by neoliberal technocrats and the IMF, is aiming at the "enclosure" of all communal holdings by assigning property rights to *ejido* plots. Together with the lowering of tariffs on U.S.-produced corn and grains, which will accelerate dramatically with NAFTA, it is no wonder the Zapatista National Liberation Army (Ejército de Zapatista Liberación Nacional—EZLN) considers NAFTA "a death sentence for indigenous peoples."[37] With the absence of state credit subsidies and constitutional protection of the *ejido,* millions of indebted peasants could be forced to sell off their "shares" of land and enter into the wage labor market by the year 2000.[38]

This dramatic escalation of neoliberal reforms has no doubt continued to alter the political landscape of Mexico, most evident in the persistent defiance by the EZLN of neoliberal capitalism and the ruling party. Rejecting the corporatist strategies of the state, the EZLN has anchored (and benefited from) a popular strategy that deemphasizes conciliatory pacts with the government while asserting the primacy of social protection from the unregulated market through democratic institutions. The left populism of *neocardenismo* gave initiative and substance to this popular break from the corporatist state, though at the same time the popular upsurge in 1988 activated an intensified effort by the ruling party to either co-opt or repress new movements and organizations.[39] Following the elections of 1988, social movements in Mexico faced two related political conditions: a heightened opportunity to link regional and sectoral concerns through a national movement, and, through a semiautonomous relationship with the PRD, to develop a politics of disengagement from the PRI. Thus, when the NAFTA debate entered the picture as a site of public conflict in 1991, many conditions for a democratic-left and environmentalist assault were in place, particularly with the struggles already entered into along the border.

CROSS-BORDER COALITIONS AND SOCIOECOLOGICAL CLAIMS ON THE MARKET

Debates over NAFTA were preceded by, and developed around, the struggle over the meaning and character of economic development. The discursive terrain on which this struggle began was at the "border," or, more precisely, the conditions of life in the "free-trade zones" and the deregulated production in the *maquiladoras*. Rapid environmental degradation, persistent declines in real wages, and highly uneven patterns of urbanization made the United States–Mexico border the site for constructing a critique of neoliberal development and trade. It was here that the moment of *neocardenismo* and the regional base of its support met the confluence of third- and fourth-wave environmentalism. In other words, the assault on national sovereignty coincided with a global environmentalism, having a common meeting place at the border. The geohistorical concurrence of political–economic crises fragmented the character of the border, making it the exemplar of free trade, environmental degradation, and the assault on national sovereignty. Thus, newly formed binational coalitions entered national and international political arenas with a rhetorical instrument grounded in an environmental science of border life. The border became the region—at least temporarily—for testing the opposed frameworks of neoliberalism and environmental justice.[40]

What is important for the purposes here is the salience of the formation of "a social system [that] creates a community of interests around the boundary [between the nations]."[41] This boundary "is characterized as a setting where the North–South dialogue takes on a distinctly environmental and spatial character, because it is here that the two nations and cultures meet physically" on a natural geophysical surface, but interrupted by the political organization of this space.[42] Not unlike the cases of Chernobyl and the Rhine River contamination in Europe during the 1980s, the extensive environmental degradation along the United States–Mexico border exemplifies the contested and shared nature of this space.

The "border" as place entered public discourse in Mexico and the United States with rapidity during the 1980s. Growing attention by the media about United States–Mexican relations and concerns "signaled the growing significance of the international boundary between" the two nations.[43] The meetings held each year by the two nations increasingly focus on topics related to "border economic and policy issues, including immigration policy, *maquiladoras*, drug trafficking, tourism, border ecology, and pollution."[44] In addition, the interrelationships of the national civil societies have increased in density, with grassroots movements, labor union networks, associations and coalitional organizations, migration patterns, and business associations proliferating.

Between 1987 and 1991 cross-border links among labor, environmental, women's, human rights, Latina/o, and consumer organizations increased significantly and began fusing civil, worker, and environmental justice concerns. In Mexico, one of the most important independent labor organizations in the country, the Frente Auténtico del Trabajo (FAT)—and a key founder of the Mexican Action Network on Free Trade (RMALC)—developed numerous links with U.S. and Canadian labor.[45] Another border group, Mujer a Mujer (MAM—Woman to Woman) collaborated with Mujeres en Acción Sindical (MAS—Women for Union Action) to form the first trinational working women's conference.[46] A broad-based community coalition, the Coalition for Justice in the Maquiladoras (CJM), formed with the intention of pressing U.S. multinational corporations to adopt socially responsible practices. The formation of the CJM subsequently helped forge important ties across a broad cross section of national organizations and local grassroots organizations. The Border Ecology Project (BEP)—linked strongly to the Enlace Ecológico (Ecological Link) and the Proyecto Fronterizo de Educación Ambiental (Border Project for Environmental Education)—has produced environmental science for border struggles with corporations and government agencies. The now defunct National Toxics Campaign (a founding member of CJM) also produced a seminal analysis of border river pollutants. Their studies were widely used and cited in congressional hearings over "fast track"[47] and NAFTA and were subsequently influential in garnering wider environmental support.[48] More generally, border area groups began to form regular and cooperative relations with the larger antitoxics and environmental justice movements. Groups such as Sin Frontera, the Southwest Organizing Project, and the Citizen's Clearinghouse for Hazardous Waste (now the Center for Health, Environment, and Justice) found the border as a site where multiple forms of social and environmental exploitation could be viewed as a systematic problem that the environmental mainstream was unlikely to address. Importantly, these and other grassroots groups and coalitions, despite their limited resources as local organizations, set an institutional tone for the formation of broader national networks involving very large movement organizations.[49]

Unlike in Canada, labor and progressive movements in the United States were relatively idle throughout the Canada–United States trade negotiations. Though popular organizing efforts in the United States began to focus on the Uruguay Rounds of the General Agreement on Tariffs and Trade (GATT) in 1985, none had achieved a significant national level coalition until January 1991, when concerns over the implementation of "fast track" legislation for NAFTA escalated.[50] In 1989, representatives from several unions, environmental groups, farmer organizations, and progressive think tanks began to develop a "Fair Trade" campaign, where worker, environmental, consumer, farmer, and human rights could be enforceably linked to trade policy. In early 1990, news of the proposed United States–Mexico Free Trade Agreement "leaked" from the Bush administration. Differences in strategy among these "Beltway" actors toward the proposed free trade agreement became explicit. One bloc, to become the Alliance for Responsible Trade (ART), emphasized the importance of involving groups in Mexico to build an international campaign. The lobbying teams formed the Citizen's Trade Campaign (CTC) and pushed the *immediate goal* of defeating the pending trade agreement. Within the year, negotiations intensified for global trade liberalization under both GATT and the United States–Mexico Free Trade Agreement (the precursor to NAFTA). The "fair trade" advocates now set forth to mobilize two uneven, and sometimes competing, strategies through the groups that began to consolidate around the two coalitions.

Reflecting the diverse political practices of U.S. movements—from the isolated left wing of the labor movement to the moderate interest-group politics of the Sierra Club—the fair trade mobilization was similarly structured. The dominance of "interest group" tactics by the most resourceful sectors of U.S. movements strengthened, indeed dictated, the formation of the Citizen's Trade Campaign. Conversely, ART's greater capacity to deploy more disruptive tactics was severely constrained by the absence of resource-rich movement organizations. ART's efforts were nonetheless supported and remarkably influential in establishing (and, even today, still maintaining) international relations with unions and human rights groups in Mexico.[51] A differentiation of political strategy among U.S. groups, unlike in Canada and Mexico, thus resulted in *two* coalitions with unequal access to movement resources.

In Mexico, out of the context of a regionally fragmented popular sector emerged the local foundations for the formation of national coalitions and international links. In 1991, the binational networks that emerged from the problems along the border (CJM, Mujer a Mujer, etc.), provided the organizational base for broader alliances. Mexico City-based efforts were linked to the border regions through the Mexican Action Network on Free Trade (RMALC), which "originated within labor unions and other NGOs [nongovernmental organizations] with prior ties to groups in the United States and Canada such as the Authentic Labor Front, Common Frontiers, People's Team, and Mujer a Mujer."[52]

The sharply repressive and co-optive nature of the Mexican state via the military and the PRI's quasi-governmental clientelistic networks, created conditions for a single, though radical, alliance of activists and movement organizations. Without the equivalent of "Capitol Hill" and the short-term effectiveness offered by lobbying and litigation, the PRI's dominance of all popular platforms, including the media and central organs of the labor movement, prevented an anti-NAFTA campaign from materializing in ways analogous to the Citizen's Trade Campaign (CTC) in the United States. Rather, the RMALC focused tactically on grassroots mobilization and education from the outset. Working consistently with a broad range of center-left organizations, the RMALC mobilized through its numerous constituent organizations. Despite the PRI-sponsored media blackout, the political reins on organized labor through the PRI-client CTM, and state repression in the countryside, the RMALC deployed an effective transnational campaign that further informed the continental challenge to neoliberalism.

Strong links between RMALC activists and several U.S. and Canadian groups reinforced ties already formed between sector-specific groups, such as Enlace Ecológico-Border Ecology Project, and United Electrical Workers and the Frente Auténtico del Trabajo (FAT).[53] The Alliance for Responsible Trade (ART) in the United States—especially through affiliations with the United Electrical, Radio and Machine Workers (UE) and Development Group for Alternative Policy, and later through Friends of the Earth (FOE)—formed a crucial node from which transnational anti-NAFTA strategy would form. Teach-ins, lobbying sessions, cross-border marches and rallies, and continental fair trade proposals were the products of the transnational networks. On the other hand, groups affiliated primarily with the Citizen's Trade Campaign, such as the Sierra Club, most large unions, and animal rights groups, remained disconnected from RMALC through most of the NAFTA fight. Some two years following the NAFTA fight, fair trade internationalism has begun to inform those groups that remained on the conventional political path of the Citizen's Trade Campaign.[54]

Despite the disunity in political strategy and resources, the national and transnational structure of the fair trade mobilization operated relatively well, and the overall effects of these broad alliances were profound. The *conventional* legislative tactic characteristic of the environmental mainstream (and the AFL-CIO) was able to operate alongside a more radical, though resource-constrained, effort to expand transnational solidarity, problematize conventional tactics, and construct an alternative vision. The duality of this mobilization was facilitated by heightened opportunities to build broader alliances through unions and environmental groups acting as linchpins within the alliance, linking various wings of the respective movements. Groups such as FOE, the Teamsters, and Clean Water Action straddled *both* U.S. coalitions in addition to active efforts at building international links alongside their more radical counterparts, such as Greenpeace and the UE. These structural "bridges" thus enabled the formation of a prototype for the broad, progressive alliance that is probably necessary to move us beyond the neoliberal project, and contributed a vision toward that end.

The basic normative claims of these networks called for democratizing international trade relationships and submitting the market and capital accumulation to social needs. For instance, the Final Declaration of the International Forum at Zacatecas opened with the following words: "We insist that trade be part of a strategy of continental development that guarantees the distribution of wealth, the elevation of living standards, and the self-determination of our peoples."[55] Cardenas's own "Continental Development Initiative" coincides with the demands of the Mexican network: "Economic liberalization is not our objective, it is but one of our tools. Development, social justice and a clean environment are our objectives."[56] The Alliance for Responsible Trade's alternative agenda makes the same appeal: "Trade and investment—the main areas addressed in the proposed agreement [NAFTA]—should not be seen as ends in themselves, but as tools toward development, social justice and a healthy environment."[57]

All the major networks, and many of the bi/trinational coalitions and grassroots organizations, thus recognize the unregulated market as potentially a common threat to society and nature. Their public literature, formal declarations, press statements, and so on, are imbued with this common understanding, though manifest within the political culture of the movements. In the United States, many oppositional networks and organizations sponsored a full-page ad in several national newspapers immediately before the first signing of the agreement in December 1992. Here, the unregulated market manifested itself as a "Sneak Attack on Democracy," of whom the "beneficiaries are multi-national corporations for whom democracy itself is an impediment to *their* free trade."[58] In Mexico, on the other hand, the concern for national sovereignty had a more radically democratic twist, making appeals to the threats to democracy as well as those to national sovereignty more effective.[59] For instance, RMALC's statement on free trade asserts that the "indispensable conditions for economic integration" should include "an ample and democratic consultation with all potentially affected sectors, through a national referendum," . . . "clear guarantees of our political independence and national sovereignty," and "true respect for the Political Constitution of Mexico and the national laws."[60]

Highlighting the contradiction between free trade and democracy, the anti-NAFTA coalitions improved their ability to forge a multi-issue language consonant with the interests of a broad range of movement constituencies. As a framework for citizen mobilization, threats to democracy carry a broad appeal. Furthermore, by linking trade and democracy in this way, activists challenged prevailing notions that capitalism is equivalent to democracy. Indeed, NAFTA opposition pressed an ideological platform (sometimes more radically than intended) that (temporarily) severed the link between liberal markets and liberal politics. For environmentalism, this helped set the tone for arguments between leading environmental groups over the role that market incentives and democratic institutions should play in the reconciliation of human economic activity with ecology. For those groups that remained opposed to NAFTA, it was much easier to learn that global ecological distress could not be resolved through free markets, only exacerbated.

In the 1990s, before the signing of the agreement, the national and international environmental community became increasingly active on the contested terrain of "sustainable development."[61] NAFTA's exhortation of "sustainable development" in the preamble was thus a perfect place for continuing the contests concerning "sustainability" and who will (or can) manage it. Indeed, most criticisms in the United States, especially those endorsed by a cross section of the environmental majors, centered on the ways in which NAFTA actually precludes "sustainable development." The focus was on issues of funding for cleanup, means of enforcement, and the mechanisms of dispute resolution.[62]

Environmentalist opposition in Mexico added and extended dimensions of environmentalism in the North (or the United States and Canada) and vice versa. This was particularly apparent in debates concerning the purpose and means of development. Not only was the problem of vast U.S. natural resource consumption sharpened in debates, but the main Mexican opposition group, the RMALC, countered the Perot-centered argument of "don't send U.S. jobs to Mexico" with "don't send Mexico's natural resources to the United States."[63] By framing environmental concerns in this way, the dichotomy between the North and the South was sharpened and globalized.

Thus, by 1992 many citizen groups—including environmental organizations—were actively constructing their understanding of NAFTA. This implied an internal engagement with one's own mission as an organization, as well as a critique of the opponents' neoliberal project, and hence unregulated market activity such as free trade. Documents such as "Look before You Leap: What You Should Know about a NAFTA," "The Citizen's Analysis of the NAFTA," and "A Just and Sustainable Development Initiative for North America" (produced in consultation with networks in all three countries) suggest a discourse that is reappropriating a critique of the market, a critique that broadens calls for environmental justice and the internalization of environmental and social costs.[64] In other words: *fair* trade.

It is not surprising on this account that Greenpeace became perhaps the most critically versed of the international environmental organizations on the ineptness of the unregulated market.[65] Greenpeace formulated a critique of free market environmentalist principles in their 1992 publication "UNCED Undermined: Why Free Trade Won't Save the Planet." Other statements and publications pertaining to NAFTA recognize and explicitly analyze the ways in which unregulated markets encourage capital to elevate profits by displacing costs onto nature and society.[66]

While a few of the larger environmental organizations linked "trade" to environmental hazards by 1992, a connection to the social consequences of free trade-induced environmental degradation emerged from the grassroots.[67] The nature of border region development and the necessity of engaging the issue of trade on social justice terms forced some of the larger environmental organizations to redirect their own internal dialogues. Local expertise and grassroots organizations sought inclusion in the fair trade campaign, while several larger coalition members (such as Public Citizen and Clean Water Action) recruited antitoxics and environmental justice groups. By early 1993, hundreds of local environmental groups participated in phone banks and rallies and signed petitions declaring NAFTA "fundamentally flawed."[68] Some of the larger environmental justice groups, especially Citizen's Clearinghouse for Hazardous Waste and the Student Environmental Action Coalition, actively coordinated grassroots mobilizations in the fight against NAFTA.

To enter a contest with the state and neoliberal capital, some of the environmental majors versed themselves in the language of grassroots dissent and consequently became familiar with some of the assumptions of free trade and its implications for environmental justice and the discourse of sustainable development. In the process, a deeper understanding of trade, and hence the market, was constructed around grassroots' concerns for environmental justice. Reception of such a critical stance on trade and the market, however, was not warmly received in the confines of the "Group of Ten" conservationists.

Yet activists within several mainstream groups worked to incorporate political economic and social justice issues. The deployment, spread, and infusion of the meaning of environmental justice exerted influence throughout the political culture of the environmental movement, enabling the concept of "fairness" in the fair trade movement to resonate with movement participants. Indeed, the environmental justice mobilization through "fourth wave" (and supportive "third wave") organizations, set an important context for the construction of a *fair* trade challenge to NAFTA, linking social justice and environmental politics. The involvement of groups representing diverse environmental constituents, such as the CCHW, Akwesasne Environmental Project, Coalition for a Liveable West Side (New York City), the Native Forest Network, the Southwest Network on Environmental Racism, and hundreds of others, brought explicitly social justice concerns to the national fair trade coalitions. Some conservationist organizations would lean toward a political co-optation of the language of environmental justice, while others, such as Friends of the

Earth and Sierra Club, struggled internally to embrace the movement from below.

Globally oriented, diverse in perspective, and broadly networked within the social movement sector, the fair trade coalitions in both Mexico and the United States transformed the understandings and capacities of those environmental groups that sustained opposition to NAFTA. Border issues, the high-level rhetoric of international trade, and the concerns of indigenous, labor, small farmers, women, and human rights groups exposed NAFTA opposition to a broad horizon of social justice concerns closely connected to the environment. These same matters were quickly interpreted, consolidated, repackaged, and distributed via daily and weekly fair trade reports to hundreds of organizations and then to millions of individuals through newsletters, electronic mail, magazines, and the press.[69] Trade, one can be sure, will remain an environmental issue, but under what terms?

NEOLIBERAL CAPITAL AND THE ENVIRONMENTAL COALITION FOR NAFTA

Political and economic elites in Mexico and the United States vigorously responded to NAFTA opposition and expended great effort to keep the environmental debate over NAFTA confined to the language of economic growth. Paramount to this strategy were tactics that brought environmentalism *into* the language of NAFTA, a process reluctantly begun under the Bush administration and elaborated under the Clinton administration with the negotiation of the labor and environmental side accords. Access to the leading third-wave environmental organizations was enhanced with the Clinton–Gore administration, and, by negotiating the side accords, this access proved to be a "clincher" for the proponents of neoliberal markets.

The environmental side agreement, or North American Agreement on Environmental Cooperation (NAAEC), acknowledges "the sovereign right of States to exploit their own resources pursuant to their own environmental and development policies."[70] Objectives of the agreement include these: "[promote] sustainable development . . . ; avoid creating trade distortions . . . ; promote transparency and public participation in the development of environmental laws, regulations and policies; promote economically efficient and effective environmental measures . . . ; and promote pollution prevention policies and practices."[71] Leading executives of pro-NAFTA environmental groups, such as the National Wildlife Federation, declared that the side accord would "encourage public participation in resolving trade and environmental disputes . . . and impose trade sanctions on countries seeking to boost trade by lowering or ignoring their environmental standards."[72] Indeed, Jay Hair with the NWF asserted that "there's going to be a near-term resolution of some incredibly difficult environmental degradation problems."[73] Several institutional mechanisms, such as the North American Development Bank (NAD-Bank) and the Commission for Environmental Cooperation (CEC), were created under NAAEC (or alongside it) to implement these proposed objectives.

Critics quickly responded that the side accords and the institutional mechanisms for enforcement and monitoring are "a lot of talk, but no teeth . . . [and] punish governments not polluters."[74] Two years after the implementation of NAFTA, the actual funding for key institutions, such as CEC and the NAD-Bank, has fallen short. Arguments in favor of the agreement by pro-NAFTA environmental organizations relied on the formation of a well-financed oversight commission and a serious border clean up project; they proposed between $30 million and $70 million to fund the CEC's oversight of NAFTA-related environmental degradation and border cleanup.[75] The Commission for Environmental Cooperation, however, has been allocated a budget of $9 million ($3 million from each country) and will likely face further constraints on funding. Furthermore, the lending capacity of the NAD-Bank has been severely limited by the effects of the peso devaluation, further constraining the feasibility of obtaining credit in poor communities along the border.[76] In terms of implementation, the NAAEC has thus far failed to realize its objectives.

Environmentalists in the United States and Mexico were the first to "test" the procedural dimensions of the NAAEC and the CEC. As of late 1996, three petitions were filed with the CEC, with the latter two representing organizations from all three NAFTA countries. Only the first case—involving a request to investigate the deaths of some 40,000 water fowl in a central Mexican reservoir—was accepted.[77] The ruling by the CEC resulted in nothing more than a report on the deaths of these birds. Much to the dismay of the Grupo de los Cien (Group of 100), the report attributed the *cause* to botulism and, while recognizing that "exposure to heavy metals, in particular chromium . . . was indicated in some of the birds," downplayed the significance of industrial toxins.[78] Moreover, the ruling by the CEC established a precedent that further narrows the scope of review under NAAEC, arguing that "failure to enforce" can only occur through administrative failure, not because the enforcement of intact environmental laws was defunded. Subsequent petitions to the CEC were met with outright rejection.[79]

Yet, during the NAFTA battle, the USA*NAFTA corporate coalition, numerous business and trade associations, and the Salinas and Clinton administrations insisted that further economic growth in Mexico and the United States (and Canada) will lead to more prosperous and sustainable ways of life. Consistent with the preemptive environmental politics of previous ruling party administrations, Salinas linked, tactically, the administrative functions for environmental protection, now as Secretary of Social Development (SEDESOL), with the politically strategic "antipoverty" Programa Nacional de Solidaridad (PRONASOL).[80] This development coincided with efforts to dispel environmental opposition through Salinas's contacts with U.S.-based international conservation groups, such as the Natural Resources Defense Council (NRDC) and the World Wildlife Fund (the latter received a contract to establish a new conservation zone). The highly visible Grupo de los Cien—which worked with the pro-NAFTA NRDC and the U.S. trade representative to draft the *Review of U.S.–Mexico Environmental Issues*—and Restauración Ambiental (a counterpart of the World Wildlife Fund), reflect clear cases where U.S.-based third-wave groups encouraged and facilitated the dissociation of several core environmental groups from prominent opposition forces in Mexico.[81] Rather than opting for a social justice critique of the environmental implications of NAFTA, as did FOE and the Movimiento Ecologista Mexicano (MEM), neoliberal environmentalism prompted a conciliatory posture toward Salinas's preemptive tactics. References were made to Salinas' plans for new conservation zones or to efforts to reconcile environmentalist concerns in Mexico by increasing funding for environmental protection and enforcement through PRONASOL.[82] Indeed, in efforts to "defang" more radical environmental challenges to neoliberalism in Mexico, the PRI leadership utilized their close connections with the Bush administration (and later the Clinton administration) and neoliberal capital in a preemptive strike to leverage support from prominent, international environmental "partnerships."[83] More effectively, however, the USA*NAFTA corporate coalition, with assistance from the Clinton administration and the environmental side accord, bought environmental advocates.

By proclaiming support for NAFTA, the "Environmental Coalition for NAFTA" charmed the corporate architects of the agreement and thus reestablished close ties to neoliberal capital. For instance, in 1992 the World Wildlife Fund (WWF) "received $2.5 million in a single donation from Eastman Kodak, whose CEO, Kay Whitmore, is cofounder of USA*NAFTA."[84] Further, corporate interests were expressed from within the large conservationist organizations. Kay Whitmore received a seat on the Board of Directors of the World Wildlife Fund, and at a more general level, quoting John Audley, "what you see is very clear, the pro-NAFTA environmental organizations have Boards that are very corporate oriented, or heavy with [corporate] foundation support, and the anti NAFTA environmental organizations are much more grassroots oriented, . . . not corporate driven. It's very clear."[85]

With less subtlety, William Reilly, the former head of Conservation Foundation and director of the EPA under President Bush, utilized his knowledge of the differences within the environmental movement.[86] In efforts to dispel fears of a unified labor–environmental opposition, Reilly organized a meeting between the U.S. Trade Representative, Ambassador Carla Hills, and representatives from several of the major environmental organizations involved in opposing fast-track legislation for the trade agreement.[87] In an exchange of memos between Hills and Reilly, Ambassador Hills raised concerns that the environmental community was a monolithic, Greenpeace-like entity, while Reilly reassured her that "there are very conservative minds in the environmental community" and that he could help develop a positive relationship.[88] By April 1991, they targeted the National Wildlife Federation, Environmental Defense Fund, Nature Conservancy, Natural Resources Defense Council, and World Wildlife Fund.

As lobbying organizations with close dependencies to the environmental policy system, they could act as effective supporters in Congress and hence neutralize the environmentalist opposition, as they did in 1993. "We broke the back of the environmental opposition to NAFTA," boasted John Adams, the head of the Natural Resources Defense Council.[89] Now eager supporters of NAFTA, the leaders of the National Wildlife Federation (NWF), Environmental Defense Fund, and Nature Conservancy used their seats on governmental trade advisory committees to "bargain" on behalf of nature in opposition to the handful of NAFTA challengers on those same corporate-dominated committees.[90] The Audubon Society's president, Peter Berle, found reward in 1994 with a seat (alongside an executive of Allied Signal, an avowed leader in the USA*NAFTA) on the Commission for Environmental Cooperation accompanying the environmental side accord.[91] Understanding the political character of the environmental mainstream, the offices of the EPA and the U.S. Trade Representative, under both Presidents Bush and Clinton, focused on winning the support of Jay Hair—the president of the National Wildlife Federation—at a very early stage in the struggle over NAFTA.

The seven environmental majors that formed the "Environmental Coalition for NAFTA" also reconfigured tactical and ideological splits among the national organizations.[92] As Dowie remarked, the "discord between pro- and anti-NAFTA enviros . . . was perhaps the nastiest internecine squabble in the movement's hundred-year history."[93] Exemplifying this rift, Stewart Hudson of the NWF adamantly argued before the Subcommittee on Trade that "the National Wildlife Federation, along with the National Audubon Society, the World Wildlife Fund, the Natural Resources Defense Council, the Environmental Defense Fund, and Conservation International, wholeheartedly supports NAFTA and . . . urges Congress to approve this vitally important agreement."[94] On the other hand, a sustained exposure of the Sierra Club to movement tendencies around grassroots tactics and environmental justice was heightened, especially by working closely with Friends of the Earth, Clean Water Action, Greenpeace, and numerous grassroots environmental groups affiliated with the Citizen's Trade Campaign.

Allying with "less accommodating enviros, like Jane Perkins, former executive director of Friends of the Earth, and Barbara Dudley . . . at

Greenpeace," the leadership of the Sierra Club received "caustic letters" from Jay Hair.[95] By aligning with groups advocating less conventional practices in the NAFTA fight, the balance of power *within* the Sierra Club shifted more favorably to factions that were urging a break with the clubishness of the G-10 in 1991.[96] Michelle Perrault, former president and chairperson of Sierra's board, helped maintain internal opposition to the Club's pro-NAFTA economists in addition to representing environmental justice concerns through her common board membership on Pesticide Action Network (PAN).[97] The former chairperson of the Sierra Club, Mike McCloskey, and Carl Pope, the executive director, were targeted by other G-10 environmental groups for "breaking the ranks" by allying with labor unions and the more left-leaning environmental groups. Jay Hair, expecting the Sierra Club to swing in support of NAFTA, as six other groups did, expressed betrayal and anger toward McCloskey and the Sierra Club, commenting, "I think the stakes are just too high for people to be into such protectionist polemics as I see coming out of the Sierra Club."[98] Further distancing the Sierra Club from previous loyalties, McCloskey affirmed their opposition: "We are not identifying ourselves in this instance with some of the other so-called mainstream groups and perhaps it does represent a *repositioning* of the Sierra Club."[99]

DECENTERING NORTH AMERICAN MOVEMENTS: TOWARD A POLITICS OF ECONOMIC AND ECOLOGICAL DEMOCRACY

In Mexico, the main formative groups in opposition to NAFTA emerged from a reorganized popular sector, which preceded and formed the populist *neocardenista* opposition to the hegemony of the Partido Revolucionario Institucional (PRI) in 1988. Amidst an increasingly repressive state, these prior mobilizations established the political space for aggregating demands for democracy and autonomy at the national and international levels, forging an alignment between a civil front and the revolutionary movement in Chiapas. Together, these movements constitute a force that has already pushed Mexico into political crisis and will no doubt force substantial changes in the workings of Mexican society. Nowhere is the evidence of a weakened PRI more clear than in the indigenous peasant and labor sectors. Most recently, the continued wage crisis induced by neoliberal strategies of accumulation has eroded the systems of control within the PRI-controlled Confederation of Mexican Workers. Much like the recent first contested election for AFL-CIO leadership in the United States, the deepening rifts within the Mexican Congress of Labor (CT) indicate that the political "protection" offered by state corporatism is unraveling still further. Torrents of state violence, the escalating militarization of the Mexican state, and widepsread control of the mass media, however, should remind the critical observer that the PRI's dual-edged saber—repression and co-optation—is swung with the sharpest blade.

In the United States, environmental organizations, at least initially, were prominent actors in constructing opposition to NAFTA. During the late 1980s, a number of events and circumstances contributed to a rejuvenation of sectors in the U.S. environmental movement and brought attention to trade issues. Importantly, the environmental justice grassroots movement challenged the environmental mainstream to more responsively address the inextricable connections between social justice and environmental quality. This urgent call for justice in environmentalism enabled the concept of "fairness" in the fair trade mobilization to resonate with the more progressive wing of large environmental groups, such as Friends of the Earth, Greenpeace, and, surprisingly, the Sierra Club. "Fair trade" thus formed a framework for applying the principles of ecological democracy to the politics of international trade.

Yet, what Dowie calls the "third wave" of environmentalism has indeed moved a substantial wing of the movement into the "boardroom," more closely connecting these organizations to both the state and capital.[100] With little more than a rhetorical receptivity to the emerging fourth wave of environmental justice, leading third-wave groups confined large capacities of the fair trade movement to the boardrooms, courtrooms, congressional lobbies, and executive advisory boards. In particular, the "Environmental Coalition for NAFTA" and its sharp display of loyalty to capital, market incentives, and the president, forced the large anti-NAFTA environmental groups to mobilize with more conventional tactics. For instance, by resigning the capacities of Greenpeace as a movement organization to its canvassing operations, a tactically innovative offensive was further limited. Third-wave environmentalism thus constrained the political practice and visions of the fair trade campaign. Coupled with an intensification of grassroots mobilizations, however, and the fissure between the Sierra Club and the rest of the G-10, tendencies to "problematize" political practices within the movement were strengthened. Indeed, the fight over NAFTA further politicized existing factions in North American environmentalism, created new ones, and exposed the limitations of an environmentalism subordinate to neoliberal markets and capital.

ACKNOWLEDGMENTS

An earlier version of this chapter was presented at the 1994 Annual Meeting of the American Sociological Association in Los Angeles, CA. I wish to thank Daniel Faber and the *Capitalism, Nature, Socialism* Boston Editorial Group in addition to Margaret Somers, Robert Adwere-Boamah, Paul Ciccantell, Ming-cheng Lo, Marina Ramirez, and Tom Waller for their comments on earlier drafts.

NOTES

1. Markets characterized as "neoliberal" are best contrasted with markets where state or community sanctions may be deployed to protect society and nature from the logic of commodification. Neoliberal markets thus epitomize the classic liberal utopia of an unregulated market society (see Karl Polanyi, *The Great Transformation: The Political and Economic Origins of Our Time* [Boston: Beacon Press, 1944]). Neoliberalism is thus defined as an encompassing perspective that asserts that the "market allocates resources to all uses more efficiently than political institutions" (Adam Przeworski, *The State and Economy under Capitalism* [New York: Harwood Academic Publishers, 1990], p. 15).

2. Michael G. Wilson, "The North American Free Trade Agreement: Ronald Reagan's Vision Realized," Executive Memorandum (Washington, DC: Heritage Foundation, 1993).

3. U.S. Council for International Business monthly newsletter, October 1994, p. 7. The USCIB is the formal representative of U.S. capital to the OECD. The OECD, or Organization for Economic Cooperation and Development, is a forum for constructing common economic agendas among leaders of the largest capitalist countries of the world.

4. See Sarah Anderson and John Cavanagh, eds., *NAFTA's First Two Years: Myths and Realities* (Washington, DC: Institute for Policy Studies, 1996); and Boyer, Gabriela, et al., *NAFTA's Broken Promises: The Border Betrayed: U.S.–Mexico Border Environment and Health Decline in NAFTA's First Two Years* (Washington, DC: Public Citizen, 1996).

5. The symbolic and political legacy of Cardenas, whose father was perhaps the most popular president in postrevolutionary Mexico, cannot be underestimated. Indeed, "Cuauhtemoc Cardenas became the symbol, the redemptive myth, capable of reversing social decay, of resuming the abandoned [revolutionary] path and promoting democratization, the defense of national sovereignty, and social equality" (Jaime Tamayo, "Neoliberalism Encounters *Neocardenismo,*" in Joe Foweraker and Ann Craig, eds., *Popular Movements and Political Change in Mexico* [Boulder, CO: Rienner Publishers, 1990], pp. 130–131).

6. Nonetheless, the labor movement in all three countries was a crucial, if not ultimately the most significant, sector to mobilize against NAFTA. See David Barkin, "Constructive Labor Strategy for Free Trade," *The Review of Radical Political Economics, 25*(4), 1993; Allen Hunter, "Globalization from Below?: Promises and Perils of the New Internationalism," *Social Policy, 25*(4), 1995; Thalia Kidder and Mary McGinn, "In the Wake of NAFTA: Transnational Workers Networks," *Social Policy, 25*(4), 1995; and Ian Robinson, *North American Trade as if Democracy Mattered* (Washington, DC: Canadian Centre for Policy Alternatives and the International Labor Rights Fund, 1993).

7. By "conventional practice," I mean political practice that follows institutionalized channels of influence characteristic of clientelism, formal litigation, legislative lobbying, and electoral aims. See Ulf Hjelmar, *The Political Practice of Environmental Organizations* (Aldershot, UK: Avebury, 1996).

8. Mark Dowie, *Losing Ground: American Environmentalism at the Close of the Twentieth Century.* (Cambridge, MA: MIT Press, 1995); Robert Gottlieb, *Forcing the Spring: The Transformation of the American Environmental Movement* (Washington, DC: Island Press, 1993); Kirkpatrick Sale, *The Green Revolution: The*

American Environmental Movement, 1962–1992 (New York: Hill and Wang, 1993); Daniel Faber and James O'Connor, "The Struggle for Nature: Environmental Crisis and the Crisis of Environmentalism in the United States," *Capitalism, Nature, Socialism,* Issue 2, 1989.

9. Dowie, *Losing Ground.*

10. Ibid., pp. 107 and 206.

11. Ibid., p. 109.

12. Ibid., p. 206.

13. For a discussion on the "legacy of ecological imperialism," see Daniel Faber, *Environment Under Fire: Imperialism and the Ecological Crisis in Central America* (New York: Monthly Review Press, 1993); and Kathryn Sikkink and Margaret Keck, "Activists without Borders: Transnational Issue Networks in International Politics" (Ann Arbor: University of Michigan, Advanced Studies Center, Working Paper Series, No. 8, 1995).

14. Most mainstream environmental organizations adopted mass-mailing and marketing strategies in the 1980s, reflecting a true crisis in their capacity to mobilize the grassroots. See Robert Mitchell, Angela G. Mertig, and Riley Dunlap, "Twenty Years of Environmental Mobilization: Trends among National Environmental Organizations," in Riley Dunlap and Angela G. Mertig, eds., *American Environmentalism: The U.S. Environmental Movement, 1970–1990* (Washington, DC: Taylor and Francis, Inc., 1992).

15. Lois Gibbs, "Foreword," in Richard Hofrichter, ed., *Toxic Struggles: The Theory and Practice of Environmental Justice* (Philadelphia: New Society Publishers, 1993), p. ix.

16. Dorceta Taylor, "Environmentalism and the Politics of Inclusion," in Robert D. Bullard, ed., *Confronting Environmental Racism: Voices from the Grassroots* (Boston: South End Press, 1993), 57.

17. The landmark First People of Color Environmental Summit in 1991 formalized the challenge by the emerging movement for environmental justice to the environmental mainstream; see Bullard, ed., *Confronting Environmental Racism;* and Bunyan Bryant, ed., *Environmental Justice: Issues, Policies, and Solutions* (Washington DC: Island Press, 1995).

18. See Terry Lee Anderson and Donald Leal, *Free Market Environmentalism* (Boulder, CO: Westview Press, 1991).

19. Gottlieb, *Forcing the Spring*, p. 317.

20. Quoted in Daniel Faber and James O'Connor, "The Struggle for Nature," p. 34, from Barry Commoner, "A Reporter at Large: The Environment," *New Yorker,* June 15, 1987. Numerous "third-wave" environmental organizations, such as the Environmental Defense Fund, helped engineer such market-based approaches to capitalizing on nature.

21. Sara Diamond, "Free-Market Environmentalism," *Z Magazine,* December 1991, pp. 18–24. For an activist presentation of industry responses to environmentalism, see Carl Deal, *The Greenpeace Guide to Anti-Environmental Organizations* (Berkeley, CA: Odonian Press, 1992).

22. Dowie, *Losing Ground*; and Gottlieb, *Forcing the Spring.*

23. William Canak, *Lost Promises: Debt, Austerity, and Development in Latin America* (Boulder, CO: Westview Press, 1989), p. 21.

24. James Cockcroft, *Mexico: Class Formation, Capital Accumulation and the State* (New York: Monthly Review Press, 1990).

25. David Reed, ed., *Structural Adjustment and the Environment* (Boulder, CO: Westview Press, 1990), pp. 83–84. See also Judith Teichman, "The Mexican State and the Political Implications of Restructuring," *Latin American Perspectives, 19*(2), 1992.

26. David Barkin, *Distorted Development: Mexico in the World Economy* (Boulder, CO: Westview Press, 1990); and Cockcroft, *Mexico*; Tamayo, "Neoliberalism Encounters *Neocardenismo*."

27. Joe Foweraker, "Introduction," in Joe Foweracker and Ann Craig, eds., *Popular Movements and Political Change in Mexico* (Boulder, CO: Rienner Publishers, 1990), p. 16.

28. Ibid., p. 16.

29. Ibid., p. 15, makes reference to Ignacio Marvan Laborde, *Tendencias actuales de los movimientos sociales en México: expresiones nacionales y regionales* (Mexico City: Instituto de Investigaciones Sociales, UNAM, 1988). Stephen Mumme suggests transformations in environmental policy during the same periods, though also a consistent "smothering" of civil society by the corporatist strategies of the state, in "System Maintenance and Environmental Reform in Mexico: Salinas' Preemptive Strategy," *Latin American Perspectives, 19*(1), 1992. For other arguments explaining the limited spaces for environmentalist mobilization and policy development, see David Barkin, "State Control of the Environment: Politics and Degradation in Mexico," *Capitalism, Nature, Socialism*, Issue 7, 1991; Daniel Goldrich and David Carruthers, "Sustainable Development in Mexico? The International Politics of Crisis or Opportunity," *Latin American Perspectives, 19*(1), 1992; Stephen Mumme, Richard Bath, and Valerie J. Assetto, "Political Development and Environmental Policy in Mexico," *Latin American Research Review, 23*(1), 1988; Richard Nuccio and Angelina Ornelas, "Mexico's Environment and the United States," in Janet Welsh Brown, ed., *In the U.S. Interest: Resources, Growth, and Security in the Developing World* (Boulder, CO: Westview Press, 1990); and Lynn Stephen, "Women in Mexico's Popular Movements: Survival Strategies against Ecological and Economic Impoverishment," *Latin American Perspectives, 19*(1), 1992.

30. McCaughan suggests a somewhat different, and not incompatible, story. During the 1980s, the prolonged and deepening socioeconomic crisis in Mexico occasioned fractures in the bourgeoisie, and in the process "the class interests represented in the old historic bloc are being politically and ideologically redefined." See Edward McCaughan, "Mexico's Long Crisis: Toward New Regimes of Accumulation and Domination," *Latin American Perspectives, 20*(3), 1993. See also Judith Teichman, "Dismantling the Mexican State and the Role of the Private Sector," in Ricardo Grinspun and Maxwell Cameron, eds., *The Political Economy of North American Free Trade* (New York: St. Martin's Press, 1993); and Diane Davis, "Mexico's New Politics: Changing Perspectives on Free Trade," *World Policy Journal*, Fall/Winter 1992.

31. Foweracker, "Introduction," p. 15.

32. Tamayo, "Neoliberalism Encounters *Neocardenismo*," p. 131.

33. Vivienne Bennett, "The Evolution of Popular Movements in Mexico Between 1968 and 1988," in Arturo Escobar and Sonia Alvarez, eds., *The Making of Social Movements in Latin America: Identity, Strategy, and Democracy* (Boulder, CO: Westview Press, 1992).

34. Tamayo, "Neoliberalism Encounters *Neocardenismo*," p. 130–131.

35. Edur Velasco Arregui, "Industrial Restructuring in Mexico During the 1980s," in Grinspun and Cameron, *The Political Economy of North American Free Trade.*

36. Adolfo Aguilar Zinser, "Authoritarianism and North American Free Trade: The Debate in Mexico," in ibid.

37. See George A. Collier, *Basta! Land and the Zapatista Rebellion in Chiapas* (Oakland: Institute for Food and Development Policy, 1994).

38. Raul Hinojosa Ojeda and Carol Zabin, "Towards a Social Charter for the North American Free Trade Area: Reconceptualizing the Role of Mexican Migration and Rural Development," unpublished paper presented at the S.A.S.E. Conference, New York, NY, 1993.

39. Judith Alder Hellman, "Mexican Popular Movements, Clientelism, and the Process of Democratization," *Latin American Perspectives, 21*(2), Issue 81, 1994.

40. In November 14, 1991, the *Tucson Citizen* reported "Grassroots Border Alliances Form: Free Trade Talks Are Said to Be Catalyst," and the *Los Angeles Times,* on August 12, 1992, reported on "A Deal That's Hazardous to Your Health." The *San Diego Tribune* of June 27, 1992, reported, "Environment Is Key to Trade Pact Issue." In addition to concern for national sovereignty in Mexico, the heightened salience of the United States–Mexico border further fueled a right-wing national populism in the United States. The right-wing opposition to NAFTA, assaults on immigrants and immigrant rights, and the progressive mobilizations to protect those rights are additional variants on the theme of "protecting" national sovereignty.

41. Lawrence A. Herzog, *Where the North Meets South: Cities, Space, and Politics on the U.S.-Mexico Border* (Austin, TX: Center for Mexican American Studies, 1990), p. 8.

42. Ibid., p. 6.

43. Ibid., p. 10.

44. Ibid., p. 11. For example, see Interagency Task Force, *Review of U.S.-Mexico Environmental Issues* (Washington, DC: U.S. Government Printing Office, 1991).

45. Clearly influenced by the old Communist Party culture, the United Electrical, Radio and Machine Workers' (UE) consciously militant style and strategy makes it a unique case within the U.S. labor movement. The UE–FAT "Proposal for Action" defined cross-border organizing as part of an effort to establish "a beachhead of democratic unionism in strategic sectors of the Mexican economy" (August 1, 1992).

46. See Mujer a Mujer, "Women Fight against 'Free Trade' Restructuring," in *Correspondencia,* monthly newsletter, February 1992. These cases are but a few, albeit important, examples of cross-border organizing.

47. Fast-track legislation is designed to expedite congressional approval of international agreements; moreover, placing NAFTA (and GATT) agreements under this rubric avoids their being treated as treaties, requiring a two-thirds majority vote in the Senate. Under "fast track," trade agreements may only be ratified without amendments and through a simple majority vote.

48. It should be noted that the organizing tactics of the Institute for Agriculture and Trade Policy (IATP), and later the Fair Trade Campaign, included a specific strategy to cultivate and deploy an environmentalist critique of free trade with the

intent of gaining broader environmentalist support (personal interview, Craig Merrillees, October 21, 1994). For examples of the use of an environmental science of border life to criticize free trade, see the testimony in Committee on Finance: U.S. Senate, *United States–Mexico Free Trade Agreement: Hearings before the Committee on Finance, February 6 and 20, 1991* (Washington, DC: U.S. Government Printing Office, 1991); and Committee on Environment and Public Works and the Subcommittee on Labor of the Committee on Labor and Human Resources, *Economic and Environmental Implications of the Proposed U.S. Trade Agreement with Mexico* (Washington, DC: U.S. Government Printing Office, 1991).

49. Several influential publications emerged from groups with border concerns during 1991 and 1992; the earliest of the critical perspectives came from the border regions. Some of these include annual publications from the Coalition for Justice in the Maquiladoras, the National Toxics Campaign Fund's *Border Trouble: Rivers in Peril* (Boston: National Toxics Campaign Fund, 1991), the Texas Center for Policy Studies' "A Response to the Bush Administration's Environmental Action Plan for Free Trade Negotiations with Mexico" (May 1991), and the National Safe Workplace Institute, *Crisis at Our Doorstep: Occupational and Environmental Health Implications for Mexico–U.S.–Canada Trade Negotiations* (Chicago: Author, 1991). More recently, the Inter-Hemispheric Resource Center and Border Ecology Project published Tom Barry and Beth Sims, *The Challenge of Cross-Border Environmentalism: The U.S.–Mexico Case* (Albuquerque, NM: Resource Center Press, 1994).

50. The role of the IATP, particularly Executive Director Mark Ritchie, cannot be underestimated. Tied into many development, small farming, and economic justice groups, the IATP was at the forefront in developing alternative conceptions of international trade within the progressive policy arena since the mid-1980s. The IATP remained one of the most important networking organizations throughout the NAFTA and GATT struggles.

51. The ART sustained itself through a close network of internationally oriented activists from several Washington, DC, groups, including the International Labor Rights Education and Research Fund, Economic Policy Institute, Greenpeace, the Development Group for Alternative Policy, and the Institute for Policy Studies.

52. Ricardo Hernandez and Edith Sanchez, *Cross-Border Links: A Directory of Organizations in Canada, Mexico and the U.S.* (Albuquerque, NM: Inter-Hemispheric Education Resource Center, 1992), p. 14.

53. The joint submissions by the Teamsters and the United Electrical, Radio, and Machine Workers to the National Administrative Office were not only a test for expanding international labor solidarity, but also a test of the side accord established under NAFTA, the North American Agreement on Labor Cooperation. Both submissions declared that Mexican officials failed to enforce relevant labor laws around incidents at two U.S.-based multinational operations where Mexican workers were fired for attempting to organize a local affiliate of the Frente Auténtico del Trabajo (FAT), independent of the dominant Confederation of Mexican Workers (CTM). The NAO filed a report that "found disagreement about the events" and concluded that it "is not in a position to make a finding that the Government of Mexico failed to enforce the relevant labor laws" (U.S. National Administrative Office, *Public Report of Review, NAO Submission #940001 and NAO Submission #940002* [Washington, DC: Bureau of International Labor Affairs, U.S. Department of Labor, 1994], pp. 30–31).

54. Two years after the agreement, large movement organizations such as the Friends of the Earth, Sierra Club, United Mine Workers of America, Teamsters, and Public Citizen remain actively committed to extending fair trade campaigns across the hemisphere, while neoliberal capital prepares for the "Free Trade Area of the Americas" (FTAA).

55. Final Declaration of the International Forum, Zacatecas, Mexico, October 1991.

56. Cuauhtemoc Cardenas, "The Continental Development and Trade Initiative," New York City presentation, February 8, 1991, published in Coalition for Justice in the Maquiladoras, *Annual Report, 1990–91* (San Antonio: Author).

57. MODTLE, "Development and Trade Strategies for North America" (Washington, DC: International Labor Rights Fund, 1992), p. 3.

58. Citizens' Trade Campaign, advertisement against NAFTA, "A Sneak Attack on Democracy," in ten national newspapers, including the *New York Times,* the *Washington Post,* and the *Los Angeles Times* (December 17, 1992).

59. This was in marked contrast to the concerns for "national sovereignty" expressed within an economic nationalist ideology and often explicitly racist subtext by the right-wing Perot and Buchanan fronts in the United States.

60. Hernandez and Sanchez, *Cross-Border Links*, p. 15.

61. Lori-Ann Thrupp, "Politics of the Sustainable Development Crusade: From Elite Protectionism to Social Justice in Third World Resource Issues," *Environment, Technology and Society, 58,* 1991.

62. For examples, see Public Citizen, "Briefing Packet for the 21st Century" (Washington, DC: Author, 1992); "The Citizen's Analysis of NAFTA"; and the October 5, 1992, "Environmental Reply to the Bush Administration on the NAFTA," written and signed by 20 environmental and consumer rights organizations.

63. Steering Committee statement from an RMALC meeting on October 4, 1993. This statement was widely used in press statements at later dates.

64. Public Citizen and Citizen's Trade Campaign, "Look before You Leap: What You Should Know about a NAFTA" (Washington, DC: Public Citizen, 1992); Development GAP, "The Citizen's Analysis of the NAFTA" (1400 I Street, NW, Suite 520, Washington, DC: D-GAP, 1992); and RMALC, CTC, ART, Action Canada Network, and Common Frontiers, "A Just and Sustainable Development Initiative for North America" (Washington, DC: Citizen's Trade Campaign, 1993).

65. While this point may not be surprising, it runs counter to the initial response by Greenpeace to NAFTA. Rather than joining the other more moderate national environmental organizations in opposition to NAFTA in early 1991, Greenpeace hesitated for nearly a year before fully opposing it, due in part to financial restraints, internal divisions, and differences between Greenpeace USA and Greenpeace Mexico. It was not until 1992, when Barbara Dudley became executive director of the USA office, that the internal political differences on the topic of NAFTA were brought under control. In an abrupt shift, Greenpeace USA then became a prominent figure in the fair trade efforts and would remain so throughout the NAFTA struggle.

66. Greenpeace activists have published numerous pieces on NAFTA. These include, among others: "Analysis of the U.S. Proposal for an Environmental Side Agreement to the NAFTA," June 1993; Barbara Dudley, "The Impact of the NAFTA on the Environment," testimony to the House Subcommittee on Environment and

Natural Resources, March 10, 1993; Trade Team, "NAFTA: Trading Away the Environment"; Carol Alexander, "The NAFTA and Energy Trade" (Washington, DC: Greenpeace USA, 1992); "NAFTA and the North American Agreement on Environmental Cooperation (NAAEC): Side-Stepping the Environment," Greenpeace Policy Briefing, September 1993; and numerous press releases. Other environmental majors have contributed to publications, undertaken numerous testimonies to house committees, and have participated in coalition statements.

67. Efforts to link trade to environmental hazards were apparent in earlier struggles over pesticide use, food imports, and deforestation. Previous movement currents thus shaped efforts to articulate the environmental and social consequences of neoliberal trade.

68. Letter and petition to Ambassador Mickey Kantor, the U.S. Trade Representative under President Bill Clinton, September 13, 1993.

69. Howard Frederick describes the role of electronic communication networks in shaping opposition to NAFTA in "North American NGO Computer Networking against the NAFTA: The Use of Computer Communications in Cross-Border Coalition-Building," unpublished paper presented at the XVII International Congress of the Latin American Studies Association, September 24–27, 1992, Los Angeles, CA.

70. Governments of the United States of America, Canada, and Mexico, *"Preamble" to the North American Agreement on Environmental Cooperation* (Washington, DC: U.S. Government Printing Office, 1993), p. 1.

71. Ibid., p. 2.

72. Jay Hair, quoted in Boyer, Gabriela, et al., *NAFTA's Broken Promises,* pp. 8–9.

73. Ibid., p. 10.

74. Andrea Durbin, "Environmental Side Agreement: Falls Short of Fixing NAFTA Flaws," Friends of the Earth Policy Brief (Washington: DC: 1993), p. 2. The most important "post-Side Accord" environmentalist critique of NAFTA and the side accords was publicly available on October 5, 1993. This 22-page document has 20 national environmental, consumer, conservation, and animal welfare groups signed-on (with memberships totaling nine million) and maintains the more critical elements of earlier anti-NAFTA, environmental "consensus" documents.

75. Boyer, Gabriela, et al., *NAFTA's Broken Promises.*

76. Ibid.

77. The first case filed with the CEC was pursued by representatives of two groups who leaned more often to the pro-NAFTA camp than the anti-NAFTA camp, the National Audubon Society and the Grupo de los Cien (or Group of 100). The staff that filed the petition was among those more skeptical of the NAAEC and the decision by its leadership to support NAFTA. The ruling by the CEC no doubt reinforced anti-NAFTA sentiments within these organizations. Both organizations opposed GATT more vigorously than they supported NAFTA (author interview with Rhona Carter, Youth and Urban-Environmental Program Coordinator National Audubon Society, October 3, 1995).

78. Commission for Environmental Cooperation, Secretariat, *Migratory Bird Mortality of 1994–1995 at the Silva Reservoir* (October 1995). The Grupo de los Cien was particularly vexed by the implications of the report, since chromium and other industrial heavy metals, which are used extensively in the cities surrounding the reservoir, were found in the tissue samples of birds and in sediment samples.

The Public Citizen, in *NAFTA's Broken Promises,* discussion of this case presents a nuanced analysis of the politics behind the report.

79. For a discussion of these petitions, see Public Citizen, *NAFTA's Broken Promises.*

80. PRONASOL, Hellman argues, is a newer version of an old strategy of PRI corporatism, enabling the current regime to deliver "very substantial material rewards to those popular groups willing to sign pacts" (Hellman, "Mexican Popular Movements . . . ," p. 136). For a discussion of state corporatism and environmental politics in Mexico, see Stephen Mumme, "System Maintenance and Environmental Reform in Mexico: Salinas' Preemptive Strategy," *Latin American Perspectives, 19*(1), 1992, and David Barkin, "State Control of the Environment."

81. U.S. Trade Representative's Office, "Review of U.S.–Mexico Environmental Issues," February 1992.

82. The Salinas administration, through the Secretaria de Comercio y Fomento Industrial (SECOFI), in anticipation of the environmental anti-NAFTA mobilization, contracted with the public relations firm Burson-Marsteller (at $5.4 million) to paint a rosy picture of Mexico's environmental reform agenda. See Bill Baldwin et al., *The Trading Game: Inside Lobbying for the North American Free Trade Agreement* (Washington, DC: Center for Public Integrity, 1996).

83. Several third-wave conservationist groups published positive spots on the Salinas administration's environmental record and concerns. A quite blatant example of neoliberal environmentalism was published in "A Conversation with Mexico's President," *International Wildlife,* September/October 1992, a publication of the National Wildlife Federation.

84. Alexander Cockburn, "The Shameful Seven," *The Nation,* June 28, 1993. Interviews with a select group of environmentalists in Washington, DC, consistently revealed Cockburn's assessment of the "Shameful Seven."

85. Author interview with John Audley, October 21, 1994, Washington, DC. John Audley worked against NAFTA as a representative with the Sierra Club for two years, and his Ph.D. dissertation analyzes the political dynamics of U.S. environmental groups in the NAFTA struggle (*Environmental Interests in the North American Free Trade Agreement* [College Park: University of Maryland, Department of Political Science, 1995]).

86. It is worth noting that the Conservation Foundation (CF) provided funds for the first Earth Day, despite a lack of attention by most of the other old conservation organizations. This was in large part due to the leadership of Sydney Howe, "one of the first mainstream leaders to take a genuine interest in urban environmental issues, which he saw as inseparable from civil rights and other social justice movements. (Howe was fired by the CF board in 1973.)" (Dowie, *Losing Ground,* p. 26). In 1972, the Conservation Foundation, still under Howe's leadership, kicked off a conference that explicitly linked the "traditional" issues of environmentalists to urban quality of life and social justice concerns. (See their 1972 report on the conference, *Environmental Quality and Social Justice in Urban America,* edited by James Noel Smith. [Washington, DC: Conservation Foundation].)

87. The information for this argument comes from interviews with representatives who attended the March meeting or were closely involved with the activities that followed.

88. These memos were referenced by John Audley in an interview on October 21, 1994.

89. Quoted in Dowie, *Losing Ground,* p. 188.

90. Dowie, *Losing Ground,* p. 188.

91. A strange irony, indeed, given that the first petition to the Commission for Environmental Cooperation was filed by dissident staff within the National Audubon Society. Perhaps this is why a "report" was released, rather than outright rejection of the petition.

92. The National Wildlife Fund, along with five other organizations, have insisted that there are "Eight Essential Reasons Why NAFTA Is Good for the Environment," October 1, 1993.

93. Dowie, *Losing Ground,* p. 187. In addition to his excellent history of the U.S. environmental movement, Mark Dowie participated in multiple forums and published several critical commentaries on "free trade and the environment."

94. Stewart J. Hudson, "Testimony before the Subcommittee on Trade of the Committee on Ways and Means" (Washington, DC: National Wildlife Federation, September 21, 1993), p. 1.

95. Dowie, *Losing Ground,* pp. 74, 187.

96. See Dowie, *Losing Ground*; also author interviews with Dan Seligman, Sierra Club trade specialist, Washington, DC, October 1995.

97. Michelle Perrault was also one of five environmental administrators invited by President Clinton to "balance" corporate representation on his "Council on Sustainable Development." She was the only one of the five to oppose NAFTA (Dowie, *Losing Ground,* p. 183). Personal interviews with Dan Seligman and John Audley.

98. Greenwire interview, September 20, 1993.

99. Ibid.

100. Dowie, *Losing Ground.*

Chapter 9

Earth First! in Northern California

An Interview with Judi Bari

Douglas Bevington

Judi Bari was an internationally known Earth First! organizer in northern California. Before her involvement with environmental issues, Bari was active in the union movement and helped lead two strikes in the Maryland/DC/Virginia area, one by 17,000 grocery clerks, and the other a wildcat against the U.S. Postal Service at the Washington, DC, Bulk Mail Center. In 1979, Bari moved to northern California, married, and had two children. After her divorce in 1988, she worked as a carpenter, building upscale houses out of redwood. This experience sparked her involvement in Earth First! In the following interview, conducted in the summer of 1993, Bari discusses her work as an Earth First! organizer. She analyzes the economics of logging and the strategies and tactics of the major logging companies and describes in detail Earth First!'s struggles in the region, which under her leadership emphasized community and worker organizing. Bari also reviews the Redwood Summer Campaign, which brought 3,000 people to northern California to engage in direct action against the destruction of the redwood forests. She discusses the attempted murder of coworker Darryl Cherney and herself, when a bomb hidden underneath her car seat exploded on May 24, 1990, as they were driving through Oakland. Bari was almost killed and was left disabled for life. The local police and FBI arrested the two activists, claiming that they had made and planted the bomb themselves. However, after eight weeks of public accusations by the police and the FBI, the local District Attorney declined to press charges for lack of evidence. No other arrests have been made. Bari and Cherney sued the FBI and the Oakland police for false arrest, presumption of guilt, and civil rights violations. Judi Bari's writings on Earth First!'s struggles with the logging industry, her campaign within EF! to disavow tree spiking, and the circumstances surrounding the bombing that nearly took her life are chronicled in Timber Wars, *published by Common Courage Press in 1994.*

Following this interview, Bari continued her lawsuit against the FBI and the Oakland police. It has survived numerous motions to dismiss and has provided an invaluable window into the FBI's campaign of infiltration and disruption against Earth First!. Likewise, Bari continued her activism in defense of the redwoods. She was a principal organizer of two massive rallies in September 1995 and September 1996 that drew thousands of people to northern California to halt the logging of Headwaters Forest, the world's last remaining stands of unprotected old-growth redwoods. During the 1996 rally, more than a thousand people were arrested for participating in nonviolent civil disobedience, making it the largest single arrest for a forest-related protest in U.S. history. Soon after, the federal government and Maxxam, the owners of the property, reached a tentative agreement to protect two of the six groves within Headwaters. Efforts continue to save the remaining forest.

During that fall, Bari was diagnosed with inoperable breast cancer. She used her public announcement as an opportunity to call attention to the rising incidence of breast cancer and its connections to environmental degradation. She spent much of the remaining months of her life preparing the lawsuit. It was her hope that she would live long enough at least to see the suit come to trial, but delaying tactics by the FBI prevented this. Despite great pain, she continued to work on the suit up until the very end. Judi Bari died on March 2, 1997, at the age of 47. In the following week, as she had requested, her community held a giant party to celebrate her life and legacy.

At the time of this writing, Bari's lawsuit continues, though it has yet to come to trial. Donations in support of the suit may be sent to the Redwood Summer Justice Project, P.O. Box 14720, Santa Rosa, CA 95402. There has also been a trust fund set up to help provide for her two teenaged daughters. Donations may be sent to the Children of Judi Bari Trust Fund, P.O. Box 2517, Sebastopol, CA 95473.

Douglas Bevington—Please describe the region in which you are organizing and that region's key players. How far do your activities and influence extend?

Judi Bari—The area is the northern California redwood region, which at this point is mostly Mendocino and Humboldt counties. It's a big, sprawling area, rugged and sparsely populated. There are only 80,000 people in Mendocino County, where I live, and there are no large towns, no urban centers. Humboldt is a little different because they have a university, Humboldt State, and a large town, Eureka. Both counties are rural impoverished areas. Our county governments are broke. They are closing down libraries and cutting back on police. Because of this, and because of the lack of urban influence, there is kind of a "wild West" mentality. There are some towns—Whitethorn and Covelo come to mind—that are virtually lawless areas, over two hours' drive from the nearest sheriff's deputy.

There are three main corporations in the redwood region. Louisiana Pacific (L-P) is the biggest redwood landowner, with 500,000 acres spread over the entire area. Tied for second are Georgia-Pacific (G-P), which owns 200,000 acres in Mendocino County, and Maxxam, which owns 200,000 acres in Humboldt.

L-P has been around since 1975, and they've been liquidating ever since they've been here. Last year, they even admitted that 90 percent of their marketable trees in this county have already been cut. They are a cut-and-run company and they are almost done. G-P has been here a little longer. G-P has just one big sawmill and it's still running. But most of their timberlands have been clearcut. They are about 70 percent done, so it's just a matter of time.

Maxxam is probably the best known of the three companies, mostly because of its outrageous financial practices. The 200,000 acres now owned by Maxxam used to be owned by Pacific Lumber Co. (Palco), a 120-year-old locally owned company, one of the oldest in the area. Palco didn't clearcut, and they had the closest thing to sustainable logging practices around. Because of this, they have the most old growth left. They have the best of what is left in the world. But in 1986, Maxxam Chairman Charles Hurwitz, a corporate raider from Texas, took over Palco in a forced junk bond buy-out scheme and then tripled the cut of old-growth redwood to pay off their junk bonds.

This area has had pretty much a singular economy based on timber. Many of the logging families here go back five generations. Everybody knows each other, everybody grew up together and often are related to each other. It's hard to understand the phenomenon of the company town until you live in one. King Timber controls all aspects of the society—the jobs, the schools, the hospital, the newspaper, and the police.

This singular economy and isolated rural culture got interrupted around 1970, when the back-to-the-land hippies moved in with their politically sophisticated, radical urban culture. What the hippies do for a living is grow the best marijuana in the world. In the mid-1980s, they actually included marijuana in the agricultural report of Mendocino County and found that it was the biggest cash crop, bigger than timber. This gives people a way to live in this area without being economically dependent on timber. Because timber is such a fluctuating market of booms and busts, during the low periods it is the marijuana money that keeps the towns going. Garberville, in southern Humboldt County, is probably the best-known local town built up with marijuana money. Many early growers used their money to open "legitimate" businesses. As the hippies gained economic influence, they also gained political influence, including passing voter referendums on local issues and electing candidates to the Board of Supervisors.

So, we basically have a bipolar social situation, with the hippies and the loggers. Of course, there are other social/cultural groups in our community, including Mexicans, Indians, fishermen, etc. But the main players in the Timber Wars are the hippies, the loggers, and, of course, the big timber corporations.

DB—What role does tourism play in the economy?

JB—Tourism is primarily on the coast in Mendocino County. It's had some effect of gentrification—yuppie owners, underpaid maids, and sky-rocketing land values. But the coast is separated from the inland by a mountain range and 36 miles of winding roads, and it doesn't seem to have as much impact inland.

DB—What are the major environmental groups in your region?

JB—The two most prominent environmental groups here are EPIC and Earth First! EPIC stands for Environmental Protection Information Center. It is based in Garberville, and they basically file lawsuits on timber harvest plans. They are the most important local group for that. They were also the sponsors of the "Forests Forever" statewide voter initiative in 1990. Earth First! takes the lead in defining the issues and also does the direct action, the logging blockades. We are one of the largest and most influential political groups here.

There are also a series of watershed groups, people living in a threatened watershed who get together, such as the Albion River Protection Association. There are scores of these groups, and they form the local base of the environmental movement here. The way that we organize in EF! is by working with the watershed groups that advise us on the issues.

Other important influences are the Mendocino Environmental Center in Ukiah and the North Coast Environmental Center in Arcata. Both are long-standing, independent local groups that maintain storefront offices, monitor timber harvest plans, lobby, and sometimes file lawsuits. They unite us all by making their physical space, office equipment, services, and political support available to all parts of the movement.

So, you can see that this is a very locally based grassroots movement. The big national groups are hardly even a factor. The Sierra Club is around, and they have funded some of the lawsuits on the timber harvest plans. Sierra Club state reps have also stepped in to negotiate with timber and government to try and compromise some of the gains we have made here. But the local Sierra Club chapters mostly focus on land use issues and not on forestry. And the other mainstream groups, like Greenpeace and Wilderness Society, just don't exist around here.

DB—How large is the active membership base of EF! in your region?

JB—EF! is decentralized and it's not a membership organization, so that question is not so easy to answer. There are three separate EF!

groupings in this area: Mendocino County, southern Humboldt, and northern Humboldt (including Humboldt State University). We all work together and support each other's actions, but we have three autonomous groups, treasuries, mailing lists, etc. The Mendocino County mailing list is about 1,500 people. In order to be on this list you need to have either given money or attended actions within a year. About two-thirds of that list is local, meaning northern California, and the rest is national.

In Mendocino, we can get about two hundred people to a demonstration, which is incredible given the small population. I don't know of another group that can get 200 people to a demonstration in this area. When we started, before Redwood Summer, we'd get 20 people. Now we get 100–200.

DB—While working with EF! to protect the forests in your region, you have also been organizing a coalition with timber workers. Where do the interests of environmentalists and the timber workers coincide and where do they conflict?

JB—The interests coincide because both the forests and the workers are exploited by out-of-town corporations, whose policy is to liquidate the forests and then leave. Cut-and-run is equally damaging to the ecology and the economy. The area that they leave behind is devastated, and to log in a manner that you are cutting yourself out of work is certainly not in the interest of the loggers. So the basis for the coalition is local people, people who have a long-term interest in the area against out-of-town corporations.

DB—Unsustainable logging practices can have short-term social benefits on a local level in terms of increased employment, etc. Are the interests of local movements capable on their own of shaping a sustainable relationship with the forest? Or is there a need for environmental movements beyond and outside the region?

JB—I disagree with the premise that liquidation logging has short-term benefits for the community. The short-term benefits are really too narrow to be considered by anybody who is thinking seriously of the area in which they live because when the timber companies do this, they bring in people from all over. When they were logging at 300 percent, they had people from Idaho, Oregon, and every place you can think of in here logging. So it is not that the local people had this great era of prosperity. The corporations had a great era of prosperity, and some of it may have trickled down into the local economies, but by and large we were not the beneficiaries of it.

Back to the coalition, some environmentalists think that there should be no logging at all, which may not be that far off, considering how little forest there is left. But this obviously represents a conflict within the coalition. What the progressive loggers have come up with instead are methods of restoration logging—restoring the damaged land while still taking out some products in order to provide a living for those forest workers. For example, when the redwoods are cut out, the tan oaks overgrow and just strangle the area. If you thin the tan oaks, then the redwoods can grow back better. And you can mill the tan oaks into flooring, which is a high-value product. That could be the basis for a commonality, creating a way for people to make a living and still live here while maintaining the forests.

Conflict arises from the fact that the environmental movement in general, certainly the large groups, are primarily urban, privileged people, whereas the timber workers are a rural industrial proletariat. So there are many real divisions between these groups—urban versus rural, white collar versus blue-collar, privileged versus nonprivileged. The timber industry has sought to exploit those differences and portray the entire environmental movement as privileged, urban people who aren't concerned about rural people. That is true to a certain extent. It is certainly true of the big mainstream groups. That is why the propaganda is effective, because it does contain some truth. We are an exception here because we are locally based.

DB—Could you give an example of a rural-versus-urban conflict in this case?

JB—The major urban-based environmental groups, the Wilderness Society, the Sierra Club, etc., basically hold the position to take wilderness areas and preserve them. They want to talk about saving particular areas, but the aspect of compensation or reemployment for displaced workers does not figure into their proposals at all. For example, Headwaters Forest is the most famous old-growth redwood forest left. It is in Humboldt County and is owned by Maxxam. It's 3,000 acres of redwood wilderness with 2,000-year-old trees and six-foot-tall ferns. It was discovered, mapped, named, and made an issue by EF! and EPIC. We've been working for years to preserve it. But how could it be preserved in a manner that the workers could support it? If it is cut, it will provide approximately twelve years' worth of work to timber workers, 2,000-year-old trees for 12 years' worth of work. But if it is preserved it will eliminate that many jobs. The solution is to combine preserving old growth with providing jobs in restoration for the devastated areas.

But when Congressman Pete Stark, under the influence of city-based environmental groups, proposed saving the 3,000 acres and a 26,000-acre buffer zone around it, the workers were not even part of the formula. Charles Hurwitz will be paid millions of dollars for the land. Nobody even questions that. But the idea of compensation for displaced workers isn't in the program Pete Stark has proposed for saving Headwaters.

As opposed to this proposal, EPIC also put forward a proposal for 79,000 acres which we hope is going to be promoted by Congressman Dan Hamburg. EF!, including myself, worked on the EPIC proposal. I was in charge of developing the worker compensation plan for this proposal. No other preservation proposals that I know of even include a worker compensation plan. What I did was that, instead of writing a proposal for the workers, I convened a committee of workers and displaced workers and they wrote it. Our basic principle was that the workers are not responsible for Charles Hurwitz's crimes and that they should not bear the brunt of them. Displaced workers need equivalent jobs at equivalent pay in their community. With this in mind, we developed a plan in which the worker component is integrally related to the biological component. The 79,000 acres includes six major stands of old growth separated by land that has been trashed out and cut over to various degrees. We proposed uniting all of these lands in an ecological preserve that will be managed for the idea of restoring wilderness qualities. And restoration efforts will be done by the displaced workers. We figured that it would create about 100 restoration jobs and would displace about 200 logging jobs. So along with that we have a workers' severance package.

This is based on what was done when Redwood National Park was created in the mid-1970s. That was back before the timber unions were busted. In fact, I think this is why the unions were busted. Back then, the unions were powerful enough to get a workers' severance plan included in the bill that preserved Redwood National Park in which displaced workers were paid two-thirds of their pay for six years. That is the final offer that we make in our Headwaters Proposal. We put this forward and found that it would only cost about $3 million a year to implement. Well, they are talking about $500 million to pay off Charles Hurwitz. But we've found an absolute closed door to our ideas among the mainstream.

DB—How responsive have the workers been to your attempts to build alliances with them?

JB—We've had some really good responses, but there's been a tremendous force trying to bring us apart. When we in EF! began trying to save the forest, I was working as a carpenter and I'd been a blue-collar worker all of my life. The reason I got involved in forestry issues was because I was building yuppie houses out of old-growth redwood, and I was appalled at it. Certainly, my job was contributing to the destruction of the forest, just as the loggers were. And I was working in the same industry, on the other end. So when we began to work to save the forest, I found that I had as much sympathy for the people in the mills, in factory conditions with which I was all too familiar from my own work history, as for the people in the trees trying to stop the cutting. I was really surprised to see the prejudice among environmentalists, that they actually blamed the loggers. Part of the reason for environmentalists' contempt for workers is that they are completely ignorant of the fact that the workers have indeed been struggling against these companies for years. So, the first thing I tried to do was to educate environmentalists about the timber workers. I taught a workshop about the history of worker organizing in the timber industry, and the International Workers of the World (IWW).

The timber workers have something to tell us that we don't know. They're out in the woods all day. They know exactly what's going on. Just giving them that respect of being intelligent, that made a lot of difference in their openness towards us. I began to meet workers by blockading them.

We'd shut down some logging operation, and there they'd be, the perfect captive audience. When we'd start talking, because the way I was talking to them indicated that I had similar experiences and sympathies and understood what their life conditions were like, they were really interested in talking to me. I began to get to know them and to know some of the issues in the workplaces and some of the working conditions. I began to advocate for those positions from whatever public forum I got from being an EF!er. I began to advocate for the conditions of the workers because the unions have been essentially busted in our area and there really isn't anybody advocating for them.

DB—What unions were busted and when?

JB—Both G-P and L P were union. There were two unions that represented them in the early 1970s: the International Woodworkers Association (IWA), an industrial union including both loggers and mill workers representing G-P workers, and the Western Council of Industrial Workers, a branch of the International Brotherhood of Carpenters and Joiners, who represented L-P mill workers. I don't believe that the L-P loggers were unionized.

In 1983 L-P decided to bust the union. What they did was institute a unilateral $2-an-hour wage cut in the saw mills, forcing a strike by the union. They managed to keep everybody out and prolong the strike. The union's reaction, because unions are so conservative these days, was to call for a nationwide boycott, which was stupid because the power of workers is in blocking production, not in calling for boycotts. And the boycott was a failure because timber is primarily bought on the wholesale level. It's not something you can successfully organize a consumer boycott of. So, the union was busted and they hired scabs. Then they took a revote and had the union decertified. With the union destroyed, the starting wage in the mill dropped from $9 to $7 per hour. When I got a divorce and needed to get a regular job, since I had factory experience, the first place I looked was the mills. I wasn't involved in environmental issues yet. I discovered that I couldn't support my kids on that salary. I had to find a better job.

G-P came down hard on the IWA, and the union quickly capitulated, agreeing to all concessions in order to prevent their union from being busted too. The concessions included an across-the-board 25 percent wage cut, bringing them down to $7 per hour also. They also agreed to give up representing the loggers. So they now only represent 550 mill workers in the county.

Anyway, I began to get a reputation for advocating for workers. I also wrote some articles, I wrote one in the *Industrial Worker,* the IWW paper, called "Timber Wars" that talks about why our interests are in common and why we shouldn't be turned against each other. I'd give it to people I knew, and they'd pass it around.

In 1989, there was a PCB spill at the G-P mill, and it dumped on some of the workers and poisoned them. The union sided with the company, and the workers didn't know where to turn. Because I had spent time as a union organizer in the straight AFL unions, I pretty much know the bureaucracy for OSHA, and I just used that knowledge to help them pursue an OSHA claim. We ended winning a willful citation against G-P, a $114,000 fine, which is of course nothing compared to what was done to them, but it was the highest thing that you could get. In the course of this, five of the G-P employees most directly affected by the spill, in order to be represented by us instead of by their union, wrote a letter to the company that said, "We do not authorize a union representative from the AFL union to represent us in this matter of the PCB spill. We authorize only Judi Bari and Anna Marie Stenberg of IWW local No. 1." That was our Wobbly union that we had formed. This must have sent a chill down G-P's spine since I was the best known EF! organizer in the region at that time.

DB—Why did you choose to work with the IWW?

JB—I never tried to organize a union, as in getting bargaining power, because this a declining industry. It was clear to me that the corporations would close up shop and run before they would give in to a union organizing drive.

The other thing was that I didn't work in the mills. I was a commercial carpenter working with the redwood that came out of these mills. And Anna Marie Stenberg, one of the principal people who organized our IWW chapter, was a childcare worker who ran a center for the mill workers' children. What the Wobblies offered us was this one big union theory by which everybody in the industry is united. No other union offered that. The Wobblies offered us the structure to address immediate issues, supply representation where none existed, and build coalitions with loggers around the larger issue of cut-and-run.

Shortly after the PCB spill, a worker for L-P named Fortunado Reyes was killed after he was ridiculed by his supervisor and told not to push the emergency stop so often at work because it was slowing down production. So he tried to clear a jam on the assembly line without pushing the emergency stop. When the line started up again, he was crushed to death behind a load of lumber. The workers were appalled. They filed their own complaint with OSHA. But because they would lose their jobs, they couldn't be as public about it as I was able to. I went to a Board of Supervisors meeting and called the head of L-P, Harry Merlo, and Fortunado's supervisor, Dean Remstedt, murderers, and I demanded that the county bring criminal charges against L-P. Amazingly, about a year later, they actually did. L-P had only been given a $1,200 fine for Fortunado's death, and they appealed it and got it reduced to $600. All that ended up happening with the criminal charges was adding a $5,000 fine from the county. Still, the people who worked there began to hear somebody advocating for their interests. That was something that they hadn't seen.

The same thing was going on at Maxxam. After the takeover of Palco, one of the things Charles Hurwitz did was raid the pension fund in order to pay off his junk bonds. He went to the company town, Scotia, and told the workers, "I believe in the golden rule: those who have the gold rule." The workers were outraged and attempted to organize.

Pacific Lumber is a company that has been nonunion for 120 years. They've resisted the Wobblies. They did it by being a paternalistic company that pays a little better than the others. It gives better working conditions and scholarships for the children of the workers to go to college. It's a company town. They actually own the town, one of the last company towns in the United States. If you work there, you can get these nice houses very cheaply. Because of that, they basically had a loyal work force. When Charles Hurwitz took over, 500 workers signed a full page ad objecting to the takeover. That was pretty radical for them. But it got co-opted by a man named Patrick Shannon who had actually done some union-busting work down in San Francisco at Yellow Cab. I think that he was just a charlatan. He came in and said he wanted to organize an ESOP, an employee stock ownership plan. Instead of this energy that workers were experiencing going into a union, this man, who was antiunion, convinced the workers that what they needed was an ESOP. The flaw in this strategy is that the company wasn't for sale. They only give workers ESOPs when a company is about to go bankrupt. Charles Hurwitz wasn't about to turn this company over to the workers. The result of it was that Patrick Shannon was discredited and the ESOP collapsed. The company reigned everybody back into loyalty and fired some of the workers from the ESOP Committee and got the workers to direct their anger at Patrick Shannon instead of Charles Hurwitz.

At this point, I met some of the workers who had been on the executive committee of the ESOP who were interested in resisting. We put out a newsletter on the workroom floor. The company paper was called *Timberline,* so this newsletter was called *Timberlyin'*. The Palco paper had a skyline of trees, ours had a skyline of stumps. This was distributed on the workroom floor, and it was a rank-and-file newsletter that lampooned management and criticized them for both their work rules and the forestry practices.

DB—How did the timber workers feel about being in coalition with EF!, given its reputation for tree spiking?

JB—Before the bombing, the coalition was beginning to grow. But it was held back by EF!'s advocacy of tree spiking, the idea of driving nails into trees to prevent them from being cut supposedly because the companies wouldn't want to risk shattering a saw blade in a mill. I have pretty much opposed tree spiking from the start. It is a pretty naive strategy made by people who have never done industrial work in their life. They don't know, first of all, how dangerous a factory really is and, secondly, how little the companies care for the lives of the workers. For them to count on the morality of the companies not to cut a tree just because it's spiked, that hasn't been the case. The companies have cut the trees anyway and risked the workers' lives. In our area a man was nearly decapitated when a saw hit a tree spike and broke up and hit him in the neck. I had been against tree spiking as an individual, but I did not have political influence within EF! to do much about it. But by 1990, I did. At the urging of timber workers that we were working with, Northern California and Southern Oregon EF! publicly renounced the tactic of tree spiking. When we did that it really opened the coalition so that it could be more public, so that more people could feel comfortable working with us.

It was in April 1990 that we renounced tree spiking, so that upped the ante. The coalition became more solid. And timber workers felt better about working with us. Right around that time we had already put out a call for Redwood Summer, for students and others to come into our area and engage in nonviolent civil disobedience to stop the liquidation of the forests. Right on the eve of Redwood Summer, L-P closed yet another sawmill and laid off yet another 200 people on the same day that they announced record profits.

We went to the Board of Supervisors in April for the first time publicly with the worker coalition. We had been working with timber workers privately for years, but now for the first time in public. EF!, IWW, and currently employed loggers and mill workers from L-P attended a Board of Supervisors meeting together. What we asked for was that the county of Mendocino use its power of eminent domain to seize all of L-P's timberlands and operate them in the public interest as the only way to save the trees and jobs. This is obviously a radical demand for Mendocino County, and it certainly caused a stir.

DB—At this point, how large was the coalition? What percentage of it were environmentalists and what percentage were workers?

JB—Everything was tiny. The number of people who came to the meeting was about 50. Of those, five were workers, although there were quite a few workers who didn't have the nerve to come to the meeting. There were about 30 workers in the coalition and maybe 100 environmentalists. Everything was amazingly tiny compared to the influence we had.

DB—How much influence would you say this coalition has had, given its small size?

JB—Because most people are cowed into inactivity, we could start with just that many active people and have a tremendous impact. We've really caused quite a bit of change in the logging methods and the rate. The logging rates are way down from what they were in 1990. Part of that is certainly caused by the economy. But it is also because they don't want to stir up that much shit again. The *Santa Rosa Press-Democrat* quoted a *Wall Street Journal* analyst after Redwood Summer on L-P's announcement that they were changing their logging practices and stopping clearcutting solely in the redwoods. It said something like, "Well, if you were a timber company and you had to worry about these people protesting all of the time, surely you would accede to some of their more reasonable demands."

DB—What kind of response was there to the coalition's demands?

JB—It was immediately after that meeting for eminent domain that I began to receive the death threats that led up to the bombing. The scariest death threat that I got was a picture of myself with a rifle scope and crosshairs drawn over my face. That picture was from that meeting. It was clear to me where I had crossed the line. Having not only advocated such radical tactics, but having done so in open coalition with workers posed a serious threat to the corporations. I think that is when they decided that they were willing to kill me if they needed to. And things certainly built from that point.

They began to organize very hard to discredit us. They had a concentrated campaign to make us appear violent, to make us appear antiworker. This campaign, when they were unable to get us to discredit ourselves or terrify us with death threats, included producing fake press releases that had our EF! logo on it.

DB—What evidence do you have of the connection between the timber corporations and disinformation about the Redwood Campaign?

JB—Maxxam filed a lawsuit against Darryl Cherney and George Shook, two EF! tree sitters, several years ago. In the course of that suit, the defense asked for all internal documents relating to EF!. One of the documents we got back was a letter from the company to Hill & Knowlton saying that they wanted to hire them to counter EF!. Hill & Knowlton is one of the biggest, most sophisticated PR firms in the world. Hill & Knowlton had an $11 million contract with the government of Kuwait. They helped persuade the U.S. Congress to go to war in 1991 based on the testimony of this poor Kuwaiti woman who talked about the Iraqis throwing babies out of incubators into the desert. She gave this testimony to Congress just before they voted to go to war. Her testimony turned out to be completely bogus. The woman was the daughter of the Kuwaiti ambassador. The whole thing was a lie. So Hill & Knowlton is this international company in the business of producing disinformation. And they were hired to produce the disinformation around the redwood issue.

We got another internal document from this lawsuit in which the Maxxam/Palco PR manager, David Galitz, acknowledges that one of the press releases allegedly from us calling for violence and chaos is fake. It had Darryl's name spelled wrong and my hometown wrong. But one week after the date on that internal memorandum, that same press release was sent to the *San Francisco Examiner, Chronicle,* and the *San Jose Mercury* by Hill & Knowlton. I believe that Hill & Knowlton wrote it as well as distributed it, but I can only prove the latter.

In another example, in April 1990, in this pulp mill in Samoa, L-P called a meeting on the clock to distribute a different fake press release, whip up hatred against us, and openly encourage workers to intimidate us.

We know about this meeting because it occurred at one of the last union mills and the Pulp and Paper Workers' Union filed a grievance against L-P for it.

DB—How do you know that press release was fake?

JB—It had our name on it and we didn't write it. It purports to be from "Arcata EF!" and there is no such group. The Arcata group is called North Coast EF!. I also know that it was fake by its content and method of distribution. Although no EF!ers received these press releases, they were distributed to the newspapers, in the mills, and in stacks in laundromats in the logging town. We know everybody in our group. This didn't come from anybody in our group. So we know we didn't write them, and we know that the method of distribution was in order to inflame timber workers and discredit us in the press.

Another example came from the Sahara Club, an antienvironmental hate group. They published a diagram of how to make a bomb, which they claimed was from an EF! terrorism manual. Of course, there is no such EF! manual, but, by printing this diagram and attributing it to us, they managed to simultaneously stir up hatred against us and distribute information on how to make bombs.

There was this rash of false information coming out in April 1990, right after the eminent domain attempt. Workers were very much riled up. Nobody, including us, really had the experience or sophistication to expect there to be fake documents distributed to the press. So most people who saw these things believed them to be real. It began to discredit us a little bit. But I still don't think that it would have worked without the bombing.

After the bombing, I was immediately arrested and blamed for bombing myself. The bomb was hidden directly under my car seat and triggered by the motion of the car. Yet, the FBI arrested me and Darryl for transporting explosives, claiming that it was our bomb and we were knowingly carrying it. Again, people aren't used to seeing blatant lies, or at least they are not used to recognizing blatant lies in the headlines of the newspaper. So, at first, a lot of people believed that it was my bomb, including some of the workers with whom we had just begun to build bridges. The bombing very much scared people. The environmentalists were terrified. Workers who were just beginning to trust us thought we must have been lying all along and began to back off. Those who didn't believe the lies, many of them were scared off. So the bombing set back the worker coalition for miles.

DB—Did the bombing destroy the coalition or has it recovered?

JB—It took years to recover. But it finally has. I don't know if we are quite up to the eminent domain stage. But years later we have finally regained much of the ground that we lost. And a lot of this is because of a man named Ernie Pardini.

Ernie is a fifth-generation logger. His family is one of the most prominent logging families in our area. Last summer, a community called

Albion decided that they didn't want to let L-P take the last of the trees in the area. So the local activists called in EF!, and in a nine-week campaign of constant actions we managed to shut down L-P in the entire watershed. Ernie Pardini's uncle was the man who was doing this cut. Partway through the struggle, Ernie came forward and denounced L-P. He did that publicly. He's probably the most articulate logger I've ever worked with. He's really smart and brave too. He stood up against his family. He stood up against his community to tell the truth about the corporation. His brother did so also. The role that I've played in this is that I gave them a forum to talk.

We have great local alternative media. We get no press coverage in the corporate media, but we have really good alternative media. I was writing a column in the *Anderson Valley Advertiser,* a local radical newspaper. I did one in which I interviewed the L-P mill workers about their working conditions. I also interviewed five loggers and wrote an exposé of their working conditions. Ernie had read these things and saw what we were doing. The other thing I have is a radio show. Ernie came on the radio show and denounced the corporations in no uncertain terms. He even denounced them for bombing me. His brother called in and joined him.

DB—Who did Ernie work for at that time?

JB—He was a logger who worked for L-P, not directly though, because loggers work through the gyppo system. He actually worked for his uncle, Mancher Pardini, one of the biggest logging contractors in Mendocino County.

DB—What do you mean by "gyppo"?

JB—It means the timber corporations contract out the actual logging to small companies that are paid a fixed rate on a logging job. The gyppos own the big heavy logging equipment, not the corporations. That's one of the real problems with this EF! idea of halting logging by sabotaging equipment. That's all very well and good if the equipment is owned by the corporation. But when the equipment is owned by a small company, that's who pays the cost for the sabotage. It doesn't hurt L-P one bit. This is also a real good way for the corporations to distance themselves. It's harder for the workers to identify the enemy when they are, say, working for their uncle. If Ernie had been working directly for L-P and he stood up, that would be one thing, but he had to stand up against his family.

Ernie has been blacklisted for it, as most workers are. We call it the Future Ex-Loggers Coalition because anyone who stands up gets blacklisted or fired. But what Ernie told us is really the next step that we needed to know. He said that loggers can't stand up against the corporations if there's nothing else for them to do, if there's no place else for them to go. We really needed to form an alternative. "Retraining" and "diversification" are just code words for dislocation to rural timber workers. So, his idea is what we've named the Mendocino Real Wood Co-op. One-third of the redwood land is privately owned, as opposed to corporate ownership, and that third is the third with the trees on it. The corporate lands are just about gone.

Ernie's idea was to unite small landowners who want their land to be logged sustainably with loggers who want to use sustainable logging methods. Because this is so small-scale that individual loggers and individual landowners can't provide enough to really have any long-term markets for these products, the idea is to do it as a co-op. What Ernie would say is that we don't need to diversify away from the timber industry, we need to diversify within timber. The idea of diversifying within the timber industry is to only use what the forest can give up as it recovers from the assault that it has sustained. If you do this on small scale with local people who are being displaced by the corporations, that provides an economic alternative for loggers so that they can stand up to the corporations. Ernie started talking to us last summer, and this is just coming into fruition now. The first jobs are just beginning to happen in the Mendocino Real Wood Co-op. There are only five or so now, but we are just starting our first timber harvest plan and have gotten a timber distributor in the Bay Area. It may be up to 20–30 jobs by the time this is printed. So that is the state of the worker coalition right now. We are working with Ernie and others to try to create an alternative economy, along with renouncing the corporations and trying to stop their cut-and-run logging.

DB—Isn't sustainable forestry in the interest of the corporations? Why do they practice cut-and-run?

JB—L-P, G-P, and Maxxam are multinational corporations. G-P is the biggest timber corporation in the world. They own rainforests in Brazil. They own timber all over the place. What they own here is only a very small part of their holdings. None of these corporations are based here. They bought these lands in order to get the most profit out of them. They operate on a basis of short-term profit. They are not interested in long-term sustainability because they have no economic interest in the long-term health of this region. Their only interest is the long-term economic health of their company. Their interest is to get as much value out of here as fast as they can and move on, in this case, to Siberia.

There are different things that they can do with the land when they are done cutting it. If they just cut the old growth and leave some decent second growth and residual old growth, they can sell it to another timber company. If they leave nothing but tiny trees, which are called pecker poles, then they call sell it to a pulp company. And if they clearcut, they can sell it for real estate. The land still has value after they take the trees off of it. That short-term value of selling the land is much greater to the corporation than the long-term investment in the ecosystem and the local economy.

Anyway, second-growth redwood isn't worth shit. It doesn't have the qualities that old growth does. It's not rot-resistant and things like that. And even by the most generous estimate, you can't cut a redwood to make boards out of it for 50 years. So if they cut down the trees and planted new ones, they'd have to wait at least 50 years to get another saw log. L-P hasn't existed for that long. The natural lifespan of a redwood is 1,000–2,000 years. Capitalism hasn't existed for 1,000 years!

A redwood tree does not even reach reproductive age, does not produce cones, for 100 years. They are cutting trees only halfway to reproductive age. So, these companies have almost ended natural reproduction of redwoods. Not only are they changing the ecosystem, but they are destroying its natural evolution.

DB—How has capital flight affected the struggles in your region?

JB—It was really a wake-up call for the timber workers in our region. When L-P opened their Mexico mill, there was a shock wave through the area. L-P employees, Walter Smith being the most notable of them, stood up and denounced L-P. Walter Smith is a second-generation local logger. Not only did he speak out, but there was no racism in what he said. He said that the issue isn't the nationality of the people who are milling the trees. It's already about one-third Mexican in our area. The issue is the exploitation that's going on by shipping it to Mexico, both the exploitation of our area and the exploitation of the Mexicans.

It was the Mexico mill that spurred people to speak out against L-P. Once they started talking, it wasn't just the Mexico mill that they were talking about. We used it as a method of showing that corporations are not working in the interests of the workers and that they really are only interested in making money for themselves. This has made it really clear for us to be able to say that the problem here is not the timber workers versus the environmentalists. It's the corporations versus the entire community. It's capital flight that has been irrefutable on that.

Ultimately, L-P has got to go. "L-P out of Mendocino County" is one of our slogans. I don't want them to do to Mexico what they've done to us, but I sure do want them to leave. And the sooner they leave, the sooner we can come up with a community based solution.

DB—How has your community been transformed by these struggles?

JB—What has been happening locally has been awesome. The movement has become really deeply community-based. Whereas it started out with just a few radical EF!ers, the concept of biocentrism is really the prevalent view among all environmental groups in Mendocino County now. A couple of years ago you couldn't have said that. The actions have also become based in the community. The corporate state threw their worst at us in Redwood Summer, with the violence, assassination attempts, and frame-ups, but EF! didn't back down. Three thousand people came to Redwood Summer. They sat in trees, blocked logging roads, chained themselves to bulldozers, and, in scenes reminiscent of anti-Klan marches in the South, stood up to hatred and intimidation by marching thousands strong through logging towns. We stood up to it all with courage, principle, and nonviolence, maintaining our attitude of not being opposed to the workers. This really left us with a legacy of respect in the community. So when the trees start going down, what has happened in the years following is that the communities have called on EF! to help save their trees.

This last summer the prime example of that happened in what has come to be known as the Albion Uprising. Albion is an unincorporated area, population 200, about nine miles south of Mendocino on the coast. The community of Albion had been working through the system to protect their watershed for years. When L-P started cutting, the local activists asked us to come in and help. We came in with our tree-sitters and our expertise in organizing, and we joined in with the local community to mount a nine-week campaign of constant direct actions. It proved to be a really powerful combination The first two tree-sitters were lesbians. We had a Native American tree-sitter. We had timber workers coming to our rallies and denouncing L-P. We cut across all of the traditional divisions in the environmental movement. And we came up with a community-based direct action movement. We just continued blockading L-P no matter what they did. They got injunctions against us. They filed a SLAPP suit against us. They arrested us, sometimes falsely. They did everything that they could to stop us, and we wouldn't stop. We maintained nonviolence. We had all of these great creative tactics. We used one tactic called "yarning," in which we take yarn and weave it in and out of the trees to slow down the logging. It is a great tactic because you have a hard time calling yarning "terrorist." Yet, you can't cut yarn with a chainsaw or a logger's ax because it's too flexible. You have to cut it with scissors. It's this aggressively unmacho thing.

We slowed them down and created so much social unrest that finally after nine weeks they gave up. After realizing that they wouldn't be able to stop us, the same courts who had ruled to let L-P cut at the beginning of the Albion Uprising ruled to stop it. So we shut them down in the entire watershed, and it has really had an impact throughout our community. What's happened as a result of that has been an even further deepening into the community of the acceptance of direct action tactics. When you file a lawsuit, they say that you first have to exhaust administrative remedies. Well, in Mendocino County, the remedies includes chaining yourself to the bulldozers.

What's happened so far this year is that the community of Albion, this time by itself, without needing to call in EF!, was able to stop a helicopter logging job by using the same kinds of combinations of direct action and filing lawsuits that we did before. They stopped it much more quickly and easily because of the legacy and the knowledge of what happened last year.

Another example involves the Cahto Indians. The Cahto Wilderness, their tribal lands, were the first thing EF! ever saved by blockading in this county. Anyway, this year when they discovered that a toxic dump was leaking onto the Cahto Rancheria and poisoning the kids, Peggy Smith, one of the Cahto tribe members, chained herself to the gate and forced the county to close the dump. So what we see here is a community which has become empowered to take this kind of action, and its effectiveness has been proved, and it has gone way beyond EF! now. To me that really measures the success of the tactics.

DB—How has the particular character of your region helped in forming alliances with labor?

JB—One difference in this area where I live and work is that we do live here. Both sides, the EF!ers and the timber workers, live here—we are neighbors, and we live in the forests. That has given our North Coast redwood region environmental movement a different character than those in Oregon, for example, where most of the environmentalists live in the city and travel out to these vast tracts of public land for their environmental activity. It is really different here because the ecosystem is so fragmented that people live right in the areas that they are trying to save. The fact that we are neighbors, the fact that we all live rural lifestyles, has really proven to be something that we have in common, that has enabled this area to develop differently than other areas as far as the possibilities of building timber worker alliances.

DB—In the past few years, there have been some important changes within EF! You mentioned that some segments of EF! have disavowed tree spiking. Also, founding member Dave Foreman has left the organization. Could you comment on these changes, your role in them, and your opinion of EF! in general?

JB—The problem with EF! is that it is subject to a COINTELPRO operation similar to what was done to the Black Panthers and the American Indian Movement (AIM). So, it's really hard to even comment on EF! because it is so disrupted that it is hard to tell what is coming from EF! for real and what is coming from the disruption campaign.

Dave Foreman quit partially because of what was going on up here. He thought that we were too radical. He thought I was a leftist and that I wasn't really an EF!er because of my social views, particularly on labor issues. He thinks that the issue should be wilderness only and preservation only. I think that is foolish. There's no point in even preserving wilderness if you don't change the society that is destroying it. Social/environmental issues are too closely linked to even begin to separate them.

What should have happened when Dave left is that those who believe that social and environmental issues are linked should have prevailed within the organization. But because the organization is being disrupted by the FBI, there has been endless infighting. Still, the character of EF! has changed in a lot of ways in that it is no longer just this kind of macho, tree spiking, eco-dude organization. It is a more broadly based group that considers more of the overall effects. We have been forming alliances all over the country with Native People and AIM in particular. This direction is definitely happening within EF!, but it's happening in the context of a disruption campaign that continues to attempt to discredit us and make it as difficult for us as possible.

DB—What forms has the disruption campaign against EF! taken?

JB—Well, the bombing and attempted frame-up of me and Darryl were certainly part of it, as well as the infiltration of Arizona EF! by an agent provocateur and the subsequent arrest and jailing of Arizona EF!ers. But there are also more subtle indications of a continuing internal disruption campaign. I, in particular, have been targeted to be discredited within EF!, including the printing of really insulting letters about me in the EF! Journal coming from people we've never heard of in areas where there is no active EF! movement. They've followed the pattern that was established in COINTELPRO that was done with the Black Panthers and AIM to foment divisions within those groups. This same technique of sending very insulting letters back and forth between the factions appears to be occurring in EF!. There's not only these letters from unknown people, called gray propaganda in FBI talk, but fake letters from known people, which the FBI calls black propaganda. One of Dave Foreman's pen names was Chim Blea. The EF! Journal printed a very insulting letter directed towards the Northeast EF! group signed by Chim Blea. The next month a letter comes in from Dave Foreman saying that he had not written that letter. A letter that was equally insulting was written back from Buck Young, one of the key people in the Northeast. Buck said he didn't write that. In fact, two letters from two different Buck Youngs were published in the same journal.

Richard Held, who is the FBI agent in charge of my case, pioneered these techniques. He has a 25-year history in COINTELPRO—he was in charge of the operations against the Black Panther Party in Los Angeles. He is one of three FBI agents named by Geronimo Pratt's defense committee as being responsible for framing and jailing him. Geronimo was isolated from the Panthers in much the same way that they are trying to isolate me within EF!

From there, Richard Held went on to reorganize the remnants of the Minutemen, a right-wing paramilitary group, into the Secret Army Organization in the San Diego area in the early 1970s to spy on, harass, and attempt assassination of anti–Vietnam War activists. In the mid-1970s, Richard Held went to Pine Ridge South Dakota to join in the reign of terror against AIM. He participated in the framing and jailing of Leonard Peltier. He then went on to Puerto Rico in 1986, where he was in charge of the Puerto Rican FBI. He directed a campaign against the Independentistas movement. So Richard Held is a major scum. But I don't want to put this all on one man. This is agency policy that we're talking about here.

Anyway, the techniques used before and after the bombing are reminiscent of techniques used throughout Richard Held's history. Richard Held personally went on television and said that Darryl and I were the only suspects in the bombing.

DB—You are suing the FBI right now because of that.

JB—Yes, I am. Actually, I really don't have any faith in the court systems at all. The reason that I sued the FBI is because it is the only way that I knew to keep this case alive at all. The FBI tried to blame me and Darryl and frame us for the bombing. So what we've sued the FBI for is false arrest, illegal search and seizure—direct civil rights violations. But we've also sued them on a First Amendment conspiracy charge, which says that they arrested us for the bombing rather than investigate it, as an attempt to sabotage our political work in the manner of COINTELPRO by attempting to portray us as terrorists so that we would be isolated, discredited, and that they could sabotage the movement. Essentially, we are suing the FBI for continuing to engage in COINTELPRO tactics 20 years after they were declared unconstitutional by a U.S. congressional investigation. To my great surprise, we survived the motion to dismiss, and we are in the discovery phase of this lawsuit. As far as I know, we are the first political group that has succeeded in getting COINTELPRO addressed in a civil court. We're demanding another congressional investigation, and we're actually getting some place with the House Committee. They recently agreed to start an investigation of the FBI's role in the case.

DB—On May 24, 1993, you held a press conference to release some of the findings of your own investigation. Could you describe what you found?

JB—Because we have the right of discovery in this case, we got released 300 photos and 270 pages of police reports from the Oakland police. I'm really amazed that they released them. I don't think they were paying attention to what they were giving out. The photos show clear as day that the Oakland police lied and knew they were lying when they arrested us. They arrested us based on their claim that the bomb was on the backseat floorboard. Therefore, we should have seen it. Therefore, we knew we were carrying it and it was our bomb. Actually, the bomb was hidden under the driver's seat. The police photos we have show that the front seat is blown through and that the backseat is intact. They have a side view of the car in which the frame is buckled under the front seat and the backseat is intact. They then removed the seat and photographed the explosion hole. That's the most dramatic photo of all. First of all, it's very horrible looking. This explosion was clearly meant to kill, not scare or maim, and it nearly did. This photo shows this huge hole in the floor with the metal curled back from a very clear epicenter. The hole is right under the front seat.

So, last week we decided to go public with this information on the anniversary of the bombing. And we put out a press release announcing that we were going to do so. As soon as we did that, two things happened. Richard Held resigned. He decided to take an early retirement at the age of 52. He announced his retirement on Friday. Our press conference was on Monday. Now, I'm not going to take credit for this retirement. He says it's because the FBI told him that he had to transfer and he didn't want to transfer. And I'm sure that his retirement has as much to do with the Democrats coming in and the fall of William Sessions as it does specifically with our case. But I can't ignore the timing of Richard Held being up in his office cleaning out his desk, while we were standing on the front steps of his office displaying these photos. The other thing is that the FBI announced that they have closed the bombing case. Also, last week, our lawyers requested a document related to our court case. The lawyers were told by the court that the document has already been destroyed. Well, the statute of limitations isn't up, and that document is specifically named in the lawsuit. So what I see going on is a cover-up.

DB—How has the rest of the environmental movement responded to yours and other EF!ers' conflicts with the FBI?

JB—Right after the bombing we got at least lip service support from most of the big groups. But, by and large, most of them aren't very aware of our struggles and many of them are fooled by the COINTELPRO operation, so they treat us like terrorists. One exception has been Greenpeace. Greenpeace has supported us in several key areas. After the bombing they paid $30,000 for us to hire private investigators and bomb experts to help us defend ourselves and investigate the bombing.

But, on the whole, we are a grassroots movement and we don't interact that well with something like the Sierra Club or these big, national, corporate-type movements that operate from the top-down, with big budgets, with underpaid canvassers, that organize themselves exactly like a corporation. The one thing that I have found to be an effective combination, though, is direct action combined with lawsuits. The mainstream environmental groups sue on timber harvest plans or various legal violations associated with overcutting. What typically happens is that as soon as they get sued, the timber companies try to cut down every tree they can as fast as they can before the judge can rule on it. What we do is put our bodies out in front of the bulldozers to prevent them from doing that. This combination of direct action and legal action has proven really effective time after time.

DB—Logging practices in your region are shaped by regulatory agencies at the federal, state, and county levels. What has been your experience with agencies at each of these levels? How responsive is each to social movement pressures?

JB—All of them are really at the beck and call of the corporations. If the laws were followed, the forest would not be being destroyed. The agencies sometimes will even admit that a cut is illegal, but still will not do anything to stop it. The agencies are really impotent. They are just part of the bureaucracy.

The real power is the corporations. That is why we take action directly against the corporations. That is why our technique is blocking production at the point of production. Although we may have an occasional demonstration at the Board of Forestry, that is not where we focus our energy. Because that is not where the power is. The power is at the corporate level. And our experience is that when we direct our actions at the point of production, towards the corporations, and ignore all of this bureaucracy, that is the only place we can really have any significant influence. Once you start blocking logging roads and once they find that they can't get these logs out and still maintain any kind of social control and order up here, then all of the sudden the regulatory agencies start scurrying to make changes. But without people chaining themselves to trees and blocking bulldozers, it does not matter how many complaints you file, they never do their job.

DB—You have said that lawsuits, combined with direct action, can be quite powerful, but you have also said that you have no faith in the court system. You say that state agencies are impotent, but you have attempted to use the Board of Supervisors to seize L-P's holdings under eminent domain. Could you clarify what role you see for the state in these struggles?

JB—I don't have any faith in the courts or regulatory agencies. They rarely enforce the regulations when that enforcement goes against the interest of the big corporations. That's why we need to use direct action to pressure them. Without the direct action, nothing happens at all. With direct action, you can win on the narrower issues. I don't think we can win on the larger issues without a revolution.

DB—But in the absence of a revolution, do you think it is effective to work to democratize the agencies of the state to make them more responsive to social movement pressure?

JB—I think we need to work on all levels, but I myself wouldn't waste my time on that particular one. They respond to the social pressures when the social pressures become so great that it is difficult for the power structure to maintain control. I don't know what you mean by working to democratize them. If you don't have bags of money, lobbying doesn't work.

DB—I was speaking more of structural changes within the state.

JB—I think they've been very adaptable. No matter what the structural change, they've managed to rearrange themselves. In 1970, there was this huge effort to do exactly what you are saying, to rearrange the bureaucracy to be more responsive and to deal with ecological issues. The Forest Practices Act was passed. This act was a great political victory with tremendous struggles around it. It called for forest management that will produce sustained yield of high-quality saw logs. That's the law in California. The Board of Forestry and California Department of Forestry were formed to oversee and enforce this. You have to file a timber harvest plan, the equivalent of an environmental impact report. There were all of these bureaucratic changes. Well, last year the Department of Forestry admitted in a report that under their oversight there has been complete depletion of the forests. Under this act, which has really strong language, and under this structure, which really appears to give some kind of democratization and control, by their own admission in the last ten years 90 percent of the remaining private old growth and 50 percent of all private forest lands have been cut. The companies have so much power that they infiltrate whatever agency you put in charge. You can put all of your effort into reforming the bureaucracy, and the corporations will just take over whatever new bureaucracy you set up.

DB—How do you see the arrival of the Clinton administration affecting logging practices in northern California?

JB—What I really see is a sense of false hope with the Clinton administration. Nothing could be worse than the Reagan–Bush years. So I do not want to discount the advantage of having Bruce Babbitt instead of James Watt, but I think that it is just a matter of degree because the basic system is still intact. Some of the larger outrages may be stopped for a while. But unless we change the underlying problems, it is not really going solve anything. The problem I see with the Clinton administration is a relaxing of some of the vigilance of some of the environmental groups and the belief that now the government is on their side. I don't think that the government is on our side.

DB—What kind of revolution do you view as necessary to create a sustainable relationship with the forests?

JB—I don't believe that it's possible to save the Earth under capitalism because I think that capitalism is based on the exploitation of the Earth, just like it's based on the exploitation of workers. But I don't believe that traditional Marxist socialism is the answer either. Marx speaks only of redistributing the spoils of raping the Earth more equitably among the class of humans. He doesn't address the relationship of the society to the Earth, and I think that is one of the principal contradictions. We need to find a new way to live on the Earth without destroying the Earth or exploiting lower classes. It needs to be socially just and it needs to be biocentric. We are calling this Revolutionary Ecology.

DB—What sort of direction do you see yourself moving in the years ahead? Do you plan to attempt the eminent domain strategy again?

JB—I wish I had the nerve to, but I'm afraid of getting killed. I think that's why they tried to kill me. That was the final impetus. Maybe one of these days I'll get up the nerve, but they did a good job of scaring me. I cannot do what I used to do—I am physically disabled from the bombing, and I think that my physical condition is declining. It certainly is not getting any better, so I see myself more in a position of writing. The FBI lawsuit is going to take a ton of my energy. So I see myself doing more writing, more working on the lawsuit and exposing the FBI and the timber companies, and less actually physically being out there on the lines as I used to.

DB—Finally, what lessons do your experiences offer about the application of class issues to environmental struggles?

JB—The people who do apply class analysis are subjected to tremendous repressive forces, including violence, blacklisting, and a sophisticated propaganda campaign to keep workers and environmentalists fighting each other. In my case, that force culminated in a car bomb assassination attempt and a FBI frame-up. That was the second assassination attempt on me in only 10 months' time. I don't think it's legitimate to try to analyze this problem of environment and class without taking this repression into account. Why aren't the timber workers speaking out? Because loggers who open their mouths get fired and blacklisted. But there have been some tremendously articulate and principled loggers and mill workers that have spoken out—Walter Smith, Gene Lawhorn, Dave Chism, Pete Kayes, Ernie Pardini—whose contributions to the struggle have been really great. And they've been unrecognized outside of the activist circle.

It needs to be recognized that the reason for this repression is that we are effective. The last thing that capitalists want is for environmental ideas to get into the hands of the workers. As long as environmental ideas are isolated among privileged people, they are not really going to do that much damage. The most that an environmental group can do is ask the government to write a law to try to get the corporations to do something different. But if radical ideas were in the hands of the timber workers, for example, they could simply refuse to cut by those methods, if they were organized. So the potential for bringing about change is greater if radical ideas are held by the workers themselves. That's why there has been such tremendous pressure to stop anybody who expresses those ideas. I think that has been largely unacknowledged. It needs to be recognized, and people who take these steps need to be supported by the rest of the movement.

Chapter 10

Racism and Resource Colonization

Al Gedicks

> The attacks on our sovereignty and treaties are really attacks on our way of life, our way of viewing things. The environment is critical to our being. The same attacks to separate us from our resources and land are being used in Brazil, Alaska, and elsewhere. It's really racism, with many different names and faces.[1]

Wisconsin has, in recent years, been the site of intense racial conflict; it is a conflict that has bitterly divided many northern communities ever since off-reservation treaty rights for Wisconsin Chippewa Indians were reaffirmed by a court decision in 1983. Chippewa seeking to exercise their lawful right to spearfish outside reservation boundaries have been met at the boat landings by angry white protesters, who have hurled rocks, fired shots, and yelled racial epithets like "Timber Nigger," at Indians.

Under the terms of treaties signed in 1837 and 1842, the Chippewa ceded large tracts of land to the federal government but retained the rights to hunt, fish, and gather on all lands within the ceded territory, roughly the northern one-third of Wisconsin. In 1983 the U.S. Supreme Court refused to hear the state of Wisconsin's appeal of the "Voigt decision," which reaffirmed Chippewa treaty rights. While the U.S. Constitution says that treaties are the "Supreme Law of the Land," the state administration of Governor Tommy Thompson has criticized the Chippewa for exercising their treaty rights. Indeed, as the Strickland report notes, "the state of Wisconsin has acted as if its 'problem' in northern Wisconsin is the result of Chippewa behavior."[2]

The racial hostility that has been directed at Chippewa spearfishers who take approximately 3 percent of the walleye fish harvest is almost beyond belief. This 3 percent, as the Strickland report notes, "is subject to more WDNR [Wisconsin Department of Natural Resources] attention and observation, monitoring, press coverage, and political manipulation than the entire other 97 percent."[3] Every study that has been done on the impact of Chippewa spearfishing, from the Wisconsin Department of Natural

Resources to the Great Lakes Indian Fish and Wildlife Commission, to the most recent report commissioned by the U.S. Congress, has failed to find any evidence that the Chippewa are threatening the fish resource.[4]

On the other hand, the state of Wisconsin, especially the executive branch, has been actively promoting plans for a mining district in the ceded territory of the Wisconsin Chippewa that has the potential to cause serious long-term damage to the resource and economic base of northern Wisconsin. The mass media has inadvertently assisted the anti-Indian movement by narrowly focusing public attention and discussion on the more sensationalistic aspects of the treaty controversy while virtually ignoring the economic and political context of the issue. Beneath all the racist rhetoric of the spearfishing controversy lies the essential and inseparable connection between the political assault against Indian treaties and the corporate assault on the environment in the 1990s. By focusing on the issue of resource control in the ceded territory, it is possible to see the convergence between the anti-Indian movement, represented by groups like Protect Americans' Rights and Resources (PARR) and Stop Treaty Abuse (STA), and the promining policy of the Thompson administration in Wisconsin. This convergence between anti-Indian sentiment and mineral interests is best understood as the most recent episode in a long history of Indian dispossessions.

TREATIES AND MINERALS

Long before Columbus set sail for the "New World," the Chippewa bands who inhabited the Lake Superior region knew of its rich copper deposits. The desire to control mineral deposits in Indian territory was a major component in the series of treaties that followed the early adventurers.

In 1826, at Fond du Lac, one of the first treaties between the United States and the Chippewa gave the U.S. mining rights to all of Chippewa country. The 1828 Treaty of Green Bay dispossessed the Winnebago, Potawatomi, and Chippewa of their lead mines. The 1842 "Miners Treaty" of La Pointe dispossessed the Chippewa of the Keeweenaw, Michigan, copper districts. Finally, the Chippewa were dispossessed of the iron wealth in northern Minnesota in the Treaty of 1854.[5]

Wholesale impoverishment of the Chippewa enriched several generations of East Coast copper- and iron-mining families, including the Aggasiz and the Rockefellers.[6] It also set in motion the great mining and lumber booms, which then went bust, leaving large portions of the Lake Superior region in a severe economic depression that still continues today.

More than a century after the first mining treaty, competition for Indian land and resources continues. The Lake Superior region of northern Wisconsin, northern Minnesota, and the Upper Peninsula of Michigan is considered a prime place for mineral deposits because it lies in the southernmost extension of the Canadian Shield (formerly known as the Chippewa Lobe). Its Precambrian glacial rock is believed to be some two billion years old. Up until the late 1960s, little was known about the distribution of metallic ores because the region was overlain by a concealing layer of glacial debris. Multinational mining and energy corporations began intensive geophysical exploration in northern Wisconsin in the wake of the revolts that swept the Third World in the 1960s. The resulting shaky investment climates—with threats of expropriation, nationalization, and higher taxes—scared many mining firms considering the billion-dollar capital outlay and the decade-long lead time to reap profits. To lessen the risks, multinationals intensified their search for "politically secure" supplies of raw materials. From the corporate viewpoint, "political stability" means the extraction of minerals at extremely high rates of profit with very little risk or interference from democratic institutions. Historically, one of their most stable investment areas has been Indian lands.

In 1975 the U.S. Bureau of Mines, under contract with the Bureau of Indian Affairs (BIA), began a systematic mineral resource evaluation of U.S. Indian reservations. In 1976, the BIA reported that copper, zinc, gold, and uranium might be found beneath various Wisconsin Indian lands.[7] In 1976, Exxon Minerals announced its discovery of one of the world's largest zinc–copper deposits near Crandon, Wisconsin. The site, claimed by the Sokaogon Chippewa under an 1854 treaty, lies a mile from the lake where they have gathered wild rice for centuries. In 1994 the Sokaogon lost an eight-year legal battle to retain treaty rights to a 144-square-mile area in northeast Wisconsin that includes the site of a proposed Exxon mine.[8] Should Exxon build a mine, acid runoff and seepage could destroy the lake, the mainstay of the Chippewa's subsistence economy and culture.[9] The tribe was not reassured when an Exxon biologist mistook the wild rice for "a bunch of weeds." Exxon's own environmental impact report blandly mentioned that "the means of subsistence on the reservation" may be "rendered less than effective."[10]

From Exxon's perspective, there was never any question of whether there was going to be a mine, only when. Robert Davis and Mark Zannis have summed up what lies behind such an attitude: "Simply stated, the difference between the economics of the 'old colonialism,' with its reliance on territorial conquest and manpower, and the 'new colonialism,' with its reliance on technologically-oriented resource extraction and transportation to the metropolitan centers, is the expendable relationship of the subject peoples to multinational corporations."[11]

Since the mid-1970s, multinational mining corporations have quietly leased the mineral rights to over 300,000 acres of land in the ceded territory of Wisconsin (see Figure 10.1).[12] The top corporate leaseholders include Exxon, Kerr-McGee, Noranda, BHP, and Rio Tinto Zinc (RTZ). The long-range planning of these corporations envisions the Lake Superior region as a new resource colony that will provide cheap raw materials for

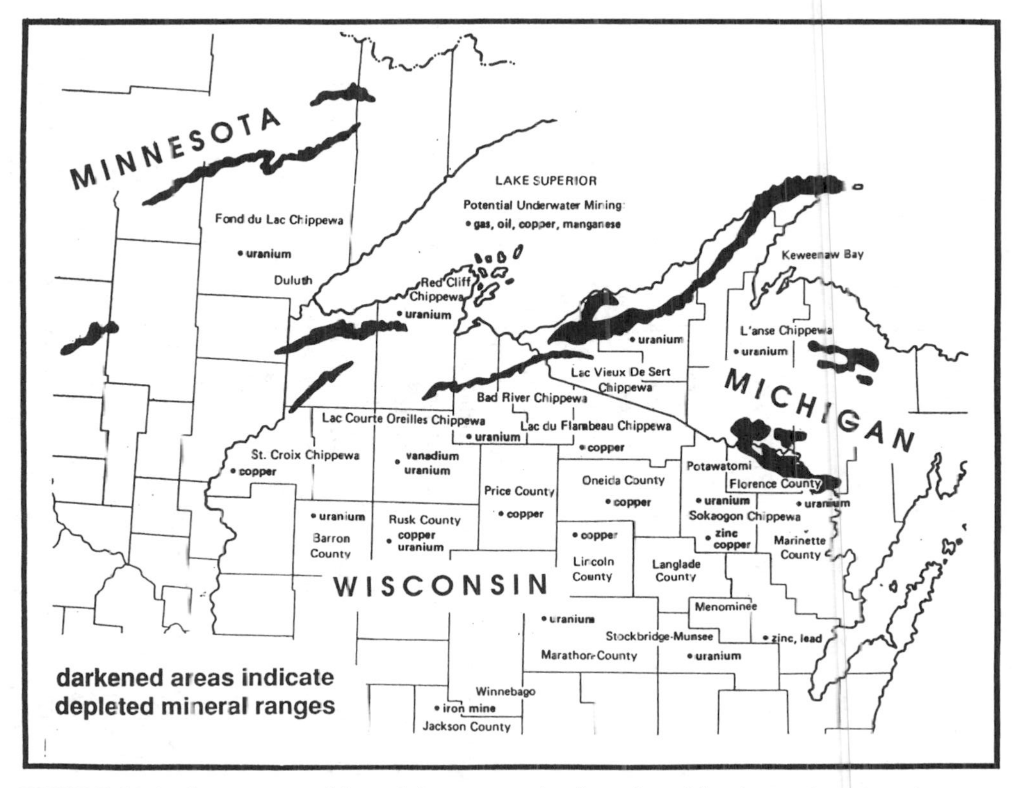

FIGURE 10.1. Areas targeted by mining companies for mineral leasing and exploration. *Source*: *Land Grab: The Corporate Theft of Wisconsin's Mineral Resources* (Madison, WI: Center for Alternative Mining, 1982). Copyright 1982 by the Center for Alternative Mining. Reprinted by permission.

corporate growth and diversification and a dumping ground for the toxic wastes left behind from the mining process. A 1976 University of Wisconsin report is explicit on this point:

> Mining waste, because of acid drainage or the discovery of potentially carcinogenic material in the waste, may have long-term effects on the natural and cultural environment. Because these effects may occur only as an act of God and long after the mining firm has left the area, repairs and compensation may become the responsibility of the public sector. In certain cases the potential for damage may be so severe as to require perpetual monitoring and maintenance similar to that done by federal authorities with radioactive waste material.[13]

SCAPEGOATING THE CHIPPEWA

While multinational mining and energy corporations were exploring, drilling, leasing, and preparing to mine the region's mineral resources, the leaders of Protect Americans' Rights and Resources and Stop Treaty Abuse were blaming the Chippewa for the economic decline of the tourist industry. Antitreaty groups found a receptive audience in northern Wisconsin, where per capita income has lagged behind the rest of the state and where unemployment rates are higher than the state and national averages.

Has northern Wisconsin tourism suffered as a result of Chippewa spearing activities, as the antitreaty groups claim? Dick Matty, Director of the Wisconsin Division of Tourism, has stated that there has been "no real negative impact" on tourism as a result of Chippewa spearfishing. Chamber of Commerce officials in northern communities like Minocqua and Boulder Junction report that tourism is booming.[14]

However, there have been significant changes in Wisconsin's tourism economy that were having negative economic effects prior to the Voigt decision and off-reservation spearfishing. A 1981 Wisconsin tourism industry study concluded: "Resort problems were shown to increase with the age of the resort. Those that appear to be having the most significant problems, however, were built prior to 1930. Twenty-five percent of these resorts were shown to have declining occupancy trends. This may be attributable to the declining quality of these resorts due to their age and the fact that over 60 percent of their owners have not made any improvements or done upkeep since the resort was built."[15] As newer and more modern resorts attract tourists away from the antiquated ones, the Chippewa have become convenient scapegoats for the failure of these mom-and-pop resorts.

The Wisconsin Department of Natural Resources (WDNR) has also contributed to the scapegoating of the Chippewa by the way it has manipulated the bag limits on non-Indian sportfishing. The WDNR's own studies have shown a steadily decreasing walleye fish population for decades due to habitat destruction and pollution. At the same time, sportfishing demands have risen. In 1979 a WDNR report recommended decreasing bag limits as an option, long before the spearfishing controversy.[16] "However," as the Strickland report has noted, "for the past two years the WDNR has only reduced the bag limit on those lakes speared by the Chippewa, making it appear that the Chippewa are responsible for the reduction in bag limits. Although the bag limit reduction has been long in coming, the state refuses to acknowledge that the lowering of the bag limit is due to other factors and not to Chippewa exercise of reserved rights. Resort owners, already feeling the pinch of changing vacation planning, fear that these lower bag limits will decrease tourism and blame the Chippewa."[17]

By scapegoating the Chippewa for the economic problems of northern Wisconsin, antitreaty groups and the state have diverted attention from the significant environmental threats to the economy and culture of both Indian and non-Indian communities in the northwoods. The Chippewa, along with the other Indian nations in northern Wisconsin, already suffer a disproportionate environmental risk of illness and other health problems from eating fish, deer, and other wildlife contaminated with industrial pollutants like airborne PCBs (polychlorinated biphenyls), mercury, and other toxics deposited on land and water. "Fish and game have bioaccumulated these toxic chemicals," according to a 1992 U.S. Environmental Protection Agency study, "to levels posing substantial health, ecological, and cultural risks to a Native American population that relies heavily on local fish and game for subsistence. As the extent of fish and game contamination is more fully investigated by state and federal authorities, advisories suggesting limited or no consumption of fish and game are being established for a large portion of the Tribes' traditional hunting and fishing areas."[18] To suggest that the treaty rights of the Chippewa are a threat to the economy of northern Wisconsin is to promote the most cynical sort of victim-blaming. "Sooner or later," says Anishinabe (Chippewa) treaty rights activist Walt Bresette, "people in northern Wisconsin will realize that the environmental threat is more of a threat to their lifestyle than Indians who go out and spear fish. . . . I think, in fact, that we have more things in common with the anti-Indian people than we have with the State of Wisconsin."[19]

RESISTING TRIBAL RESISTANCE

In their attempt to gain control over low-cost Indian resources in the ceded territory of Wisconsin, multinational corporations have contributed to a growing tribal nationalism on the reservations. "It is not surprising," says one legal scholar, "that tribal nationalism should reemerge most dramatically in the management of reservation resources. Tribal occupancy of land has always been at the very foundation of the unique existence of America's Indian tribes."[20] Since the 1950s, tribes have gone from being "politically stable" resource colonies to sovereign governments trying to assert and defend their treaty rights.

Both the Lac du Flambeau Chippewa and the Lac Courte Oreilles Chippewa have objected to mines proposed by Noranda and RTZ, respectively, because the pollution from these mines would degrade the habitat for hunting and fishing in Chippewa ceded territory. Federal courts have consistently found that treaty Indians have "an environmental right" to preserve fishing habitats.[21] And in 1986, in Wisconsin, the Exxon mine project was put on hold—partly due to stiff resistance from a coalition of Indians and environmentalists.[22] Seven years later, Exxon and Canada-based Rio Algom formed the Crandon Mining Company (CMC) and announced plans to pursue mining permits once again. Indians, environmentalists, and sportfishing groups have vowed to stop the project.

Promining interests have responded with renewed calls for Congress to terminate treaties. Often, abrogation has been portrayed as needed to "free" the Indian.[23] At the same time, the increased exercise of tribal authority in zoning, taxation, construction, and land use ordinances has spurred non-Indian resentment.[24] Organized reactions to the exercise of tribal authority began in the western states of Washington, Montana, and Wyoming with the formation of such groups as Quinault Property Owners Association, Montanans Opposed to Discrimination, and Wyoming Citizens for Equality in Government. In 1976 many of these groups came together in the Interstate Congress for Equal Rights and Responsibilities (ICERR). This organized anti-Indian network "linked on-reservation non-Indian landowner opposition to tribal governments with off-reservation non-Indian sport and commercial fishermen opposed to tribal treaty protected fishing rights."[25]

In Wisconsin, antitreaty groups organized after the 1983 Voigt decision reaffirmed the off-reservation treaty rights of the Chippewa. These groups took names like Wisconsin Alliance for Rights and Resources (WARR) and Equal Rights for Everyone, and tried to convince the public that the Chippewa were out to "rape" the resources. Their members used slogans like "Save a deer, shoot an Indian" and "Save a fish, spear an Indian."

In 1985, WARR founder Larry Peterson started yet another antitreaty organization, called Protect Americans' Rights and Resources. The new organization has tried to avoid some of the overt, blatant racism of its predecessors. According to PARR, the problem is not so much the Indian as the federal government, which keeps the Indian separate from the rest of society. The solution, according to Peterson, is to make Indians like all other citizens.

PARR's literature emphasizes the national concerns of its members. In 1987, PARR hosted a national meeting of the anti-Indian movement in Wausau, Wisconsin, and representatives from almost every active antitreaty organization in the country attended. Out of this meeting came a national effort "to push the U.S. Congress to study and change federal Indian policies."[26]

PROMINING/ANTI-INDIAN

With the election of Tommy Thompson as state governor in 1986, the antitreaty movement went from the margins to the mainstream of Wisconsin politics. Thompson, in a campaign speech to PARR, had said: "I believe spearing is wrong, regardless of what treaties, negotiations or federal courts may say." And he appointed, as secretary of the Wisconsin Department of Administration, a man who had been Exxon's chief lobbyist while the company was seeking state permits to mine near Crandon—James Klauser. Klauser was also a mining consultant to the Wisconsin Association of Manufacturers and Commerce (WMC), the state's most influential business lobby. Speaking before a 1981 Democratic Party meeting in Madison, Klauser said, "Wisconsin has world-class mining potential. . . . It's really hard to appreciate the mammothness and the potential impact of this industry."[27] Some of the most active members of WMC are Milwaukee-based mining equipment firms like Harnischfeger Industries, the world's leading supplier of underground and surface mining equipment.[28]

One of Klauser's first acts as the governor's top policy adviser was to arrange a meeting in the governor's office for Exxon vice-president Ray Ingram to explore the possibilities of the company's reapplying for mine permits.[29] Although Exxon did not reapply at that time, Kennecott Copper Corporation announced its decision in May 1987 to seek permits for its Flambeau River site near Ladysmith, Wisconsin. But the authorities of Rusk County (where the site lay) had adopted a tough mining code to protect the environment. According to reports in the *Wisconsin State Journal,* Kennecott developed a sophisticated strategy to override this "extremely onerous" code and "neutralize" the opposition.[30]

Integral to this strategy was the role played by the governor's ad hoc task force on mining, created in May 1987. This body recommended that legislation be passed so that so-called local agreements between communities and mining companies could bypass county ordinances. This legislation was drafted with the help of a Kennecott attorney and added to Wisconsin's 1988 budget bill. Thompson signed the amendment into law in May 1988. It was the single most important piece of legislation for promoting mining in Wisconsin, and it was passed without public hearings or public discussion of any kind.

THE "INDIAN PROBLEM"

The other potential legal obstacle to opening the northwoods to mining was the assertion of treaty rights. Governor Thompson's response to racist violence against the Chippewa for exercising their lawful treaty rights was to propose buying or leasing such treaty rights from them in exchange for cash and government services. Klauser was the governor's personal representative in the treaty negotiations.

But most tribes rejected such negotiations outright, and two voted down proposals in referendums. One of the major issues of concern to those arguing against such proposals was the possibility that to lease out treaty rights would provide a way for mining companies to acquire mineral rights on reservation lands. In fact, although rarely mentioned, after such buyouts, Congress is more likely to consider changing treaty rights. U.S. Representative David Obey (D–WI) told Douglas County Democrats on February 16, 1990, that such an agreement, once ratified by Congress, would be a "new deal; the old deal is gone."[31]

When the anti-Indian movement increased its agitation, the state was able to intervene and suggest that some form of de facto treaty abrogation was a reasonable way of resolving the conflict. Throughout the controversy, both Thompson and Klauser continued to meet with representatives of PARR and the more militant Stop Treaty Abuse. The political legitimacy extended to these racist groups should not be underestimated.

Both PARR and STA are member groups of the Citizens Equal Rights Alliance (CERA), a national alliance of anti-Indian organizations. Fred Hatch, a former Bureau of Indian Affairs attorney who is now general counsel for STA and its delegate to CERA, describes CERA as "a political lobby of ranchers, doctors, lawyers, businessmen, and large corporations like Burlington Northern and Exxon, everybody's friend. . . . All of these companies are having problems with the Federal Indian policy."[32] In 1988, CERA joined other right-wing extremist groups like the American Freedom Coalition and corporations like Exxon at a "Multiple Use Strategy Conference" in Reno, Nevada.[33] This conference was called in response to attempts by a broad environmental coalition to reform the 1872 Mining Law, which gives mining companies access to federal land for as little as $5/acre without having to restore it or pay royalties. By preaching the "continued multiple use" of public lands, the industry has appealed to ranchers, loggers, and anti-Indian groups in a backlash against emerging environmental reforms.

THE COUNTY CONNECTION

While the conference was important in establishing links between the anti-Indian movement and the antienvironmental movement, CERA's ties with the Wisconsin Counties Association (WCA) are crucial for the legitimation of anti-Indian politics in mainstream national political debate.

The backlash against Chippewa treaty rights has extended to many Wisconsin county governments, who fear that the Chippewa, as well as practicing off-reservation hunting and fishing, might exercise their right to harvest timber on public (county) lands. That treaty claim had been rejected by a federal judge, but the counties continue to be concerned about questions of jurisdiction as between tribe and county.

WCA Executive Director Mark Rogacki has taken a leadership role in organizing counties "from all over the United States, which are on or near Indian reservations, into a National Coalition on Federal Indian Policy."[34] According to Rogacki, the exercise of treaty rights is "not in tune with contemporary society or the needs of local government."[35] In 1990, Rogacki brought together county association representatives from nine other states to examine how treaty rights affect local government and to prepare a strategy to lobby Congress for change.

Klauser was there, too. Rogacki explained his participation reflected the fact that "the [Thompson] administration and the county governments share the perspective that the federal government has to settle these matters."[36] Also among the speakers at the conference was the WCA general counsel, Milwaukee attorney Robert Mulcahy. Mulcahy has led a court battle, on behalf of most Wisconsin Counties Forest Association members, against the rights of the Chippewa to harvest timber. In February 1991, Mulcahy told a WCA meeting in Madison: "Timber is not the only issue. . . . Not many understand. Mining is the real issue. Serious claims are being made against mining in the Ladysmith area. Anti-mining groups are surfacing and doing leafletting. There is even State legislation introduced for a moratorium on mining."[37] At that meeting, a Wisconsin Counties Forest Association representative told WCA members that 56 percent of the public land in the state is in the County Forest system and that the counties "would realize big profits if they could open these lands to mining." Once again, we see the convergence between the anti-Indian movement and pro-mining interests at both the state and county levels of government.

TREATY RIGHTS AND ENVIRONMENTAL PROTECTION

"What is perhaps most important about Indian treaty rights," says Anishinabe (Chippewa) treaty rights activist Winona La Duke, "is the power of the treaties to clarify issues which would otherwise be consigned by nation-state apologists to the realm of 'opinion' and 'interpretation.' The treaties lay things out clearly, and they are matters of international law."[38] One of the landmark cases that clearly demonstrated the inseparable connection between treaty rights and environmental protection was the widely publicized Northwestern U.S. treaty rights conflict over fish in the late 1960s.

The case attracted considerable attention when Marlon Brando, Dick Gregory, and Jane Fonda stood alongside Native American fishermen and women who were exercising their treaty rights to fish in defiance of sportsmen and state authorities. Indian fisher-people in the Northwest had been arrested during similar "fishing wars" since the turn of the century, despite the 1854 Medicine Creek Treaty, which affirmed their right to fish "as long as the rivers run."[39] In 1974 U.S. Judge George Boldt of the Federal District Court, Tacoma, Washington, ruled that Indians were entitled to catch as many as half the fish returning to off-reservation sites that had been the "usual and accustomed places" when the treaties were signed. When Judge Boldt made his ruling in *U.S. v. Washington,* he realized that these rights were meaningless if there were no fish to catch. The issue was addressed by Judge William Orrick of the Federal District Court, Tacoma, Washington, in Phase II of *U.S. v. Washington* in 1980. The judge ruled that "the most fundamental prerequisite to exercising the right to take fish is the existence of fish to be taken."[40] If the state of Washington were allowed to destroy the fishery habitat by licensing dams and logging operations, it would amount to an abrogation of the Indians' right to fish. The court ruled that treaty Indians have an implicit right to have the fishery habitat protected.

While the ruling did not give the Northwest tribes absolute power to veto development projects, it did give them legal standing to challenge actions or policies that may have a detrimental effect on their right to a fishery. The burden of proof was divided between the state and the tribes. Since Judge Orrick's decision, the Indians of the Skagit River—the Upper Skagit Tribe, the Sauk–Suiattle Tribe, and the Swinomish Tribal Community—have used their treaty rights to protect the fish in the Skagit River from proposed dams and a nuclear power plant. "Without the treaty case law," said Russ Busch, attorney for Upper Skagit and Sauk–Suiattle, "more than half of our arguments against nuclear plants would have no teeth."[41]

While the boundaries of the tribe's right to protect the environment are still being reviewed in the courts, the tribes have continued to develop their management capabilities to such an extent that state and federal agencies have been forced to acknowledge the region's tribes as comanagers of the fishery resource.

REGIONAL RESOURCE COMANAGEMENT

Yet another possible model of resource comanagement has developed out of the intense treaty rights conflict over Chippewa spearfishing in northern Wisconsin. Once the Chippewa's treaty-guaranteed rights were affirmed by the courts in 1983, the tribes had to develop their managerial capabilities at the same time that they were confronting racist mobs at the boat landings. To counter the racist violence and hostility directed against Chippewa spearfishers, a number of religious, environmental, and treaty rights support groups came together to provide witnesses for nonviolence at the boat landings and to conduct public education about Chippewa treaty rights. These groups included Witness for Nonviolence, HONOR (Honor Our Neighbors' Origins and Rights), Midwest Treaty Network, the Wisconsin Greens, Northern Thunder, and the American Indian Movement (AIM). By 1991, almost 2,000 witnesses had been trained in the philosophy and practice of nonviolence, treaty history, and environmental and cultural issues in northern Wisconsin.[42] As a result of their nonviolent activism on behalf of Chippewa treaty rights, many witnesses have joined Anishinabe Niijii (Friends of the Chippewa), an Indian–environmentalist coalition, in defending Indian lands against unwanted mining development.

After years of racial backlash to off-reservation spearfishing, the anti-Indian movement in Wisconsin has suffered a number of setbacks that have resulted in much smaller crowds protesting Chippewa spearfishing. In March 1991, Federal District Court Judge Barbara Crabb of Madison, Wisconsin, issued a permanent injunction against protesters interfering physically, whether on the lake or at the boat landings, with Chippewa fishermen and women.

Now that the immediate threat of anti-Indian violence has subsided, the important long-range resource management work of the tribes will assume greater visibility and importance. One of the great ironies of the Chippewa spearfishing controversy is that the Chippewa and their white neighbors were fighting over the right to harvest contaminated fish. A recent Department of Natural Resources (DNR) study shows that increased walleye consumption is leading to higher levels of mercury in Chippewa blood, posing a health risk, especially for pregnant women and young children.[43] The walleye is a large gamefish that is a staple in the Chippewa diet; it also has the most mercury-tainted flesh. The DNR's most recent "Health Guide for People Who Eat Sport Fish from Wisconsin Waters" lists 217 bodies of water with varying consumption limits for mercury-polluted fish. As more lakes are tested, more lakes are added to the list. Since 1982 the DNR has tested 700 of the state's 15,000 lakes, with usually about one in three making it to the advisory list.[44]

In the spring of 1992, for the first time, Chippewa spearfishers actually refused to spear certain lakes because of Department of Natural Resource (DNR) fish advisories. Anishinabe (Chippewa) treaty rights activist Walt Bresette says, "I refuse to eat the fish and I refuse to let my family eat the fish because of the toxins." If new mines are constructed in northern Wisconsin, they will be a major new source of mercury contamination. Mercury is released during mineral processing and is leached into waterways from mine wastes (tailings). In the nearby Upper Peninsula of

Michigan, the National Wildlife Federation and the Michigan United Conservation Clubs sued the White Pine copper mine and smelter for emitting mercury, lead, and arsenic over the waters of nearby Lake Superior at five times the legal limit. The U.S. EPA had listed Copper Range, the mine's owner, as Michigan's "most prolific polluter."[45] In 1995, the company paid $1.8 million in penalties for air pollution violations in the largest settlement since the passage of the Clean Air Act.[46]

The prospect of a new resource colony in northern Wisconsin, with multiple mines at the headwaters of the state's major rivers, and mountains of toxic mine wastes, is a nightmare scenario for Walt Bresette, a member of the Red Cliff Chippewa Tribe and a founder of the Lake Superior Greens, an environmental organization. Nor is this what the Lake Superior Chippewa tribes had in mind when they formed the Great Lakes Indian Fish and Wildlife Commission (GLIFWC) in 1984. GLIFWC's role was to provide coordination and services for the implementation of their treaty rights in the ceded territory and to represent tribal interests in natural resource management. In addition, GLIFWC sought to provide ecosystem protection, "recognizing that fish, wildlife and wild plants cannot long survive in abundance in an environment that has been degraded."[47] In place of the resource colony being imposed by multinational mining corporations and the state, Bresette wants to declare an "environmental zone" to be jointly managed by the state and the tribes. "The ceded territory (treaty lands) would be the first 'toxic free zone' in the state if not the country," says Bresette. Rather than mines and toxic waste dumps, the zone would be a haven for vacationers and people seeking a healthy place to live.

"Under this plan," writes Bresette and Winnebago coauthor James Yellowbank, "the 42 percent of Northern Wisconsin which is ceded land would be phased into a comanagement program that ultimately would prohibit pollution. It would simultaneously develop jobs and rural community reinvestment opportunities through a ten year environmental cleanup program."[48] By proposing the general outlines of an alternative economic development plan for the ceded territory, Bresette and Yellowbank, along with the Wisconsin Greens, want to reopen and broaden the public debate about the economic and environmental future of northern Wisconsin. At the same time, the serious consideration of such a plan would pose a direct ideological challenge to the traditional export-based models of economic development. Since the promise of job opportunities is the major, if not the only, appeal of the mining industry to depressed northern Wisconsin communities, the treaty-based alternative economic development plan for northern Wisconsin holds much promise as an ecologically preferable way of creating sustainable jobs for both Indians and non-Indians.

Indian control over reservation air and water quality is long overdue. Tribal lands were overlooked in the original versions of many federal environmental laws of the 1960s and 1970s, including the Clean Air Act and the Clean Water Act.[49] In 1984 the U.S. EPA announced that it would pursue government-to-government relations with tribes.[50] In 1994–1995, four Wisconsin Indian tribes asked the U.S. EPA for greater regulatory authority over reservation air and water quality. The Forest County Potawatomi Tribe asked for tougher air pollution standards on its reservation under the federal Clean Air Act.[51] Meanwhile, the Sokaogon Chippewa, Oneida, and Lac du Flambeau Chippewa Tribes were granted independent authority from the EPA to regulate water quality on their reservations. Under amendments to the Clean Water Act, the EPA can designate tribes as independent regulators of surface water quality in the same way the EPA can give authority to states. Tribal regulatory authority would affect all upstream industrial and municipal facilities, including Exxon's proposed mine in the Swamp Creek watershed. Because Swamp Creek flows into the Sokaogon Chippewa's Rice Lake, the tribe has to give approval for any physical, chemical or biological upstream activity that might degrade their wild rice beds.[52]

At public hearings on the Sokaogon Chippewa request, local citizens, lake associations, and the Wolf River Watershed Alliance testified in support of tribal regulatory authority. Many of the local lake property owners associations expressed extreme dissatisfaction with the way in which Republican Governor Tommy Thompson and his chief aide, James Klauser, paved the way for mining by making the DNR Secretary a political appointment and eliminating the Public Intervenor's Office. The experts hired by the Public Intervenor had raised serious questions about the scientific adequacy of Exxon's groundwater studies and their waste disposal plans. Many citizens applauded the tribe for trying to preserve clean water for everybody.[53]

Some local business people testified in opposition, charging that the regulations would "shut down northern Wisconsin." This was the same kind of misinformation used by those who opposed Chippewa off-reservation spearfishing during the turmoil lasting from 1984 through 1992. The Wisconsin Mining Association, representing some of the largest mining equipment companies in Milwaukee, warned that tribal water quality authority "could be the most controversial and contentious environmental development affecting the state in decades."[54]

Both Governor Thompson and the Wisconsin DNR had urged EPA denial of tribal regulatory authority over air and water quality standards.[55] Within a week of EPA approval of Sokaogon Chippewa and Onedia water quality authority, Wisconsin Attorney General James Doyle sued the EPA in federal court, demanding that the federal government reverse its decision to let Indian tribes make their own water pollution laws. Several Republican state legislators called upon Congress to change the Clean Air Act to disallow tribal authority over clean air standards.[56] Once again, mainstream politicians are using scare tactics to suggest that Indian sovereignty over reservation resources is an economic threat to small business owners, while they ignore the serious potential for long-term damage to the resource and economic base of northern Wisconsin from large-scale mining and waste disposal.

In response to the state's challenge of EPA's delegation of water quality authority to the tribes, the Wolf River Watershed Alliance filed an *amicus,* or "friend of the court," brief supporting EPA's approval of tribal authority. "If the state is stupid enough to appeal this thing, we'll certainly write a brief detailing all the instances where the state has been derelict in its authority or abdicated its responsibility," said Robert Schmitz, president of the alliance.[57] Meanwhile, a federal court ruling in Montana has upheld the right of Indian tribes to set water quality standards on their reservations.[58]

SHOWDOWN AT LADYSMITH

The Kennecott/RTZ open pit copper mine in Ladysmith, Wisconsin, lies just 140 feet from the Flambeau River, one of the state's most pristine waterways and a prime area for walleye fishing. The site is also within the ceded territory of the Lake Superior Chippewa. In 1989, the Lac Courte Oreilles (LCO) Chippewa Tribe opposed RTZ's mine application, arguing that the state of Wisconsin "cannot issue a permit (to mine) unless and until RTZ can prove that its mining activities will not degrade the plant or animal resources in the ceded territory." Wisconsin's five other Chippewa tribes, along with treaty-rights support groups and grassroots environmental organizations, soon joined the LCO Chippewa in their battle against RTZ.

The Ladysmith battle has a significance far beyond this community of 3,800. Kennecott/RTZ's ability to overcome an Indian–environmentalist opposition at Ladysmith was seen as a hopeful sign by at least six companies planning to mine identified ore deposits across northern Wisconsin. The mine's permits were granted in January 1991 after studies claimed that "no threatened or endangered species are known to exist at the mine site."[59] In June 1991, two endangered mussels (the purple wartyback and the bullhead) were discovered in the river, well after the 30–day limit had expired for citizens and tribes to appeal a ruling that had upheld the permits. Meanwhile, an internal memo had been leaked from the Wisconsin Department of Natural Resources (DNR) that revealed that agency scientists were aware during the permitting process that endangered species were probably present at the site. In August 1991, the LCO Chippewa Tribe and the Sierra Club won an injunction against further mine construction until the Wisconsin DNR prepared a supplemental environmental impact statement on the endangered species. After the DNR issued a whitewash study on the endangered species, the LCO Chippewa Tribe and the Sierra Club filed a new lawsuit, asking the DNR to reopen the hearings on the mine permits. The suit was thrown out, in part because the tribe had 30 days to appeal the January 1991 decision to grant the permits and failed to do so. Larry Leventhal, a lawyer for the Lac Courte Oreilles, said the ruling "essentially says if an agency hides information from the public, it won't be held accountable for it."[60]

Resistance to the mine reached a new level when Anishinabe (Chippewa) treaty rights activist Walt Bresette climbed a ten-foot-high security fence carrying a war club once used by Black Hawk and clubbed several earth movers at the Ladysmith mine site. It was an act of defiance but did no physical damage. "It's obviously an illegal permit," said Bresette,

> so I don't recognize the authority of this mine to move forward. There's no proper environmental impact statement completed on this. The State of Wisconsin has, in this permitting process, failed to protect the legal interests of the Lake Superior Chippewa. It's as though the Voigt Decision never happened. Having secured the boat landings in defense of treaty rights, it's as though those battles were waged for nought. Wisconsin, under Governor Tommy Thompson, has a legal responsibility to protect the interests of the Lake Superior Chippewa. Instead he acts like a modern day Andrew Jackson who said, after the Cherokee of the 1830s won their rights, "Chief Justice Marshall has made his ruling, now let him enforce it." Shortly thereafter, despite being armed with the protections of the U.S. Constitution, the Cherokee were marched from their homelands; many died in what has been termed the "trail of tears." There will be no trail of tears from northern Wisconsin. It is time to stay and fight for our rights. This mine as proposed is bad for democracy, bad for the economy and ultimately a threat to the environment. I call upon other tribal members to defend their homeland.[61]

While the Chippewa and their allies in the environmental and treaty rights movements lost the battle to stop the mine, they nevertheless established themselves as a political force to be taken into account in the ongoing resistance to future mining projects. Robert Wilson, the chief executive of RTZ Corporation, recently paid a backhanded compliment to the Chippewa–environmental alliance in Wisconsin when he noted that the greatest political risk for new mining projects no longer comes from the developing countries, but from the United States and Australia, where opposition movements, as at Ladysmith, have resulted in costly project delays.[62]

The determination of Native nations to protect their environment from pollution threats was dramatically illustrated in the summer of 1996 when the group Anishinabe Ogitchida ("Protectors of the People") successfully blockaded acid supply trains en route to the infamous White Pine copper mine in Michigan's Upper Peninsula. Chippewa protesters said that a spill from tankers carrying sulfuric acid through the Bad River Chippewa reservation would poison their water and scar their land. Besides the threat of a spill, the Chippewa were protesting plans by the Copper Range

Corporation to use the acid to extract ore from the White Pine copper mine. The company was going to inject a sulfuric acid solution into old mines in order to leach out the remaining ore. The project involved injecting 550 million gallons of acid into underground tunnels only five miles from Lake Superior.

When the EPA granted a permit for the project without holding a hearing or drafting an environmental impact statement, Anishinabe (Chippewa) activist Walt Bresette decided to form Anishinabe Ogitchida and camp out on the railroad tracks to prevent shipments of sulfuric acid from passing through the Bad River Chippewa Reservation on their way to the mine. "Sovereignty is not something you ask for," says Bresette. "Sovereignty is the act. This blockade has drawn public attention to the largest ecological threat in our region's history."[63] In addition to the blockade, the Bad River and Keweenaw Bay bands of the Chippewa filed a federal lawsuit challenging EPA's decision not to require a permit for the project or conduct an environmental impact statement. After the suit was filed, EPA decided to do an environmental analysis that could take as long as two years. In October 1996, Copper Range announced it was suspending its solution mining operation because of the uncertainty posed by the new EPA regulatory review process.

LESSONS FROM THE WISCONSIN EXPERIENCE

Historically, the anti-Indian movement is linked to the economic expansionary needs of American capitalism. When the dominant European American society needed land and raw materials, Indians were defined as a problem or threat, and their lands and resources were taken. Now, having been left with land nobody else wanted, it turns out that some of the last remaining energy and mineral resources are located on Indian lands or on off-reservation lands in the ceded territory of Wisconsin. Once again, national hysteria surrounds "outdated Indian treaties" and the so-called tribal misuse of resources. Much of this hysteria is orchestrated by anti-Indian groups, but increasingly the state itself has played an important part. However formidable this corporate–state alliance may be, it has not been able to roll back Chippewa treaty rights or overcome a grassroots resistance movement of Indians and non-Indians to create the kind of "stable investment climate" that would allow multinational mining corporations to proceed unhindered with their resource extraction schemes.

The Wisconsin Chippewa have not only defended their treaty rights at the boat landings in northern Wisconsin but have also interpreted the treaties to include their rights to protect the habitat for their hunting, fishing, and gathering activities. Chippewa tribal leadership on this issue has attracted both grassroots and mainstream environmental organizations and has led to a growing alliance among Indians, environmental groups, and citizen action groups. And just as important, in terms of future legal action, is official EPA recognition that off-reservation developments pose serious pollution problems for the tribes that necessitate some kind of government-to-government cooperation between the tribes and the states.[64]

Whether one looks at the Northwestern U.S. fishing rights conflict, or the recent victory of the Cree and Inuit to save their lands from being submerged by the James Bay hydroelectric projects in far northern Quebec,[65] or the Chippewa treaty controversy in Wisconsin, there is an integral connection between the defense of treaty rights and the defense of the environment.[66] Once the Chippewa of Wisconsin or the Crees and Inuit of Canada asserted their sovereignty and their rights to control the natural resources within their respective territories, the focus of the debate shifted from *how* this project will be developed to *who* will be involved in the decision-making process.

This shift in the framework of the debate is most significant. Multinational corporations and prodevelopment governments rarely, if ever, make explicit provisions for real public participation in these resource decisions. For the most part, the extensive planning for these megaprojects is done in secret and presented to the public as a fait accompli. This helps to promote a "psychology of inevitability" about these projects and discourages any potential opposition from arising until it is too late to stop the project. However, once Native groups are able to assert their rights to participate in the decision-making process, the momentum of the corporate–state machine is slowed down, at least temporarily, as Natives and their environmental allies share their concerns about the wide-ranging social, economic, and environmental impacts of these projects with a larger audience.

ACKNOWLEDGMENT

This is a revised and abridged version of a chapter from Al Gedicks, *The New Resource Wars: Native and Environmental Struggles against Multinational Corporations* (Boston: South End Press, 1993).

NOTES

1. Ed Bearheart, St. Croix Chippewa tribal council member, quoted in Sharon Metz, "A Legacy of Broken Promises," in Bob Hulteen and Jim Wallis, eds., *America's Original Sin: A Study Guide on White Racism* (Washington, DC: Sojourners, 1992).

2. Rennard Strickland et al., *Keeping Our Word: Indian Treaty Rights and Public Responsibilities* (Unpublished monograph, University of Wisconsin, Madison, WI, 1990), p. 10.

3. Ibid., p. 24.

4. See Thomas R. Busiahn, *Chippewa Treaty Harvest of Natural Resources: Wisconsin 1983–1990* (Odanah, WI: Great Lakes Indian Fish and Wildlife Commission, 1991); and *Casting Light Upon the Waters: A Joint Fishery Assessment of the Wisconsin Ceded Territory* (Minneapolis, MN: U.S. Department of the Interior, Bureau of Indian Affairs, 1991).

5. Robert H. Keller, "An Economic History of Indian Treaties in the Great Lakes Region," *American Indian Journal, 42*(2), 1978.

6. William G. Gates, Jr., *Michigan Copper and Boston Dollars* (New York: Russell and Russell, 1951); David Allan Walker, *Discovery and Exploitation of Iron Ore Resources in Northeastern Minnesota: The Formative Years, 1865–1901* (PhD dissertation, University of Wisconsin, Madison, 1973).

7. U.S. Bureau of Indian Affairs, *Status of Mineral Resource Information for the Bad River, Lac Courte Oreilles, Lac du Flambeau, Mole Lake Community, Potawatomi, Red Cliff, Public Domain and St. Croix, and Stockbridge–Munsee Indian Reservations of Wisconsin. Administrative Report BIA-20* (Washington, DC: Author, 1976).

8. "Mole Lake Band Loses 8-Year Court Battle over Land Claim," *Milwaukee Sentinel,* March 22, 1994.

9. Robert P. W. Gough, "A Cultural-Historical Assessment of the Wild Rice Resources of the Sokaogon Chippewa," in *An Analysis of the Socio-Economic and Environmental Impacts of Mining and Mineral Resource Development on the Sokaogon Chippewa Community* (Madison: University of Wisconsin, COACT Research, 1980).

10. Exxon, *Forecast of Future Conditions: Socioeconomic Assessment, Crandon Project,* 1983, p. 316.

11. *The Genocide Machine in Canada* (Montreal: Black Rose, 1973), p. 37.

12. Al Gedicks et al., *Land Grab: The Corporate Theft of Wisconsin's Mineral Resources* (Madison, WI: Center for Alternative Mining Development Policy, 1982).

13. Michael D. McNamara, *Metallic Mining in the Lake Superior Region: Perspectives and Projections* (Madison: University of Wisconsin, Institute for Environmental Studies, Report No. 64, 1976), p. 51.

14. Jim Thannum, "1990 Chippewa Spearing Season—Conflict and Cooperation," Great Lakes Indian Fish and Wildlife Commission, Odanah, WI, 1990, p. 15.

15. "The Wisconsin Tourism Industry Study—Second Home Ownership, 1981 Data," Wisconsin Department of Development, Madison, 1987.

16. "Long Range Management Plan: Muskellunge Management Plan at 7-1 to 7-7; Walleye Management Plan at 9-1 to 9-7." Wisconsin Department of Natural Resources, Madison, 1979.

17. Strickland, *Keeping Our Word,* p. 20.

18. *Tribes at Risk: The Wisconsin Tribes Comparative Risk Project* (Washington, DC: U.S. Environmental Protection Agency, 1992), p. ix.

19. *Wisconsin Treaties: What's the Problem?* (Madison, WI: Midwest Treaty Network, 1991), p. 1.

20. Daniel H. Israel, "The Reemergence of Tribal Nationalism and Its Impact on Reservation Resource Development," *University of Colorado Law Review, 47*(4), 1976, pp. 617–618.

21. Fay G. Cohen, *Treaties on Trial: The Continuing Controversy over Northwest Indian Fishing Rights* (Seattle: University of Washington Press, 1986).

22. Al Gedicks, "Exxon Minerals in Wisconsin: New Patterns of Rural Environmental Conflict," *The Wisconsin Sociologist, 25*(2–3), 1988.

23. Carl L. Burley, "Indian Lands—An Industry Dilemma," *Journal of the Rocky Mountain Mineral Law Institute, 28,* 1982. There are 18 mining and oil and gas associations on the board of the Rocky Mountain Mineral Law Foundation, including the powerful American Mining Congress.

24. The federal government's General Allotment Act of 1887 provided for the division of communally held reservations into individual parcels or allotments to be transferred to individual Indians, with remaining "surplus" land to be made available to white settlers. The result was the movement of non-Indians onto Indian reservations and the creation of "checkerboard land ownership" patterns on every "allotted reservation."

25. Rudolph C. Ryser, *Anti-Indian Movement on the Tribal Frontier,* Occasional Paper 16 (Kenmore, WA: Center for World Indigenous Studies, 1991), p. 4.

26. Ibid., p. 27.

27. Don Behm, "Northern Environmental Stewards Oppose Consensus Trade-Offs," *Wisconsin Academy Review, 28*(1), 1981, p. 19.

28. Lee Bergquist, "Mining Gets New Backing," *Milwaukee Sentinel,* January 23, 1995.

29. "Governor Can't Sway Exxon to Keep Mine Process Alive," *Milwaukee Journal,* January 18, 1987.

30. Jeff Mayers, "Sophisticated Lobbying Neutralizes Mining Foes," *Wisconsin State Journal,* March 24, 1991.

31. "Indians Criticize State Officials," *Milwaukee Journal,* February 25, 1990.

32. Scott Kerr, "The New Indian Wars," *The Progressive, 54*(4), April 1990, p. 23.

33. Ryser, *Anti-Indian Movement,* p. 48.

34. Sharon Metz, "A Legacy of Broken Promises," *Sojourners,* June 1990, p. 1.

35. Scott Kerr, "Conference on Treaties Sparks Anger," *Milwaukee Journal,* January 14, 1990.

36. Ibid.

37. "WCA Wants to Promote Mining on County Lands," *Honor Digest,* February/March 1991 (Milwaukee: Honor Our Neighbors' Origins and Rights, Inc.).

38. Winona La Duke, "Prelude: Succeeding into Native North America," in Ward Churchill, ed., *Critical Issues in Native North America* (Copenhagen: International Work Group for Indigenous Affairs, Document 62, 1989), p. xi.

39. Bruce Johansen and Roberto Maestas, *Wasi'chu: The Continuing Indian Wars* (New York: Monthly Review, 1979), p. 187.

40. Cited in Cohen, *Treaties on Trial,* p. 142.

41. Michael Garrity, "The Pending Energy Wars: America's Final Act of Genocide," *Akwesasne Notes,* Early Spring 1980, p. 10.

42. Rick Whaley, "Witness for Nonviolence: Green Politics in Action," *Green Letter* (San Francisco, CA), Fall 1991. See also Whaley and Walter Bresette, *Walleye Warriors: An Effective Alliance against Racism and for the Earth* (Philadelphia: New Society, 1993).

43. Ron Seely, "Toxic Fish Taint Spearer's Victory," *Wisconsin State Journal,* May 17, 1992.

44. Will Fantle, "Fishing for Trouble," *Isthmus Newsweekly* (Madison, WI), *17*(23), June 5, 1992.

45. "Environmental Groups Sue Mine over Air Emissions," *Milwaukee Journal,* August 18, 1992.

46. Joe Williams, "U.P. Company OKs Record Settlement," *Milwaukee Sentinel,* February 1, 1995.

47. Great Lakes Indian Fish and Wildlife Commission, *1990 Annual Report* (Odanah, WI, 1991).

48. Walt Bresette and James Yellowbank, "Regional Resource Co-Management: Saving the Land for the Next Seven Generations," *Indian Treaty Rights Newsletter, 2*(1), Winter 5, 1991, p. 2.

49. Margaret Knox, "Their Mother's Keepers," *Sierra, 78*(2), March/April 1993, p. 54.

50. Steve Fox, "Taking Us Down to the River: An Indian Pueblo Challenges Upstream Polluters," *The Workbook, 17*(4), Winter 1992, pp. 153–154.

51. Steven Walters, "Tribe Seeks Air Pollution Protection within 60 Miles of Reservation," *Milwaukee Sentinel,* December 13, 1994.

52. Don Behm, "2 Tribes Hope to Control Reservations' Water Quality," *Milwaukee Journal,* February 5, 1995.

53. Mike Monte, "Indian Tribes: Our New Environmental Conscience?" *Wisconsin Outdoor News,* July 21, 1995.

54. James Buchen, "Delegation of Federal Clean Water Act," *Badger State Miner,* October/November 1995.

55. Jeff Mayers and Nathan Seppa, "Thompson Vexed by Tribe's Move," *Wisconsin State Journal,* December 13, 1994.

56. Steven Walters, "Tribe's Request Could Jeopardize Current, Future Jobs, Groups Say," *Milwaukee Journal/Sentinel,* April 27, 1995.

57. "Tribal Water Plan Supported," *Daily Journal* (Antigo, WI), February 1, 1996.

58. "Indian Lawsuit Will Proceed," *Wisconsin State Journal,* April 5, 1996.

59. Wisconsin Department of Natural Resouces, *Final Environmental Impact Statement, Flambeau Mining Co., Copper Mine, Ladysmith, Wisconsin* (Madison, WI: Author, 1990), p. 67.

60. "Judge Dismisses Lawsuit to Block Open-Pit Mine," *Saint Paul Pioneer Press,* June 13, 1992.

61. Walt Bresette, "Why I Must Act to Help Stop the Ladysmith Mine," *Ceded Territory Newsletter, 1,* May 1992.

62. "Managing Political Risk," *Institutional Investor, 26*(12), November 1992, pp. 23–24.

63. Meg Jones, "U.P. Copper mining project to close," *Milwaukee Journal-Sentinel,* October 15, 1996.

64. *Tribes at Risk,* p. 41.

65. Zoltan Grossman, "Chippewa Block Acid Shipments," *The Progressive, 60*(10), October 1996, p. 14.

66. See, for example, Harvey A. Feit, "Protecting Indigenous Hunters: The Social and Environmental Protection Regime in the James Bay and Northern Quebec Land Claims Agreement," in Charles Geisler et al., eds., *Indian SIA: The Social Impact Assessment of Rapid Resource Development on Native Peoples* (Ann Arbor: University of Michigan, Natural Resources Sociology Lab, Monograph 3, 1982).

Chapter 11

Ecological Legitimacy and Cultural Essentialism

Hispano Grazing in the Southwest

Laura Pulido

"The Rio Grande drainage area north of El Paso offers a more complete example of regional suicide than most people ever imagined."[1]

Governmental and other official and quasi-official bodies commonly assume that the poor of the world are a major threat to the environment, and hence lack what might be called "ecological legitimacy." Such legitimacy attaches to a group when it is seen as a valid environmental actor, when its commitment to preserving the environment is not regarded as suspect. Ecological legitimacy is associated with environmental stewardship, that is, the practice of caring for the land in a sustainable manner.

In particular, ecological legitimacy often eludes poor rural populations because officialdom has long assumed that landless and land-poor groups do not care about protecting their environment.[2] However, the rise of the environmental justice movement in the North and the spread of peasant and indigenous struggles in the South have sharply challenged the ideology that the poor are incapable of caring for their own environmental conditions of life.[3]

Ecological legitimacy may be drawn from different sources and crafted in various ways. One source that nonwhite and indigenous communities have exploited is what might be called (for lack of a better expression) "romanticized cultural heritages."[4] Such discourses are often predicated on cultural assumptions: first, that definition of cultural differences is the principal determinant of intergroup conflict and, second, that some cultures are inherently more sensitive to nature than others, whether or not this is in fact true. Romanticized cultural heritages are a form of essentialism, in that the characteristics of a particular group are regarded as unitary and fixed or eternal. This kind of essentialism has been common with respect to gendered identity[5] and is growing in indigenous environmental discourses. Increasingly, the struggles of some U.S. minorities and "Third World" peoples to assert their environmental legitimacy are cast within this kind of culturalist framework.

A number of writers—who insist that differences among various national and ethnic groups do in fact exist—have noted the dangers of cultural essentialism. One problem is that this type of cultural explanation typically sidesteps any analysis of socioeconomic (material and power) relationships and hence serves to essentialize ethnic differences. Cultural essentialism denies or obfuscates the whole problem of social or historical agency, obscuring dominant power dynamics such as the struggles between rich and poor and landowners and tenants, thus reifying cultural differences. Instead of examining how and why various constellations of wealth and power result in different environmental practices, cultural essentialism tends to view variations in environmental practices as originating in "natural" ethnic or cultural differences.[6]

Despite these and other problems associated with cultural essentialism, a seldom noted fact is that the grant of ecological legitimacy via one form or another of culturalism may serve at least three important purposes. First, the construction of an alternative narrative positing local peoples as capable ecological stewards is a form of resistance, as it affirms a historically denigrated ethnic or national group. At the same time, it critically scrutinizes dominant modernist approaches to socioeconomic development and resource use.[7] In this sense, culturalism offers a counterhegemonic discursive framework that is essential to the success of any oppositional struggle and alternative development path. Second, culturalism helps to consolidate the moral authority of the group in question. Moral authority, after all, is a form of power and legitimacy that arises from the belief that the relevant actors act in ethically sound or correct ways and therefore are deserving of popular support. Examples of morally authoritative, ecologically legitimate struggles include the Chipko movement in India[8] and the Brazilian rubber tappers.[9] Third, resistance by definition develops within the context of socioeconomic or political oppression, and since culturalism is a readily available resource, it may be an effective or even indispensable strategy in the struggle for ecological legitimacy.

In this chapter these general observations are applied to the process whereby a Hispano community development group in northern New Mexico, Ganados del Valle (Livestock of the Valley), sought to establish ecological legitimacy.[10] I examine the context in which romanticized, culturalist nature–society narratives emerged, and explore some of the advantages and drawbacks of their use. My thesis is that culturalism has served Ganados del Valle well, despite some serious theoretical and political problems. Ganados is a good example of the use of culturalism because it has been struggling to gain access to grazing land using various tactics, while being opposed by mainstream environmentalists and state resource managers who have denied Ganados's ecological legitimacy by claiming that Hispanos are poor resource managers.[11] To counter this opposition and to challenge the historical vilification of Hispano resource practices, Ganados has often relied on culturalist arguments in its claim to ecological legitimacy. At the same time, supporters of Hispano grazing rights have employed both structural and culturalist arguments to account for the phenomena of overgrazing, emphasizing in particular the harmonious ecological relations of indigenous peoples in general.

I first situate this case study by presenting a brief overview of northern New Mexico and then explore the ways that scholars have characterized the links between Hispano resource use, poverty, and culture. In particular, I focus on those commentators who blame Hispanos for local poverty and environmental problems and those who try to explain this poverty and soil erosion in structural terms. Both "blame the victim" and "blame the structure" approaches suggest why Ganados has employed romanticized culturalist arguments in its efforts to establish ecological legitimacy. I then describe Ganados del Valle's struggles to win grazing rights. Finally, I explore how culturalism has in fact been used by Ganados to assert its ecological legitimacy, and I also consider some of the larger political and theoretical issues associated with this strategy of "discourse of struggle."

ECOLOGICAL LEGITIMACY: HISPANO RESOURCE USE, POVERTY, AND CULTURE

The Chama Valley, the site of Ganados del Valle, is located in the northern Rio Grande watershed. It was inhabited by Pueblo Indians until the late 1600s, when Spanish explorers and mestizo settlers began permanent settlement of the region.[12] Hispanos initially settled the region through a system of land grants (*mercedes*) and developed an agropastoral system based on vertical transhumance grazing and subsistence agriculture.[13] Communities were organized to include private land for the home, garden, and feed production, and collective ownership of the highlands for grazing, timber, and other resources. Water was furnished by *acequias*, a gravity-based irrigation system well suited to arid environments.

Established as frontier outposts, Hispano villages were always oriented to subsistence, but their economic and social marginality escalated when the United States conquered New Mexico through the Treaty of the Guadalupe Hidalgo (1848).[14] Anglo control led to the loss of Hispano land and water rights through a variety of mechanisms, including the U.S. government's inability to recognize communal land ownership, high legal fees, Hispanos swindling one another, and, last but not least, outright fraud by Anglos. Regardless of the means, the end result was the commodification of land and Anglo encroachment, which brought more intense patterns of grazing, hunting, logging, and other forms of resource extraction.[15]

As Hispanos' ability to make a living was eroded through the loss of grantlands, the villages slid into deep poverty. Faced with declining economic opportunities, Hispanos pursued seasonal work strategies, outmigrated, and occasionally rebelled.[16] Although the region has been the site of numerous studies and development projects, such efforts rarely addressed the fundamental problem of the loss of land. Instead, development agents sought to build a crafts economy, teach job skills, or promote the Americanization of villagers—all focused on changing the individual while ignoring the fact that a thriving rural economy is impossible without an adequate land base.

Poverty is highly racialized in the region. As retirees, telecommuters, and tourists (most of whom are Anglo) flock to "the land of enchantment," the resulting land speculation destroys Hispanos' dreams of a viable rural economy. The newcomers drive up the cost of land, and as ex-urbanites they have values and goals different from those of low-income rural people, which further erodes Hispanos' efforts at community autonomy. In the face of seasonal unemployment of 18.9 percent in the Chama Valley,[17] the state has pushed tourism, based on wealthy outsiders' desire to experience the landscape, cultural diversity, and natural resources of the area.[18] The result has been the creation of tourist towns (Santa Fe and Taos), exclusive game ranches, luxury ski resorts, and a booming secondary home market. Such a strategy ensures the continued poverty of Hispanos as it is predicated on seasonal, low-wage tourist jobs.[19] The 1990 per capita income in Rio Arriba County was $11,979 for whites, as compared to $7,496 for Latinos. The result is a highly polarized economy geared toward wildlife and wilderness production for the enjoyment of urbanized middle-class residents.

Economic polarization partly rests upon Anglo romanticization of the nonwhite population. This continues a long tradition of Anglos desiring the landscape, artifacts, and sense of place associated with northern New Mexico Indians and more recently, Hispanos. Because Anglos have such contradictory ideological and material relationships with Hispanos, a rich and conflictual set of narratives has developed to explain persistent Hispano poverty. Poor resource management—in particular, overgrazing—has been central to explanations of poverty. Not only do such narratives deny the ecological legitimacy of Hispanos, but they exonerate Anglos and capitalism for the region's deep poverty.

CULTURE, RESOURCE USE, AND MORAL AUTHORITY

Although a large body of literature seeks to explain the "decline" of northern New Mexico, few scholars have critically examined the political subtext of the dominant arguments. Studies focusing on resource management typically posit Hispanos as poor resource managers.[20] This literature has significantly shaped dominant perceptions of Hispano ecological relations. Certainly, whether Hispano resource management is considered environmentally sound or damaging has enormous implications for establishing the legitimacy of current Hispano grazing claims for and efforts toward local control. Two perspectives are presented here: one that seeks to deny the ecological legitimacy of Hispanos, attributing persistent Hispano poverty to overgrazing; and one that seeks to affirm the ecological legitimacy of Hispanos, and as a result locates overgrazing and poverty within the context of Anglo and capitalist domination.

The first perspective, which attributes Hispano "decline" to overpopulation and overgrazing,[21] is largely Malthusian in nature and clearly results in the denial of Hispanos' ecological legitimacy. In particular, it identifies a number of specific causes of regional poverty, including general ignorance, partible inheritance (resulting in small farms), high birthrates, and a generally poor environmental ethic:

> Production per acre in much of New Mexico is less than half of what it formerly was. The difference is due to loss of soil fertility through erosion and the *practice of taking all from the land and returning nothing to it.* . . . *The smallness of farms* in northern New Mexico and the practice of subdividing land into strips perpendicular to rivers or irrigation ditches make conservation practices difficult to apply[22]
>
> Continued overstocking and overgrazing have resulted in the deterioration of the land. *Agricultural productivity is low because of poor farming practices*; the quality of product is frequently poor, resulting in low prices.[23]
>
> For Spanish-Americans to blame their deficient economy upon the many rejected land-grant claims and loss of communal lands is an effort to divert attention from the many other problems that stem from the region's physical incompatibility with agriculture, as well as with its settlement and demographic patterns.[24]

Because overgrazing leads to poverty, and because Hispanos are charged with overgrazing, they are viewed as responsible for their own poverty and also as bereft of ecological credibility. Such a perspective is held by a wide variety of New Mexicans and has real consequences. For example, one member of the Audubon Society who resisted Ganados' grazing efforts explained, "The population here [Hispanos] overdid it [overgrazed]. And to a large extent have been saved by outfits like Los Alamos who come in and hire 12,000 people."[25] Clearly, for this "conservationist," Hispanos have no ecological legitimacy.

Other important elements in this argument pertain to history and culture. Emphasizing both the long *duration* of these problematic environmental practices, as well as associating these practices with an amorphous

"Hispano culture," the perceived lack of Hispano stewardship is portrayed as natural and inevitable.

> Many of the . . . ranges are yielding less than 20 percent of their potential. Undesirable plants have either increased or invaded most ranges. Accelerated erosion has occurred on many depleted ranges particularly on sandy and steeply sloping areas. . . . The problem is further complicated because the numerous small operators have had too few acres to support their livestock. *Constant heavy grazing for over 100 years* has greatly depleted the grazing resources on lands owned by small operators. . . . The problem is further complicated because producers commonly graze seasonal ranges at the wrong time, thereby critically overusing the plants during the growing season.[26]
>
> If there is a flaw in the relationship of the villagers to their environment, it is that they, like the people of pioneer and subsistence cultures everywhere, have consistently underestimated their capacity for injuring the land. The mesas and mountains may indeed be the *alma,* the soul of the village culture, but their elevated status has not protected them from abuse.[27]

Clearly, considerable energy has been spent on debating the grazing practices of Hispanos. Because this literature blames Hispanos for environmental degradation, they are "delegitimized" as successful resource managers. This delegitimization is based on specific grazing practices (overgrazing, grazing at the wrong time, creating farms that are too small), as well as a general moral shortcoming, as evidenced by the failure to practice an appropriate environmental ethic. Delegitimization occurs because of the political need to clarify responsibility and to impose accountability on the agents of environmental degradation. Great weight is attached to environmental degradation/stewardship because it provides a necessary, but rarely articulated, moral subtext to environmental/development initiatives. Accordingly, those who are linked to ecological stewardship garner moral authority (whether the International Monetary Fund or the Moskito Indians). In the case of Hispanos, because they are not considered environmentally valid actors, their claims can be dismissed on the basis of their "proven track record" and their general moral shortcomings. Thus, Hispanos are denied "standing," as it were, in a discursive arena controlled by mainstream environmentalists, scientists, and resource professionals.[28]

In response to such interpretations, others have developed critiques far more sympathetic to Hispanos and rooted in structural analyses. Many of these arguments are drawn from political ecology, which suggests that colonialism, capitalism, and modernization outweigh individual agency, culture, or even sheer population numbers in accounting for environmental change.[29] Recognizing the consequences of conservative analyses, left scholars have depicted the economic problems of Hispanos as emanating from structural shifts, thereby leaving intact their ecological legitimacy, as "traditional" practices are associated with ecological stewardship.[30] Bobrow both characterizes and illustrates this practice:

> Some academics and policy-makers . . . have criticized the practice of grazing on open common land . . . as they were carried out by both Native Americans and Hispano pastoralists in northern New Mexico. . . . These critics point to the scenario of land degradation laid out in biologist Garrett Hardin's "tragedy of the commons"—open commons frequently become severely eroded, there is a decline in the quality and quantity of forage and there is serious long term ecological damage and accompanying economic decline.
>
> In the case of New Mexico's Hispano land grants, the tragedy of the commons was not a result of the absence of a vested interest (the vested interest was individual and community survival), it was the tragedy of the foreign land tenure system which treats land as a commodity *inflicted* on a culture which views land as a common resource.[31]

Peña, in a critique of deBuy's cultural ecology analysis, also suggests that if Hispanos in fact contributed to environmental degradation, it was because of economic pressures:

> Land degradation in Chicano community grants was the result of the partitioning and contraction of land holdings combined with the industrialization and commercialization of former common property resources. Evidence overwhelmingly demonstrates that environmental degradation in New Mexico was initiated in the 1870s after the arrival of the railroad, land speculators, and capitalist ranching, mining and lumber interests.[32]

This position clearly seeks to refute the "blame the victim" approach of previous analysts. In his detailed study of Hispano cultural ecology, Van Ness argues that "from an ecological perspective the superiority of the Hispanic land tenure for a subsistence economy is clear."[33] In a similar vein, Stoller suggests, "This pattern of vertical transhumance for livestock raising is common among pastoral peoples in mountainous, high altitude areas around the world; it was well adapted to the topography of northern New Mexico and southern Colorado. . . . The Culebra river valley and its settlers are excellent examples of a group of people who developed a culture that was environmentally sound, sane and satisfying."[34] Here, both authors seek not only to challenge dominant stereotypes but to vindicate Hispano resource use patterns by emphasizing their sustainability.[35]

Similar to conservative analyses, political projects and moral positions underlie these arguments. But, unlike the former, these are used for oppositional purposes, with several consequences. First and foremost, these scholars seek to validate the ecological practices of historically marginalized and dispossesed people by positing them as able and conscious ecological stewards. This validates both their grazing practices as well as Hispano culture in general, as this culture is associated with good morals and ecological knowledge. A second consequence is to enhance Hispano moral authority by reintroducing the notion of victimization. Stressing Hispano dispossession (as opposed to blaming the victim) casts them in a light conducive to public support. Victimization coupled with ecological legitimacy is a powerful combination in achieving moral authority. Finally, this scholarship sets the stage for activists to develop romanticized discourses and culturalist arguments. It provides the necessary intervention to create a new political space.

Thus, it should not surprise us when members of Ganados make conscious links between Hispano culture and sound resource management. As one local activist asserted, "We've been here for hundred of years and the land is still here. We know how to take care of it, that's why they [environmentalists] want it so much."[36] I will now turn to Ganados itself and its efforts to assert ecological legitimacy.

ROMANCING THE LAND: GANADOS DEL VALLE AND OPPOSITIONAL ENVIRONMENTAL DISCOURSES

Ganados del Valle was formed in the early 1980s when a few Hispanos decided to cooperatively manage their flocks.[37] Recognizing the need for meaningful economic development that was culturally and environmentally appropriate, activists created vertically integrated businesses based on grazing, lamb, and high-quality woven products. Although many Hispanos still owned a few head of sheep or cattle in the 1980s, these were not economically viable operations due to the small livestock numbers and limited range availability.[38] Both land enclosure and land speculation have made the cost of land prohibitive.[39] Accordingly, it was hoped that what one could not do independently—run a viable operation—could be done collectively. By 1995, Ganados had five businesses, including weaving, sheep and grazing cooperatives, several community projects, and was grossing over $300,000 annually.

Ganados is considered to be an economic success, as it has increased the income of over forty rural households. It is also considered to be environmentally responsible in that it promotes sustainable development. Despite severe regional poverty, for example, Ganados has consistently opposed polluting manufacturing activities and environmentally damaging resource extraction, such as mining and luxury ski resorts. Moreover, Ganados has worked to develop land-use and zoning regulations to protect the county's environmental quality. This environmental consciousness is also apparent in Ganados's businesses. In addition to a recycling business, Ganados uses guard dogs instead of poison to protect its flock against predators, produces organically grown lamb, and practices sustainable grazing.

In the late 1980s, Ganados was attempting to expand but was limited by a lack of grazing land. Given the shortage of grazing opportunities, Ganados asked the New Mexico Department of Game and Fish (NMDGF) for permission to graze on Wildlife Management Areas (WMAs) in Rio Arriba County. The WMAs were initially acquired by the NMDGF to increase elk habitat and hunting opportunities.[40] Ganados proposed a project that would allow the cooperative to graze while conducting research in conjunction with New Mexico State University. Ganados argued that the WMAs had a dense matte covering, preventing further plant growth, and that use of a short-term grazing system could improve the grasses.[41] However, Ganados's plan was opposed by both the NMDGF and local environmental groups.[42] Along with hunters, these groups constitute the agency's primary constituencies.[43] Throughout the struggle, Ganados' ecological legitimacy was questioned, both in terms of its ability to conduct sustainable grazing and the sincerity of its environmental commitment. Ganados struggled for access to the land by participating in research projects, mediation, civil disobedience, and lobbying. At one point, in order to dramatize its plight, members of Ganados, along with 2,000 sheep, trespassed onto one of the WMAs. Despite attracting significant attention, the NMDGF eventually prohibited grazing on WMAs throughout the state.

The right to graze was fought for in other arenas, as well, including a lawsuit against the Sierra Club Legal Foundation filed decades after an aborted collaborative project. In the 1960s, prompted by renewed Hispano land grant struggles, the Sierra Club Legal Foundation embarked on a joint project to buy land with La Cooperativa Agricola del Pueblo de Tierra Amarilla.[44] "The preservation of land and the perpetuation of the economic and social values of an ethnic minority [are] goals which are central to the philosophies of the Sierra Club Foundation and La Cooperativa."[45] The Foundation asked Albuquerque businessman Ray Graham to contribute $100,000 toward the project, but, despite these efforts, no land was ever purchased.

In 1989, Graham learned of Ganados' struggle with the NMDGF, and, seeing the connection between his gift and Ganados' need for grazing land, inquired how his donation had been spent. When the Foundation did not respond to Graham's inquiries, he sued. Meanwhile, Ganados persuaded the New Mexico Attorney General to investigate the matter. After several years of investigation, audits, and legal wrangling, a settlement was reached requiring the Foundation to make $800,000 available to Ganados. Ganados then set up a nonprofit land trust to purchase grazing land for its members to use.

The WMA struggle and, to a lesser extent, the lawsuit represent two discrete episodes in a much larger grazing conflict, one that goes back over a century and demonstrates how ecological relations are the site of both material and discursive power struggles. The land grant struggles of the 1960s were crucial not only because they inspired collaborative projects but also because they, along with left scholars, set the stage for contemporary oppositional discourses. Sylvia Rodriguez has shown, for example, how recent struggles in the Taos area have led to a rearticulation of land and Hispano identity.[46] Thus, Ganados' oppositional discourse, specifically its use of culturalism to promote ecological legitimacy, is but one example of many ongoing romanticized resistant discourses.

CULTURALISM AND ECOLOGICAL LEGITIMACY

Ganados challenged the entrenched belief that Hispanos were responsible for regional soil erosion and poverty in many ways, using scientific, economic, legal, and, of course, culturalist arguments. The basis of Ganados' ecological legitimacy was the cooperative's self-definition as an ecological steward. From stewardship would flow both standing as a valid environmental actor as well as authenticity in terms of its environmental commitment. Culturalism proved a useful way of establishing ecological stewardship because of the general exoticization of Hispanos by the dominant society, the regional reification of cultural differences, and the larger swirl of Hispano oppositional discourses.

Culturalist arguments served not only to validate Hispano resource practices but, by extension, Hispano culture. Specifically, it was argued that, because Hispano culture was associated with ecological stewardship, any grazing program involving Hispanos would be a success. As a consequence, cultural preservation, in addition to grazing rights, emerged as a goal. According to Ganados' narrative, Hispano culture should be preserved because it is rich, associated with stewardship, and threatened. This last point was important not only to Hispanos but also potentially to other development interests, since the region is built on cultural tourism. Recognizing the extent to which the current fascination with Hispano culture is a function of tourism, one local explained, "You know, they are trying to make us into a colonial Williamsburg of the Southwest. . . . Tourism is so altered by . . . gas prices, tourist preferences, etc. It's cool to be Hispano now, but it may not be in ten years. Coyotes are in now, where are they going to be in ten years, or furniture making?"[47] Aside from arguing that Hispano culture should be preserved because it is distinct, Ganados drew on other moral concerns, in particular the theme of dispossession/victimization first raised by left scholars. The subtext to the argument was that because Hispano culture had been victimized by the loss of grantlands, this was all the more reason to support the project. Thus, access to grazing land would preserve an "endangered culture" and was morally desirable.

> Elk and deer are not endangered in northern New Mexico. But the survival of New Mexico's Hispanic pastoral culture is endangered. Ganados del Valle's proposal to graze the wildlife refugees is an oppor-

> tunity to strengthen one of the United States' richest cultures, improve the wildlife habitat and raise the standard of living in one of the nation's poorest rural counties.[48]

Having established a moral and economic basis to support its grazing claims, Ganados emphasized that Hispano culture is one of ecological stewardship. "Our fathers knew how to take care of the land. You get out of the land what you put into it."[49] This local resident, like many activists, firmly believed that Hispano culture was inherently more careful with natural resources than Anglo culture. "Respect for water and land . . . is transmitted from generation to generation and has become a cultural characteristic of Indians and Indo Hispanic people."[50] Building on general romanticized images of Hispano history, landscape, folklore, and material culture suggests an environmental relationship that is "sound, sane and satisfying."[51] This has enabled Ganados to argue that their proposed grazing effort would be a success precisely because it was rooted in Hispano history and culture:

> We are a pastoral people. Our pastoral history goes back thousands of years spanning from the Iberian Peninsula to this continent. This pastoralism is reflected in the way the Spanish and then Mexican governments organized land use. . . . The common lands were used for moving large herds and flocks of livestock around the ecosystem which revitalized rangeland with each season. Under this system there was ample forage for domestic livestock and wildlife.[52]

There is no doubt that culturalist arguments helped Ganados achieve ecological legitimacy. Moreover, ecological legitimacy, in addition to the moral authority cultivated on other fronts, contributed to very real gains. For one, culturalist arguments provided a means for Hispanos to challenge certain institutions and practices that were otherwise unassailable, such as private property. In effect, one subtext to Ganados' claims was an affirmation of communal land ownership. Besides linking communal land ownership to sound environmental management, Ganados offered an alternative to the dominant capitalist market ideology and private property relations that subsume our lives and yet resonated with some Anglo supporters.[53]

Perhaps more importantly, the use of culturalist arguments has led to real changes in the social formation. The cultivation of ecological legitimacy coupled with cultural distinctiveness has been crucial in attracting public awareness and support. It was partly because of Ganados' compelling image that Graham and Ganados connected in the 1990s, eventually resulting in the beginnings of a community land trust. This is a clear example of how social practices are engaged in a dialectic with structures of inequality and are able to transform the prevailing material relations.

Nevertheless, using culturalism to establish ecological legitimacy, as a general matter, is of great concern, as the culture, morals, and everyday practices of poor and nonwhite people are currently under attack. This does not mean that such strategies should not be used, only that as academics and activists we should be conscious of their implications. One problem with romanticized ecological discourses is that they are often predicated on a unitary view of culture, one in which all Hispanos are thought to share the same culture, values, and practices. Though there is strong evidence that Hispanos did, in fact, produce far less environmental degradation than Anglos, emphasizing this position overlooks not only the considerable variation that exists within any social group but also the complexity of cultural evolution.[54]

Another consequence of culturalism is the reification of cultural differences between Anglos and Hispanos without paying sufficient attention to the social relations in which each group is embedded and the social practices that arise from those relations. Both Anglos and Hispanos suggested that cultural differences were an important source of the conflict, as different cultures (in this case, Anglo and Hispano) approached land in entirely different ways. For instance, it was routinely argued that Hispano culture recognizes neither "nature" nor "wilderness/wild."[55] In contrast, the Anglo land-use tradition embodies a nature–human dualism that sees humans as conceptually separate from nonhuman nature. A New Mexican environmentalist has cogently summarized the perceived distinctions:

> In the eyes of land-based people, the environment is an ecosystem in which the people exist as one part of a harmonious whole, deriving food and materials, as needed, for their continued social, cultural, and economic existence. In the eyes of environmentalists, the same land may represent an area that should be protected for its own sake, for its beauty and pristine qualities, for wildlife habitat, or for recreation.[56]

Cultural explanations such as these often account for a range of differences that more likely arise from racism, imperialism, or class exploitation. Cultural essentialist arguments have important political consequences. For one, they compel us to forfeit the opportunity to make explicit the economic relationships between various groups. Overlooked is the fact that "culture" is partially constituted by a group's location within the historical trajectory of capitalism and the particular position within a geographically specific set of social relations. Even though Anglos and Hispanos may indeed conceptualize land differently, this practice is at least partially due to the fact that Hispanos were at one time part of a precapitalist empire (Spain), while contemporary Anglo land-use and environmental traditions were formed within the context of industrial capitalism.

The reification of cultural differences that seems to exist beyond or independent of economic structures also has the potential to reproduce the existing social formation. The extent to which mainstream environmental practices serve to actively *reproduce* inequalities of wealth and power is ignored by attributing conflicting land-use plans to vague cultural arguments. Consequently, mainstream environmentalists and professional resource managers are encouraged to continue their project of land preservation and other environmental practices that lead to the further marginalization of oppressed groups. Arguments of cultural difference may lead to a "multicultural" initiative of some type, which do not begin to address the structural and ideological relations that initially engendered such inequalities. An example of this lack of analysis of power relations comes from one Anglo writer's depiction of local ethnic relations: "As for Chama's personality, its special character springs from a blend of strong midwest Anglo and Southwest Spanish heritages blended so well within the last quarter century that neither culture dominates or struggles to dominate."[57]

Another difficulty with culturalist arguments is that they are undergirded by a static conception of cultural change. The issue of "cultural preservation" is particularly important in that it assumes that Hispano culture can or will die. By casting Hispanos as soon-to-be-extinct people, it overlooks the fact that Hispano culture will continue to develop and evolve as long as there are people who identify themselves as Hispano, regardless of where they live and work.

Anthony Bebbington has pointed out the complex ways in which "tradition" serves to inform the present, particularly in the quest for ecological legitimacy and cultural authenticity.[58] By pointing out how "tradition" can be used to overturn prevailing power relations, for example, we can avoid seeing minority cultures as either static and unitary (some activists' interpretation) or as embarked on unidirectional assimilation (modernist interpretation). Yet, by consistently stressing "traditional" culture, Ganados obscures both the phenomenal job it has done and also how the success of the project itself testifies to contemporary Hispano culture and its environmental ethic. Ganados is a remarkable example of cultural change within the context of inequality. It blends new and old by creating a viable alternative. It has successfully reclaimed a maligned cultural heritage and identity, but its operation draws upon the latest marketing innovations and technologies, scientific resource management, and academic arguments.

The fact that Ganados' dominant discourse is rooted in a series of problematic cultural formulations illustrates the extent to which structures of inequality set the terms of resistance. However, it is also worth noting that all the critiques of this strategy are largely theoretical, with *potentially* adverse consequences. In contrast, the use of culturalist arguments has resulted in real material gains. Finally, while I believe that all of the critiques are valid to varying degrees, inherent in them is the belief that the actors are not aware or conscious of the strategies they are using. Conventional critiques of strategic essentialism almost assume that marginalized groups are defined by romanticized discourses, regardless of other practices. Perhaps it is time to reconsider strategic essentialism in light of the goals and ambitions of marginalized communities. In some instances, it is difficult to imagine a strategy of resistance that does *not* use the master's tools.

CONCLUSION

In this chapter I have tried to show not only the need for poor rural populations to establish ecological legitimacy but also the role of culturalism in developing this legitimacy. Ecological legitimacy is, I believe, a useful framework to understand romanticization, or other strategic tactics used by those fighting for environmental justice. I have pointed out that it was useful in the public relations arena, but it was absolutely essential in order to challenge the dominant interpretation of Hispano grazing. Even though Ganados was not successful in changing WMA grazing policy, it waged an important battle and challenged conventional ideas about resource use, poverty, culture, and social justice—one that could not have occurred if Ganados had not established its ecological legitimacy.

The use of culturalism in the development of ecological legitimacy is situationally specific and reflects a particular form of oppression, exoticization. Culturalism is not a strategy available to those of despised and denigrated cultures, such as inner-city African Americans or recent Mexican immigrants. Other groups may develop moral authority based on other aspects of their experience. Given the material and ideological forces shaping Hispanos' lives, including their insertion into a tourist economy, their relatively long history in the area, their attachment to a particular landscape, and the nature of Hispano poverty, it is hard to see how culturalism would *not* have emerged as an important element in the formation of an oppositional discourse. Nevertheless, a remaining political task is to devise nonessentialist bases for moral authority when there is an option to do so.

While some instances of romanticization and strategic essentialism result in disaster,[59] Ganados is not such a case; overall, Ganados has been able to use strategic essentialism in a careful and responsible way. Moreover, it has not prevented the cooperative from working closely with other communities, such as the Navajo, in creating other rural cooperatives. Perhaps of greatest concern is the short-changing of contemporary Hispano culture. It is true that Ganados springs from a long tradition of resistance, and its vision of development draws from the past—but neither of these facts should reduce Ganados to an historical preservation project. The leadership required to undertake such an initiative, the level of commitment among cooperative members, their desire to protect a beloved landscape, and members' eagerness to build a viable economy for themselves and their children are all striking features of contemporary Hispano culture. Yet, much of this gets lost in culturalist arguments. Conversely, perhaps this is a relatively small price to pay when, by actively reshaping their material world and relations, Ganados is building a rare example of social and environmental justice, which in turn is creating a new Hispano culture.

ACKNOWLEDGMENTS

This paper has benefitted from conversations with Bill Lynn and the comments of Melissa Gilbert, Mike Murashige, Jim O'Connor, Devon Peña, Miguel de Oliver, Margaret Villanueva, and three anonymous reviewers. I remain responsible for all interpretations and shortcomings.

NOTES

1. J. Russell Smith, quoted in A. Harper, A. Cordova, and K. Oberg, *Man and Resources in the Middle Rio Grande Valley* (Albuquerque: University of New Mexico Press, 1943), p. 28.

2. For an excellent discussion, see K. Zimmerer, "Social Erosion and Social (Dis)courses in Cochabamba, Bolivia: Perceiving the Nature of Environmental Degradation," *Economic Geography, 69*(3), 1993, pp. 312–327.

3. A. Bebbington, H. Carrasco, L. Peralbo, G. Ramon, J. Trujillo, and V. Torres, "Fragile Lands, Fragile Organizations: Indian Organizations and the Politics of Sustainability in Ecuador," *Transactions of British Geographers, 18*(2), 1993, pp. 179–196; R. Bullard, *Dumping in Dixie: Race, Class, and Environmental Quality* (Boulder, CO: Westview Press, 1990); J. Friedmann and H. Rangan, *In Defense of Livelihood* (West Hartford, CT: Kumarian Press, 1993); A. Gedicks, *The New Resource Wars: Native and Environmental Struggles against Multinational Corporations* (Boston: South End Press, 1993); R. Guha, *The Unquiet Woods: Ecological Change and Peasant Resistance in the Himalaya* (Berkeley: University of California Press, 1990); R. Peet and M. Watts, "Introduction: Development Theory and Environment in an Age of Market Triumphalism," *Economic Geography, 69*(3), 1993; L. Pulido, *Environmentalism and Economic Justice: Two Chicano Struggles in the Southwest* (Tucson: University of Arizona Press, 1996); and V. Shiva, *Staying Alive: Women, Ecology, and Development* (Atlantic Highlands, NJ: Zed Books, 1989).

4. A. Bebbington, "Modernization from Below: An Alternative Indigenous Development?" *Economic Geography, 69*(3), 1993, pp. 274–292.

5. For an overview of gender essentialism in nature–society relations, see C. Nesmith and S. Radcliffe, "(Re)mapping Mother Earth: A Geographical Perspective on Environmental Feminisms," *Environment and Planning: D, 11,* 1993, pp. 379–394. For specific critiques, see J. Seager, *Earth Follies* (New York: Routledge, 1993); and R. Schroeder, "Shady Practice: Gender and the Political Ecology of Resource Stabilization in Gambian Garden/Orchards," *Economic Geography, 69*(4), 1993, pp. 349–365.

6. For general critiques and examples of strategic essentialism and cultural reification, see D. Fuss, *Essentially Speaking* (New York: Routledge, 1989); A. Kobayashi and L. Peake, "Unnatural Discourse: 'Race' and Gender in Geography,"

Gender, Place and Culture, 1(1), 1994, pp. 225–244; and K. Anderson, "Constructing Geographies: 'Race,' Place and the Making of Sydney's Aboriginal Redfern," in J. Penrose and P. Jackson, eds., *Constructions of "Race," Place and Nation* (Minneapolis: University of Minnesota Press, 1994), pp. 81–99.

7. Shiva, *Staying Alive.*

8. Guha, *The Unquiet Woods*; H. Rangan, "Romancing the Environment: Popular Environmental Action in the Garhwal Himalayas," in Friedmann and Rangan, *In Defense of Livelihood.*

9. A. Cockburn and S. Hecht, *The Fate of the Forest* (New York: Verso, 1989).

10. "Hispanos" refers to the Spanish-speaking population of northern New Mexico and southern Colorado, which is a subset of the larger Chicano population.

11. For a detailed analysis of this struggle, see D. Peña, "The 'Brown' and the 'Green': Chicano and Environmental Politics in the Upper Rio Grande," *Capitalism, Nature, Socialism, 3*(1), 1992, pp. 79–103, and Chapter 12, this volume. See also Pulido, *Environmentalism and Economic Justice.*

12. R. Dunbar-Ortiz, *Roots of Resistance: Land Tenure in New Mexico, 1680–1980* (Los Angeles: University of California, Chicano Studies and American Indian Centers Publication, 1980).

13. Transhumance grazing is an extensive grazing system that takes advantage of changes in temperature and grasses by grazing over a range of environments over the seasons. See D. Gomez Ibañez, "Energy, Economics and the Decline of Transhumance," *Geographical Review, 67*(3), 1977. For a complete discussion of Hispano cultural ecology in the region, see J. Van Ness, "Hispanic Land Grants: Ecology and Subsistence in the Uplands of Northern New Mexico and Southern Colorado," in C. Briggs and J. Van Ness, eds., *Land, Water and Culture: New Perspectives on Hispanic Land* (Albuquerque: University of New Mexico Press, 1987).

14. N. Gonzalez, *The Spanish Americans of New Mexico* (Albuquerque: University of New Mexico Press, 1969); and G. Sanchez, *The Forgotten People* (Albuquerque: University of New Mexico Press, 1940).

15. D. Peña, "An American Wilderness in a Mexican Homeland," paper presented at the Western Social Science Association, Reno, 1991; Harper, Cordova, and Oberg, *Man and Resources.*

16. S. Forrest, *The Preservation of the Village* (Albuquerque: University of New Mexico Press, 1989); S. Deutsch, *No Separate Refuge: Culture, Class and Gender on an Anglo-Hispanic Frontier in the American Southwest, 1880–1940* (New York: Oxford University Press, 1987); R. Rosenbaum, *Mexicano Resistance in the Southwest* (Austin: University of Texas Press, 1986); P. Bell Blawis, *Tijerina and the Land Grants* (New York: International Publishers, 1971); and R. Gardner; *Grito! Reies Tijerina and the New Mexico Land Grant War of 1967* (Indianapolis, IN: Bobbs-Merrill, 1970).

17. United States Census Bureau, *Income in 1989 of Households, Families, and Persons by Race and Hispanic Origin, New Mexico: Summary of Social, Economic and Housing Characteristics* (Washington, DC: U.S. Government Printing Office, 1989); United States Census Bureau, *Income and Poverty Status in 1989, New Mexico. Summary of Social, Economic and Housing Characteristics* (Washington, DC: U.S. Government Printing Office, 1989). Maria Varela has calculated that 50 percent of the households in the Tierra Amarilla census district made less than $10,000 in 1990 (M. Varela, Testimony of Maria Varela, Co-Director of

Ganados del Valle, before the U.S. Senate Committee on Energy and National Resources on Proposed Rangeland Reforms, Albuquerque, NM, May 14, 1994).

18. A. Richardson, State of New Mexico Economic Development and Tourism Department, interview with author, Santa Fe, July 1990. For historical perspectives, see M. Weigle, *Hispanic Villages of Northern New Mexico: A Reprint of Volume II of the 1935 Tewa Basin Study with Supplementary Materials* (Santa Fe: The Lightening Tree, 1975); and M. Works, *A Place for Things: Material Culture and Socio-Spatial Processes in Northern New Mexico*, paper presented at the Annual Meeting of the American Association of Geographers, San Diego, April 1992.

19. For the early roots of this pattern, see J. Bodine, "A Tri-Ethnic Trap: The Spanish-Americans in Taos," in *Spanish-Speaking People in the United States: Proceedings of the 1968 Spring Meetings of the American Ethnological Society* (Seattle: University of Washington Press, 1968); and S. Rodriguez, "Art, Tourism, and Race Relations in Taos: Towards a Sociology of the Art Colony," *Journal of Anthropological Research, 45*(1), 1989, pp. 77–79.

20. Archdiocese of Santa Fe (1947), *Proceedings of the Archdiocesan Rural Life Conference on Rural Problems of New Mexico,* Bioregions Archive, Hulbert Center for Southwest Studies, Colorado College, "Agricultural Studies and Agroecology File"; W. Denevan, "Livestock Numbers in Nineteenth-Century New Mexico, and the Problem of Gullying in the Southwest," *Annals of the Association of American Geographers, 57*(4), 1967; W. deBuys, *Enchantment and Exploitation* (Albuquerque: University of New Mexico Press, 1989); Division of Research, Department of Government, University of New Mexico, *The Soil Conservation Problem in New Mexico* (Albuquerque: University of New Mexico Press, 1946); Harper, Cordova, and Oberg, *Man and Resources*; Soil Conservation Service, Division of Regional Planning, Southwest Region, *The Sociological Survey of the Rio Grande Watershed* (1936); in Bioregions Archive, Hulbert Center for Southwestern Studies, Colorado College, "Environmental History File"; M. Weigle, *Hispanic Villages.*

21. K. Weber, "Necessary but Insufficient: Land, Water, and Economic Development in Hispanic Southern Colorado," *Journal of Ethnic Studies, 19*(2), 1991, pp. 127–142; Richardson, interview with the author; G. Libecap and G. Alter, "Agricultural Productivity, Partible Inheritance and the Demographic Response to Rural Poverty: An Examination of the Spanish Southwest," *Explorations in Economic History, 19,* 1982; and W. Scott, "Spanish Land-Grant Problems Were Here before the Anglos," *New Mexico Business, 20,* 1967.

22. L. Redman, "Soil Conservation and Its Relation to the Community and the Family," in Archdiocese of Santa Fe, *Proceedings,* pp. 17–18; emphasis added.

23. P. W. Cockerill, "Rural Economic Problems in Low Income Areas in New Mexico," in ibid., pp. 5–6; emphasis added.

24. A. Carlson, *The Spanish-American Homeland: Four Centuries in New Mexico's Rio Arriba* (Baltimore, MD: Johns Hopkins University Press, 1990), p. 110.

25. T. Jervis, Audubon Society, interview with author (Los Alamos, NM, July 1990).

26. Northern Rio Grande Resource Conservation and Development Project, "Amendment of the Northern Rio Grande Resource Conservation and Development Project Action Plan," 1969, p. 10, in Bioregions Archive, Hulbert Center for Southwest Studies, Colorado College, "Environmental History File"; emphasis added.

27. deBuys, *Enchantment and Exploitation,* p. 297.

28. I do not mean to imply that this is the *intention* of such actors, but I do argue that this is one *result.*

29. L. Thrupp, "Political Ecology of Sustainable Rural Development: Dynamics of Social and Natural Resource Degradation," in P. Allen, ed., *Food for the Future: Conditions and Contradictions of Sustainability* (New York: Wiley, 1993); M. Watts, *Silent Violence* (Berkeley: University of California Press, 1983); and A. Wright, *The Death of Ramon Gonzalez* (Austin: University of Texas Press, 1992).

30. Van Ness, "Hispanic Land Grants"; Harper, Cordova, and Oberg, *Man and Resources.* This practice is common throughout the political ecology literature. Ashish Kothari notes: "Poverty or the lack of adequate economic opportunities often force people to degrade their own environment: for instance, firewood collection is a serious threat to forests in some places. What needs to be understood, however, is the genesis of this situation: more often than not, it lies in state policies which deprive the poor of their meager resources, and do not provide adequate alternative avenues for economic and social security" (quoted in J. Martínez Alier, "Ecological Struggles in India: Interview with Ashish Kothari," *Capitalism, Nature, Socialism, 4,* 1993, p. 113).

31. S. Bobrow, *The Community Land Trust: A Strategy for Ganados del Valle to Acquire and Secure Land for Agro-Pastoral Development.* Master's thesis in the Community and Regional Planning and Rural Development Concentration, University of New Mexico, Albuquerque, 1992; emphasis added. See also C. Eastman and J. Gray, *Community Grazing: Practice and Potential in New Mexico* (Albuquerque: University of New Mexico Press, 1987).

32. Peña, 1991, "American Wilderness," p. 3.

33. Van Ness, "Hispanic Land Grants," p. 194.

34. M. Stoller, "La Tierra y la Merced," in R. Teeuwen, ed., *La Cultura Constante de San Luis* (San Luis, CO: The San Luis Museum Cultural and Commercial Center, 1985), p. 13.

35. For a more nuanced discussion of the complexities and transgressions of Hispano resource management, see D. Peña, *Pasture Poachers, Water Hogs, and Ridge Runners: Archetypes in the Site Ethnography of Local Environmental Conflicts,* paper presented at the 36th Annual Conference of the Western Social Science Association, Albuquerque, New Mexico, April 1994.

36. S. Martinez, Ganados del Valle, interview with author (Los Ojos, NM: August 1991).

37. For the full history of Ganados and its connection to previous resistance struggles, see Peña, " 'Brown' and 'Green' "; and Pulido, *Environmentalism and Economic Justice,* Chapter 4.

38. P. Torres, Rio Arriba County Extension, interview with author (Española, NM, August 1990); and P. Kutsche and J. Van Ness, *Cañones: Values, Crisis and Survival in a Northern New Mexico Village* (Albuquerque: University of New Mexico Press, 1981), p. 45.

39. Only 28 percent of the county is privately owned. Over 50 percent is held by the U.S. Forest Service in the Carson and Santa Fe National Forests, both of which have been reducing their stocking rates. It is important to realize that, in order to qualify for a USFS permit, "base land" is required, which Ganados does not own. Another 17 percent of the county is owned by the Jicarilla Apache, who have in the past entered into leases with Ganados. Although the WMAs constitute

only 2 percent of the county, this obscures the fact that they comprise 20 percent of the rangeland in the Chama area.

40. New Mexico Department of Game and Fish, *Bill Humphries Wildlife Management Plan* (Santa Fe: NMDGF, 1984); *Rio Chama Fish and Wildlife Area Management Plan* (Santa Fe: NMDGF, 1984); and *Edward Sargent Fish and Wildlife Area Management Plan* (Santa Fe: NMDGF, 1980).

41. Ganados drew heavily on Savory's Holistic Resource Management in formulating their proposal (A. Savory, *Holistic Resource Management* [Covelo, CA: Island Press, 1988]).

42. The Nature Conservancy, Audubon Society, National Wildlife Federation, and Sierra Club individual members.

43. W. Evans, New Mexico Department of Game and Fish, interview with author (Santa Fe, NM, July 1990).

44. This was an initiative associated with Reies Lopez Tijerina and *la alianza*. See Blawis, *Tijerina*; and Gardner, *Grito!*.

45. La Frontera proposal in M. Varela, "Ganados Wins Sierra Club Foundation Settlement," *Noticias Nuevas de Ganados del Valle,* Spring and Summer 1995, p. 1.

46. S. Rodriguez, "Land, Water, and Ethnic Identity in Taos," in Briggs and Van Ness, eds., *Land, Water and Culture.*

47. M. Valdez, interview with D. Peña (San Luis, CO, September 1990); in Bioregions Archive, Hulbert Center for Southwestern Studies, Colorado College, "Oral Histories Collection."

48. Ganados del Valle, "The Grazing Proposal and the Issues," mimeographed paper (Los Ojos, NM: Ganados del Valle, n.d.).

49. M. Morales, Canjilon resident, interview with author (Canjilon, NM, August 1990).

50. T. Atencio, "Cultural Philosophy: A Common Sense Perspective," *Upper Rio Grande Waters: Strategies.* A Conference on Traditional Water Use, The Upper Rio Grande Working Group (Santa Fe, NM, October 1987), p. 11.

51. M. Stoller, "La Tierra y la Merced."

52. M. Varela, Testimony of Maria Varela before the U.S. Senate Committee on Energy and Natural Resources on Proposed Rangeland Reforms, Albuquerque, NM, May 14, 1994.

53. K. Cassutt, Sierra Club member, Interview with author (Santa Fe, NM, July 1990).

54. Peña, "Pasture Poachers"; Peña, "An American Wilderness"; and R. MacCameron, "Environmental Change in Colonial New Mexico," *Environmental History Review, 18*(2), 1994, pp. 17–39.

55. Peña, "An American Wilderness"; and Atencio, *Cultural Philosophy.*

56. L. Taylor, "The Importance of Cross-Cultural Communication between Environmentalists and Land-Based People," *The Workbook, 13,* 1988. One could also question the way "land-based people" and "environmentalists" are posed as mutually exclusive categories.

57. E. Daggett, *Chama, New Mexico: Recreation Center, Its History, Industries, Recreations* (Albuquerque: Starline Corporation, 1973), p. 42.

58. Bebbington, "Modernization from Below," 1993.

59. For an interesting example, see H. Rangan, "Romancing the Environment," pp. 155–181.

Chapter 12

The "Brown" and the "Green" Revisited

Chicanos and Environmental Politics in the Upper Rio Grande

Devon Peña
María Mondragon-Valdéz

Over the past five years, many students of political ecology have been tracking the problematic issues of race and class politics within an environmental context.[1] The relationship between Anglo environmentalists and communities of color is a contradictory feature of the larger debate surrounding what has been appropriately defined as environmental racism. One of the best arenas to examine the tensions generated by this conflict is the Upper Rio Grande bioregion of the American Southwest, in particular, rural Chicano communities in northern New Mexico and south central Colorado, which are the focus of this chapter. The battles over land use and resources that have frequently arisen between Indo-Hispanics residing in rural enclaves and Anglo environmentalists provide an important vantage point from which to study the evolving political dilemma between the "brown" and the "green."

The foremost national environmental organizations, or "Group of Ten," have been criticized by environmental justice activists and theorists for failing to integrate issues of race and class in their political discourses and actions.[2] Critics often characterize "mainstream" or institutional environmentalism as narrowly focused bureaucratic structures that promote white middle-class interests through lobbying and litigating for "citizen" participation in the politics of regulatory regimes and nature conservation. A 1992 exposé on diversity typified the Group of Ten as facilitating their reformist strategies through "professionalized . . . top-down . . . quasi-corporate organizational structures." These findings specifically highlighted the lack of race and class diversity within Group of Ten executive boards, staff, and predominantly white middle-class constituencies.[3] Environmental justice theorists posit that, because the Group of Ten organizations are so homogeneous, they tend to be overly preoccupied with wilderness preservation, wildlife conservation, global warming, and overpopulation.[4] Issues related to exploitation, poverty, and inequity in the experiences of poor, ethnic–minority, and working-class communities are thus conspicuously absent from their reform agendas.

Until very recently, mainstream environmentalists have been blind to the fact that working-class, ethnic–minority populations suffer disproportionately from environmental degradation in both urban and rural settings. It is puzzling that mainstream groups seem not to perceive how social injustice is often associated with environmental injustice, considering that three out of every four toxic waste sites are located in low-income communities of color. As early as 1983, data from the General Accounting Office indicated that Latinos constituted 75 percent of all rural Americans residing in the Southwest and West who were exposed to pesticide-contaminated drinking water.[5] A decade later, research on industrial pollution in the Midwest demonstrated the "skewed social distribution of toxic waste disposal sites" as indicative of an "environmental regime that discriminate[s] along lines of race and class."[6] Despite evidence of environmental crisis facing minority working class communities, mainstream environmentalists persist in overly emphasizing wilderness ideals. Ignoring the debilitating environmental conditions that surround racially segregated urban barrios, rural villages, and reservations, mainstream ecoactivists seemingly prefer being "enslaved to the defense of feathers and furs for the sake of wilderness appreciation"[7] to becoming embroiled in the complex ecological and political predicaments that plague communities of color.

While mainstream ecoactivists tend to ignore environmental justice struggles of people of color in the United States, they inadvertently contribute to the displacement of indigenous communities on a global level. The Group of Ten environmental organizations tend to promote an international policy that "sacralizes wilderness" without understanding how this action can result in the removal of humans from nature through ill-conceived conservation strategies. Some critics claim that this emphasis on strictly biological conservation is now a strategy to "integrate" indigenous communities into an ecotourism global market. At an international conference held in Hawai'i in 1990, the globalization of all forms of tourism was charged with "exacerbating racism, prostitution, the gap between rich and poor, environmental degradation, cultural destruction, and economic dependency."[8] Ecotourism is touted as more socially responsible than mass tourism, but the backpacker and conservationist remain wealthy intruders in the native's homeland.

Operating in this mode, the ecotourists as much as the ecoactivists simply do not see the impoverished villages for the trees. Clearly, it is not in their interest to concede that resource-dependent communities should continue to use "wilderness" space. In places like Nepal, villagers must adjust to the enclosure of traditional use areas not just for tree plantations, dams, and mines but for trekkers, campers, climbers, and other ecoadventurers. In this regard, the ecotourism industry works with NGOs, governments and corporations to jointly develop areas to accommodate the growing number of tourists demanding First World amenities.[9]

Thus, a major unresolved contradiction among wilderness advocates has been the pervasive inability to reconcile human interests with protection of nature, which has often led to an antagonistic relationship between indigenous communities and environmentalists. Unspoiled wilderness is a deeply entrenched nineteenth century ideal inherent to European and American Romanticism. Both British naturalists and the mainstream American conservation movement have often clashed with local cultures that live outside their acceptable norms, despite the fact that many traditional practices are sustainable and conducive to maintaining a balance between ecosystem integrity and biodiversity.

Like the rain forest struggles in the Third World, the so-called American Southwest is a hot bed of activity for a variety of struggles between indigenous local cultures and intrusive political economic forces acting under the sway of multinational corporate interests. In the context of the Southwest, struggles for cultural survival and political autonomy by Native Americans are a widely recognized and celebrated cause of progressive environmentalists. This is because much of our popular literature has long posited that indigenous communities have *never* caused environmental degradation. Environmental historians are now documenting how—through time—preindustrial societies have in many places overused local resources. However, the levels of exploitation, while not entirely benign, pale in comparison with capital-intensive overproduction.

Despite the popularity of Robert Redford's forgivably romantic film adaptation of John Nichols's *The Milagro Beanfield War*, struggles by Chicanos in the Southwest for cultural survival and social justice are less well known, poorly understood, and often treated in an unsympathetic manner. This is especially the case in some Southwestern regions where *mexicanos* predominate numerically and racial bias has historically contributed to the subjugation of Latino communities. The issue here is not who has been championed the most but, rather, how the largely unexamined attitudes of Amerindian and Chicano cultures have sustained important traditions regarding stewardship of land and water and right livelihood. As these rural communities become educated about the long term consequences of ecosystem destruction, they organize their own powerful challenges to environmental degradation by extractive industries. In this respect, the "red" and the "brown" are often the better role models for environmentalists, as these activists are highly motivated to protect the ecosystems their communities depend on. Yet, most Group of Ten organizations continue to overlook the ecological implications of the struggles waged by rural land-based Chicano communities. This criticism extends to some environmental historians who have been slow to acknowledge that intersections of class and race are defining factors in the making of the region's contradictory and multicultural green politics.

CRITIQUE OF RACIALIZING DISCOURSE IN SOUTHWESTERN ENVIRONMENTAL HISTORY

Why have Chicano environmental issues been generally dismissed as being unimportant? Perhaps because rural Hispano (for a definition, see Chapter 11, note 10, p. 308) villagers have often been stereotyped as being disinterested in ecological issues and specifically lacking a conservation ethic? Some environmental historians, government land and resource managers, and ecoactivists have tended to promote this notion of rural people in general. However, some critics focus on Chicanos in particular. For example, Hal Rothman, an environmental historian, argues that in the early part of the twentieth century, "Few people understood the concept of conservation. Anglo cattlemen and timbermen and Hispano and Native American pastoralists alike in northern New Mexico were not among those who did."[10] William deBuys, a wilderness advocate in New Mexico, suggests that the conservation ethic as it developed in the Southwest is a distinct by-product of a twentieth-century American conservation movement inspired by Thoreau, Muir, and Leopold. Aldo Leopold, in particular, is seen as the savior of an "American wilderness" in a (New) Mexican homeland that had been subjected to abuse by all cultures and classes. In this argument, "frontier" peoples (including *mestizo* villagers) presumably had no knowledge of ecology prior to the arrival of the refined and better-educated class of Anglo naturalists and environmentalists.

According to deBuys, contemporary villagers cannot legitimately criticize the U.S. Forest Service (USFS) for taking possession of the Spanish–Mexican common land as most of these areas had presumably been badly overgrazed and damaged by the villagers' sheep-herding ancestors. deBuys denies the expropriation of common land by a powerful class of capitalists as the root cause of environmental deterioration despite evidence indicating that the Anglo-dominated market system mandated overdevelopment and export of regional resources. In this instance, land loss transformed small *rancheros* into an oppressed colonial labor force. To disregard that colonial conditions favored absentee owners conveniently overlooks the destructive waves of large-scale industries, rail technology, and mineral rushes that were prevalent throughout the late nineteenth and early twentieth centuries. deBuys chooses to emphasize the "real third-world problems" of poverty, lack of knowledge (education), and overpopulation.[11] The USFS is cast as the heroic savior of the forests of the Sangre de Cristo Mountains through the scientific management of natural resources badly damaged by all those in pursuit of profit *and* sheep-herders of the past.

deBuys would have us ignore class and (not ideologically misconstrued) ethnic differences in land tenure and land use practices. From a Chicano perspective, the USFS brought ecological ruin to the watershed commonwealth, displaced smallholders from communal lands, and created social disharmony in a region that had been more or less sustainably managed for generations before the arrival of industrial and speculative capital. deBuys's glaring omission of preferential treatment accorded to corporate interests (specifically logging, mining, and ski industries) by the Forest Service, regardless of their impacts on the environment, can only be deemed as denial.[12]

The racist construction of Chicanos as quaint, but violent and lawless, land thugs is expressed in a more recent book, *Colorado's Sangre de Cristo Mountains*, written by Tom Wolf, a self-described "environmental entrepreneur":

> After centuries on the nethermost fringes of many different empires, local Hispanics have evolved their own unique folkways. These include a long-memoried sense of justice and a porous wall between church and state. San Luis's *predominantly Roman Catholic culture harbors violence and lawlessness* as well as a profound love of land and community.[13]

Echoing Rothman and deBuys, Wolf argues that the "tragedy of the commons" damaged the environment well before the arrival of Anglo settlers and that Chicanos must share blame for abusing the land:

> All men rushed toward ruin, when everyone exploited, shot, gouged, and burned everything. . . . Right next to the village of San Luis, separating it from the Taylor Ranch, lies the San Luis Vega, the only communally—rather than publicly owned—commons left in the United States (unless you count Boston Common). You do not have to be a range scientist to see that the people of San Luis have badly overgrazed the San Luis Vega. . . . No one escapes whipping when it comes to land abuse in the Sangres.[14]

Wolf states that it was only during "the Progressive Era [that] a truly heroic early Forest Service managed—emphatically *managed*—the forested watersheds of the Sangres back to their potential."[15]

This perspective overlooks the legacy of tragic mismanagement of the so-called public domain by the USFS and BLM (Bureau of Land Management) and it also trivializes well-documented environmental regulations imposed by the Spanish Laws of the Indies and Iberian–Moorish water allocation customs. As Malcolm Ebright, a land grant historian, has observed, conservation ethics and explicit environmental regulations were woven into Spanish–Mexican water and land laws. For example, Governor Cuervo y Valdéz of Spanish colonial New Mexico outlined the principles for the management of Santa Fé's *la cienéga* wetland commons in 1705:

"[livestock] shall not trample or eat the grass that grows there so that anyone who needs it [the grass] can mow it to feed their horses."[16] Following this tradition, in 1717, Lt. Governor Juan Paéz Hurtado stated:

> At this time every year [after the peak of spring snow melt] a *bando* [ordinance] is promulgated ordering that all the pigs running loose in this Villa of Santa Fé be rounded up, so that no damage is done either to the planted fields, or to the Ciénega of this said Villa, also that the same is done for the horses, cattle, sheep, and other animals, so that these [animals] will not trample or eat their [the Villa's common] grass.[17]

This and other evidence of Spanish colonial era and Mexican period environmental regulations is apparently cited for naught, as many environmentalists and federal resource managers stubbornly cling to poorly recast versions of Garrett Hardin's tragically inaccurate theory of the commons.[18] Hiding behind the biased constructions of "primitive" Hispanos, Wolf obscures the dominant source of contemporary environmental and economic crises in market forces unleashed by capitalism and federal land use policies favoring corporate interests.

BIOREGION AND CAPITAL

Before proceeding with the details of our two Rio Arriba case studies, we would like to point out deficiencies in the accepted bioregional paradigm, particularly as outlined by so-called deep ecologists and radical environmentalists.[19]

Bioregions are principally defined by watersheds. The rivers that flow within the watershed are a result of the incessant interaction between geology and hydrology. Ultimately, rivers create ecosystemic niches that support diverse biotic communities. The biota and water define the nature and extent of the boundaries of individual bioregions. But bioregions are also differentiated by their cultural distinctiveness. What we call "material culture" emerges out of the historical interaction between humans and their local environment.[20] In the creation of culture, humans over time engage in a "discourse" with their local environment, a process which endows the particular locale with a "place-centered identity." The embodiment of humans in nature characterizes what ancient Romans called Genus Loci, or spirit of place. Cheney describes this place-centered ontological stance as "storied residence."[21] It is this form of cultural distinctiveness that defines the human dimension of bioregions, including the types of materials and designs used to construct shelters, the way place names originate, the types of legends and lore that evolve, and differences in poetry, dance, music, and ritual. As Wendell Berry puts it, "the pattern of reminding implies affection for the place and respect for it." Sense of place conveys the idea that "local culture will carry the knowledge of how the place may be well and lovingly used, and also the implicit command to use it only well and lovingly."[22]

This approach omits two other important qualities of bioregions: first, the local ethnoscientific knowledge bases; second, the conflict between small-scale, regional economies and the predatory mode of advanced capitalism. The limitation of the bioregional paradigm is that the reality of power, conflict, and inequality is often overshadowed by overly romantic notions such as "earth-bonding ritual" and "spiritual ecology." Many deep ecologists and bioregionalists recognize the destructive impact of "industrialism," but they often fail to acknowledge the *capitalist* character of the forces responsible for the increasing pace of environmental degradation. Many ecoactivists acknowledge the struggles of indigenous people in the Third World (the Chipko "treehuggers" being the most popular cause). But, generally, both mainstream (shallow) and radical (deep) environmentalists remain practically disengaged from actual community-based struggles to unify the conservation of biological and cultural diversity. The great irony of this stance is that the usurpation of land and the impact of industrial capitalism are largely responsible for eroding local sensibilities of place. This apparent insensitivity toward the plight of rural communities has created much misunderstanding and animosity between communities of color and many environmentalists.

To understand the destruction of these bioregional communities, one must acknowledge that racialized class differences play a major role in the environmental history of many places. Capital, understood in bioregional context, represents a powerful intrusive force altering the flow of rivers, reducing the diversity of biotic communities, and degrading the complex mosaic of native natural and cultural landscapes. The power of capital to undermine the material position of traditional subsistence economies derives from its ability to enlarge the scope and increase the intensity of resource extraction through breakthroughs in reductionist science and technology.[23] In this process, the moral (and legal) authority of local cultures is greatly marginalized through land loss and the suppression of economic diversity and autonomy—all features that are inextricably linked to ecological and cultural destruction. While ecocide and cultural genocide come into the orbit of capitalist imperatives, this does mean that capital goes unchallenged, especially in communities that maintain a fiercely militant sense of place against the expansion of extractive industries and gentrification within the bounds of the reservation and village enclaves.

The fundamental problem with the enigmatic positions of deep ecological discourse is the tendency to fragment and ultimately misrepresent the irreconcilable claims of identity politics in a manner that distracts us from the more pressing issue of the sublation (*aufheben*) of political economic contradictions. "Sacred places" and "earth-bonding" rituals are the romantic contrivances of privileged outsiders who seem afflicted with a propensity to "exoticize" and "museumize" the native peoples and landscapes of the Southwest. This dilettante-like celebration of a sense of place is symptomatic of a long-standing literary and philosophical tradition of searching for earthly paradises on other people's lands. The problem with this new mystique of "eco-consciousness" is that it contributes little to the resolution of the myriad conflicts that resulted from the expropriation of land and water resources by corporate capital and the state. For how can society have any hope of maintaining or revitalizing a bioregional sense of place when the landscape has been violently appropriated and mutilated and the autochthonous cultures have been displaced?

CULTURAL AND ENVIRONMENTAL HISTORY OF THE UPPER RIO GRANDE

The Rio Grande runs 1,900 miles from its headwaters in southern Colorado's San Juan Mountains to the flat, sandy coastal plains and salt water marshes surrounding the "Rio de las Palmas." It is at this south Texas border, between Mexico and the United States, that the river empties into the Gulf of Mexico. In its course from headwaters to mouth, the Rio Grande corridor is a complex, changing mosaic of every major life zone in North America: from alpine tundra and krumholtz, subalpine and montane forests, and piñon–juniper foothills, to cold deserts, cañon lands, prairies, and finally salt marshes and estuaries.

When the *mestizo pobladores* arrived in New Mexico to establish the first Spanish colonial settlements under Oñate in 1598, they encountered a marginal environment for sustaining human inhabitation. The settlers adapted to these conditions by developing a hybridized irrigation technology whose origins can be traced to both *arroyo* (flood-plain) irrigation of the Pueblos and Islamic customs transferred from the Iberian Peninsula to the northern Mexican frontier by the Spanish. The gravity-driven earthen ditch, or *acequia*, system allowed compact village clusters within the tributary valleys of the watershed to establish sustainable agropastoral livelihoods. Over generations, *acequias* created vast riparian corridors lined with native cottonwoods, alders, willows, and grasses. This botanical profusion of plant life was dependent on water leakage from the *acequias*. In time, the arid upland valleys were transformed into veritable puzzle parts of the natural landscape mosaic as cottonwood *bosques* became wildlife habitat and movement corridors for elk, deer, antelope, and a rich variety of other species. In the *acequias*, we have a clear case demonstrating how anthropogenesis can support and extend native biodiversity.[24]

Many scholars have found that the Chicano agropastoral village culture of the Upper Rio Grande is uniquely well suited to maintain a balanced human presence in arid and semiarid environments.[25] But tremendous environmental, demographic, economic, and cultural changes have occurred in the Upper Rio Grande during the past 120 years. Scholars of the bioregion have not overlooked these changes. However, the cultural and the environmental history of the Upper Rio Grande has not been sufficiently incorporated into analysis of the impacts of an advanced multinational extractive economy.

The processes of social change brought about by the intrusion of the global market economy have not merely undermined old sustainable patterns but have ensured that there will be a seemingly endless cycle of disputes between the original settler communities and external political economic forces. The case studies we examine here indicate the existence of contradictory relationships between and among villagers, extractive development interests, conservation groups, and the state. Clark Knowlton captures the underlying causes of such conflict in a pioneering study of land grant struggles:

> Under American law [land] was and is a commodity to be bought and sold in the market place. . . . To most Spanish-Americans [*sic*] land has never been a commodity to be bought and sold. The land providing a living for the family has always been as much a part of the family as the home or children. To sell it was equivalent to selling a family member.[26]

The conflict between the market, individual capitalists, and agropastoral systems was a clash between different cultural and legal traditions and modes of production. As such, it involved a violent transition from a sustainable, subsistence-oriented economy to a globalized export-oriented and extractive economy. It is through the commodification of land and water that capital unleashes ecological destruction of bioregional resources. Sarah Deutsch notes that in 1860 sheep stocking rates in the Rio Arriba averaged about 12 head per square mile, a figure that was well within the carrying capacity of the land. By 1900 stocking rates exceeded 100 to 120 per square mile, an increase she attributes to the arrival of the railroad and the transformation of the largely subsistence sheep and cattle industries into large-scale commercial operations by outside Anglo speculators.[27] This is affirmed by a variety of studies showing how environmental degradation in the Upper Rio Grande was associated with the export market and industrializing capitalist interests.[28] The most striking evidence of this emerging pattern is outlined in a classic study of overgrazing in northern New Mexico's public domain. Unequivocably, the carrying capacity of New Mexico's montane grasslands was overwhelmed only after the intrusion of Anglo large-scale commercial cattle and sheep ranching that rapidly expanded in response to the growing demand for meat in the American West.[29] In general, clearcut logging, large-scale mining, and accelerated dam-building by capital led to the degradation of the montane forests, grasslands, and watersheds of northern New Mexico between the late 1870s and into the 1920s. This process of degradation was accompanied by the dispossession of Spanish–Mexican-era land grants, the enclosure of associated common lands, and the loss of water rights. The disruption of the sustainable agropastoral pattern was thus associated with incipient rural proletarianization as Chicanos turned to wage labor to support themselves.[30]

We want to avoid the misguided and romantic contrivances of deep ecology. Therefore, we juxtapose the study of environmental and cultural history with a critical appreciation of the political economy of this bioregion. We posit that the processes of environmental change intersect with the "race" and class politics of new social movements. We also argue that the study of what can be defined as "multicultural ecopolitics" reveals serious differences in land ethics and land use practices; this is especially true in rural Hispano enclaves of the Rio Arriba. Our bioregional approach requires that we evaluate ecopolitics in light of the biogeographic and economic conditions in given localities. In other words, we expect that differences in land tenure patterns—as much as class and race—play critical roles in shaping environmental politics in the Rio Arriba.

A COMPARATIVE POLITICAL ECOLOGY OF ENVIRONMENTALISM IN THE RIO ARRIBA

Our study focuses on two rural historic village cores in the Rio Arriba or Upper Rio Grande watershed: specifically, the Tierra Amarilla Land Grant in northern New Mexico's Chama basin and the Sangre de Cristo Land Grant in the Culebra basin of south central Colorado's San Luis Valley (see Figure 12.1). Many of the inhabitants of the Tierra Amarilla and Sangre de Cristo villages are descendants of the original *pobladores*, or land grant settlers. Spanish and Mexican settlement of the Rio Arriba evolved from a system of land grants. Initially, land grants (or *mercedes*) were given to individuals in the military as a compensation for duty; to civil authorities as a reward for services; and to elite for grazing range. Grants were also made to private *empresarios* (entrepreneurs) to encourage the development of commerce and trade. Later, community groups (composed of several extended families) were granted land to expand the settlement frontier. The sedentary Pueblo Indians also received grants to confirm their aboriginal land and water rights.[31] In both cases, complex struggles over land and water rights, economic development, and cultural survival provide a unique opportunity to examine the intersections of class and race dynamics in the context of environmental politics.

The historic roots and patterns of development of these communities closely parallel each other. Both are Mexican-era land grants, and in each case the communities were later organized by land grant activism beginning in the 1960s. Both communities experienced similar patterns of land loss, imposition of waged labor, out-migration, and environmental degradation.

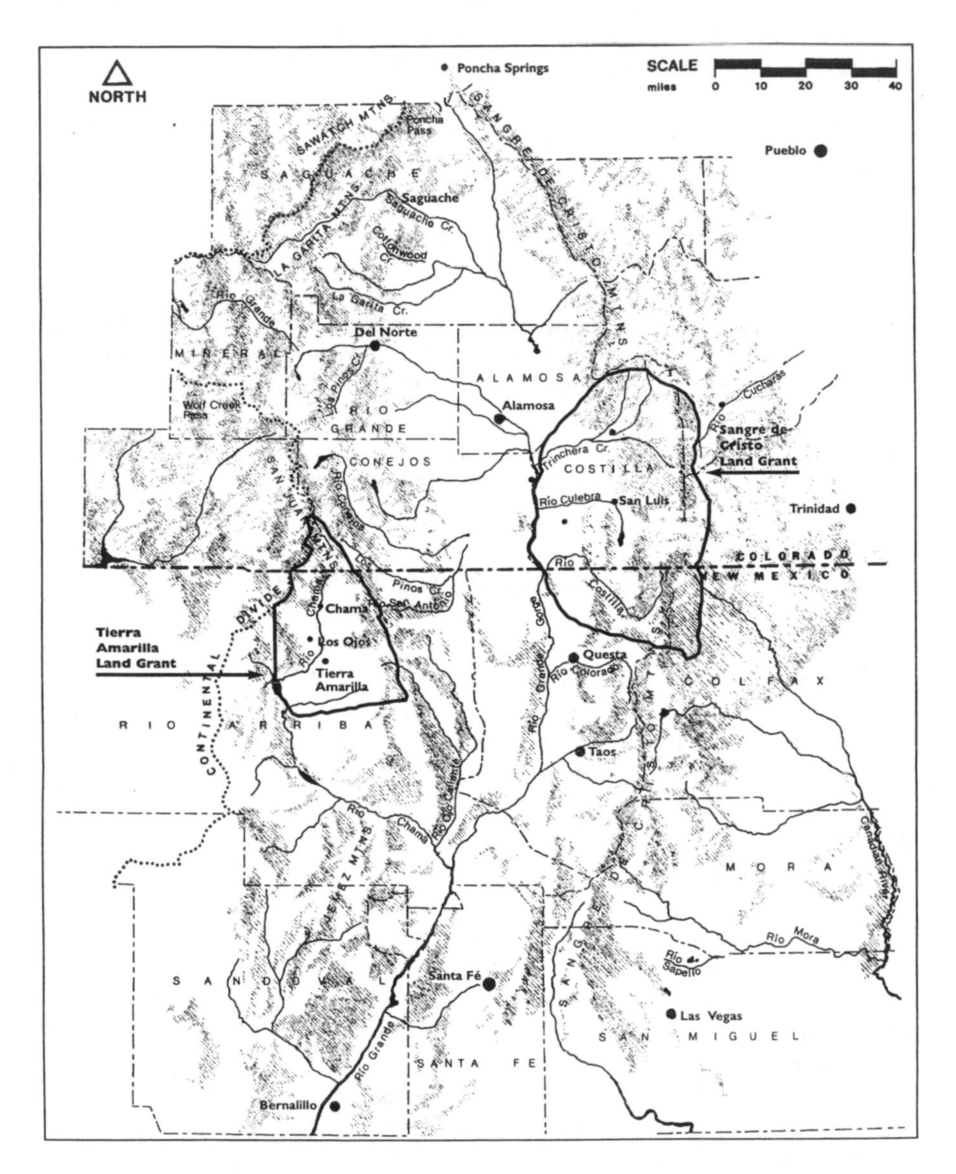

FIGURE 12.1. Map of the Rio Arriba bioregion, highlighting the Tierra Amarilla and Sangre de Cristo Land Grants.

In each case, the collective social and economic profile is characterized by a predominant multigenerational Chicano population with high rates of poverty, unemployment, and out-migration by youth and displaced farmers. Both areas lost their traditional communal land base through privatization or federal confiscation.

With the advent of the "War on Poverty" of the 1960s, our case study sites were within a seven-county area in northern New Mexico and southern Colorado designated as "the Appalachia of the West" and defined as experiencing "long-term economic deterioration." In each case the federal government supported a series of development projects to revitalize the economic status of these rural village areas.[32] The long-term failure of government programs to deal with changing patterns of land tenure and market-dominated control of resources has resulted in more poverty, more displacement from the land, and greater control of natural resources by absentee landowners on private land and the public domain. The new political relationships between environmentalists and Chicanos have produced different outcomes and contemporary asymmetries for each of these communities.

In the Tierra Amarilla case, the local people organized opposition to the loss of grazing range on common lands beginning in the 1960s with the formation of La Alianza Federal de Mercedes (Federal Alliance of Land Grants). Today, a nationally recognized livestock enterprise, Ganados del Valle, and its handweavers' counterpart, Tierra Wools (now known as Los Ojos Handweavers), are by-products of this land grant activism. Together, these contemporary, member-owned enterprises provide locals with educational and economic opportunities to revitalize agropastoral traditions and artisan skills. These self-development projects have been successful, but they have had to wage a battle with wildlife conservation, hunting, and recreational interests and their allies in the New Mexico Department of Game and Fish (NMDGF). While the outcome of this struggle did not return grazing rights to communal land that was surreptitiously converted to the public domain, the conflict captures the essence of class and racial contradictions that divide many environmentalists from other progressive social sectors. However, the deeper implications of the Tierra Amarilla case are intrinsic to the meaning and nature of sustainable development and the place of traditional *ranchero* economies in the future of land use management on the public domain.

A parallel to the Tierra Amarilla struggle is the Sangre de Cristo Land Grant. The Rio Culebra villages (established in 1851) have the oldest water rights in Colorado and the state's sole community commons. The overarching similarity of the Tierra Amarilla and Culebra villages is that both have self-managed their water rights through historic *acequias* (communal irrigation associations). The corollary to Tierra Amarilla's Alianza was the Culebra villagers' Asociación de Derechos Cívicos (Association for Civil Rights) and the Land Rights Council (LRC). The difference between these communities is that, while La Alianza failed in its litigation, a new generation sought to redress social injustice by focusing the movement on economic revitalization. In contrast, the Asociación (active during the 1960s) and the LRC (active since 1979) have waged a 37-year-long legal and political battle to restore historic use rights. Remaining in a reactive stance as perpetual litigants, the Culebra villagers have never been able to fully focus their energies on organizing meaningful long-term economic development programs at the level of Ganados.

The Culebra villagers have had to address loss of control over wood gathering by promoting energy self-reliance and resisting mining and logging exploitation in the communal uplands. At the forefront of the Sangre de Cristo environmental justice movement are Peoples Alternative Energy Services (PAES, founded in 1977), Costilla County Committee for Environmental Soundness (CES, established in 1988), and La Sierra Foundation (LSF, organized in 1993). A complement to these various community groups is the Costilla County Conservancy District (CCCD), a quasi-governmental entity with a focus on protecting the watershed through tax allocations from the villages. Each of these groups has worked at some level with mainstream and radical environmentalists.

GANADOS DEL VALLE: WILDLIFE PRESERVATIONISTS AGAINST THE GRASSROOTS

"The Forest Spirit drove the man out of the forest because he chopped down too many trees for, indeed, he was full of greed, without shame."[33] The "Forest Spirit" folktale is a good example of the ethnopoetic oral tradition in rural northern New Mexico. Like other local cultures of the Southwest, the Chicano compact village communities of the Tierra Amarilla Land Grant region have a long history of struggle against intrusive development which endangers not just the health of the land but the ability of the community to sustain itself.

The Tierra Amarilla Land Grant was extensively settled by the 1860s. However, during the 1870s and 1880s the Chicano settlers lost their common lands to the infamous "Santa Fe Ring," a notoriously corrupt and devious group of lawyers, politicians, and entrepreneurs who engaged in widespread land speculation and title fraud throughout northern New Mexico.[34] Thomas B. Catron, a member of the Santa Fe Ring, played a major role in the theft and enclosure of the Tierra Amarilla Land Grant. The community continues to remember how:

> In the first half of the century our timber and grazing resources were stripped by large corporations; leaving us, a pastoral people, without adequate land or water to support our families and way of life. In the second half of this century, poverty deepened in northern New Mexico.

> Much of this was due to loss of traditional grazing lands that were part of the land grant.[35]

Over the years, the Tierra Amarilla Land Grant heirs were involved in a series of expensive, time-consuming, and complex litigation procedures to regain their lost common lands. All litigation ultimately failed. In the late 1960s, Reies López Tijerina led a struggle to restore the land grant commons that culminated with the famous "Tierra Amarilla Courthouse Raid."[36] Finally, in 1988 and 1989, El Consejo Land Grant Association supported the Amador Flores family to successfully reclaim and establish residence on a 600-acre parcel of the land grant.

The people of the Tierra Amarilla region have a long history of self-directed community organizing efforts that derive from turn-of-the-century mutualist traditions. For example, in the late 1960s the villages organized the Cooperativa Agrícola de Tierra Amarilla (an agricultural cooperative) and, later, La Clínica del Pueblo (a rural health clinic). In the community's words, these grassroots initiatives evolved because "we were tired of being neglected by the government, of seeing our agricultural land go idle and seeing people die on the highway."[37] The contemporary nonprofit counterparts, Ganados del Valle and Los Ojos Handweavers, are the most recent examples of ecologically sustainable and culturally appropriate self-reliance projects created by the people of the Upper Chama basin. The artisan skills of Los Ojos Handweavers involves a revitalization of both Rio Grande blanket and Saltillo *tapete* (tapestry) traditions and the creation of a line of contemporary clothing. The return to high-quality craftwork at Los Ojos Handweavers is a classic example of a "free association" of skilled artisans. One significant aspect of Ganados' work is the regeneration of the *churro* sheep, a rare Iberian breed that had nearly disappeared when the group reestablished a herd.[38] The cultural and economic revitalization goals of these member-owned enterprises are all the more significant given the context of a community that faces the depletion not only of its land base and water rights but also of its increasingly mobile, mostly unemployed, and "unskilled" youth. By the 1990s, Ganados and Los Ojos Handweavers (and several spin-off agricultural and cultural-based businesses) were the largest nonprofit, nongovernmental employers in the Chama Valley.[39] These businesses are successful because they draw on the strengths of cultural traditions embedded within family and friendship networks and sustainable agropastoral practices.[40] Despite these achivements and the history of land grant struggles, the community became embroiled with a new emerging political force—a coalition of mainstream environmental organizations that included the New Mexico chapters of The Nature Conservancy (TNC) and the Sierra Club with their allies in the New Mexico Department of Game and Fish (NMDGF).

In Rio Arriba County over half of the land is in National Forest or state-owned public domain. The remaining 45 percent of the land base is held by Indian reservations and private owners. As a rule, public grazing permits and private leasing in Rio Arriba is dominated by large cattle operations, many from southern New Mexico and Texas.[41] Potential public domain grazing in close proximity to Ganados is divided into three "Wildlife Management Areas" (WMAs), all under the control of the NMDGF. While the state-administered WMAs can accommodate sheep grazing, these areas are formally defined as elk sanctuaries (which many see as little more than wildlife plantations managed for the benefit of out-of-state big-game hunters).

From the beginning, members of Ganados struggled to gain access to USFS and BLM grazing allotments and private grazing ranges to pasture their herds. The search for grazing land is very difficult, as most private land has been subdivided for residential and recreational development. Eventually, Ganados obtained a limited grazing lease from the neighboring Jicarilla Apaches. Prior to expiration of the Jicarilla lease, Ganados repeatedly approached the NMDGF for a legal grazing permit in the WMAs. Notwithstanding the fact that other New Mexico WMAs have been consistently grazed by private livestock and that a sustainable grazing management plan was submitted by Ganados, and despite repeated attempts to find an equitable solution that would accommodate elk and sheep, the permit requests were rejected. With the termination of grazing permits on the Jicarilla land, Ganados faced imminent collapse. In desperation, Ganados decided to trespass as a direct action strategy to press their case in public and force the NMDGF to reconsider their grazing plan.

In August 1989, Ganados drove the members' flocks onto the W. A. Humphries WMA, an act calculated to demonstrate the collective political will of the community and force the NMDGF to resolve this dilemma. This move was deemed necessary, as the return of the flocks to family long-lot pastures would have led to the grazing of home-grown winter feed. In turn, this would have forced the sheep herders to purchase increased amounts of commercial feed, a prospect few could afford. The only choice would have been to sell or slaughter the flocks and re-endanger the *churro* breed.[42] Ganados issued a press release that restated its desire "to be a partner with the NMDGF" and reiterated its intent "to work out an ecologically sound management plan which would improve wildlife habitat." While the "trespassing" incident finally succeeded in bringing all parties to the negotiating table, ultimately, all proposals for grazing on the Humphries WMA were rejected.[43]

An alternative seemed to emerge during discussions over possible changes in the future status of another nearby elk sanctuary, the Sargent WMA. Ed Sargent, who made his fortune with a huge sheep-grazing operation between 1910 and the 1930s, had come into possession of this 22,000-acre section of the Tierra Amarilla common lands after the arrival of the railroad. In the 1970s, TNC purchased the land from Sargent's granddaughter. Eventually, the state of New Mexico acquired the tract with taxpayer funds and hunters' fees. Once the state acquired the land, it turned management over to the NMDGF for elk hunting and winter recreation.[44] The Sargent land transfer did not explicitly prohibit livestock grazing, and in fact TNC allowed the former owner to graze cattle on the land before it became a WMA. Dismissing this contradiction as irrelevant, TNC, the elk lobby, and the NMDGF held firm on the grazing ban.

The dispute between TNC and Ganados was perhaps unavoidable. Tom Wolf, formerly with the New Mexico chapter, explained that TNC's professed conservation philosophy was based on the "fundamental separation of man and nature."[45] Ganados did not share the TNC's view of the former common lands as merely "elk habitat." This position ignores the bioregion's environmental and cultural history whereby grazing and weaving are as much a part of the landscape as the elk herds. Despite evidence suggesting that grazing could contribute to the restoration of vegetation quality,[46] and the absence of endangered or threatened species, Ganados was "politically out-maneuvered by stronger, more influential, wildlife, outdoor recreation and hunting interests" and their powerful lobby.[47]

Just as this dispute seemed destined for obscurity, a coalition of New Mexico environmental organizations (including TNC, the Sierra Club, and the Audubon Society) agreed to mediation with Ganados in an attempt to find a more just solution to the conflict and avoid the appearance of ethnoracial bias.[48] During the fall of 1990, Ganados met with the coalition to find common ground. While the discussions sensitized environmentalists to the needs of the community, the grazing ban remained intact. Nevertheless, Ganados is totally dependent on private grazing contracts, as they have been denied equitable access to public domain pasture.

As poetic justice might have it, another twist in this plot was revealed in 1989 after Ganados put the flocks on the Humphries WMA. In this concurrent event, a 25-year-old donation made to the Sierra Club Foundation by Firestone heir and philanthropist Ray Graham III for the benefit of Tierra Amarilla surfaced. In 1970, Graham had funneled $100,000 through the Sierra Club Foundation to help the Tierra Amarilla community purchase its own grazing range. After several attempts to work with the Foundation, the community was told that the donor was no longer interested in purchasing grazing land. When Graham read about the trespassing incident, he phoned Ganados to ask why the community had not used his donation to buy grazing land. It would be an understatement to say the community was not angry that the Foundation purposefully misinformed them of the status of Graham's donation. When all the dust cleared, it was apparent that this donation and other endowments earmarked for New Mexico conservation projects had been diverted into an interest-bearing account to buy land for the Foundation's San Francisco headquarters and to pay staff salaries. Not only had there been financial malfeasance and commingling of funds, there were missing financial records and correspondence that were central to this dispute.[49]

An angry Graham sued the Foundation in California, citing misappropriation of funds. The Foundation survived Graham's challenge on a technicality. Ganados took the matter to the Attorney General in an attempt to resolve the problem in New Mexico. As the Foundation continued its foot-dragging, the New Mexico Attorney General threatened to sue to return the money to the community with Ganados acting as sole representative for the Tierra Amarilla community. A court-ordered settlement conference convened from 1994 to early 1995. Late in February 1995, the court criticized the Foundation's disingenuous behavior after it refused to sign a court-mediated final agreement. In September 1995, the Sierra Club Foundation grudgingly agreed to release $900,000 to Ganados. After legal costs, this bittersweet victory was not enough for Ganados to purchase adequate land. Perhaps this is why Ganados leadership felt that their attempts to pursue a sustainable grazing ecology had been blocked by a "Green Wall," whose politics of "pride and prejudice" ultimately promoted a situation where locals had to battle "the environmentalists who should be and have been our friends."[50]

LA SIERRA DE LA CULEBRA: RESTORING A WATERSHED COMMONWEALTH?

The Sangre de Cristo Mountain Range extends from southern Colorado into northern New Mexico. The focus of this case study is centered around the Culebra Peak region: a 30-mile-long spur that is geographically situated within the central portion of the range. What makes this particular area of the Sangre de Cristo Mountains an interesting site to study is that the highland habitats are monopolized by wealthy absentee landlords, while the resource-dependent villages of the lowland valleys are economically marginalized *mexicano* enclaves. With the exception of 565 acres of encroaching national forest lands (at the upper edge of the range), Costilla County is a land use anomaly as it is entirely privatized. The two largest baronial ranches within this section of the range are owned by the Malcolm Forbes Corporation (267,368 acres) and the Taylor Family Partnership (77,000 acres). However, there are several other large and small parcels belonging to a variety of absentee landowners from throughout the United States and around the world.[51]

The Culebra Range lies within the boundaries of the nearly one-million-acre Sangre de Cristo Land Grant issued to two applicants by the Mexican government in 1844. By 1849, the first of several villages was established by *mestizo* agropastoralists who were recruited as settlers under the promise of perpetual usufructuary rights to the Sangre de Cristo Mountains to hunt, fish, gather wood, and recreate. As was a custom within Spanish–Mexican land grants, villagers were given a lowland communal pasture and allocated individual *ranchos* and water rights. A necessary complement to the village was the community-managed *acequia* irrigation system. This type of settlement pattern created an agroecological landscape that allowed the early settlers to maximize productivity, conserve water, and maintain self-reliance in an arid uplands environment.[52]

As a consequence of the size of the Sangre de Cristo Land Grant and its disputed contemporary claims, this is perhaps the most complex and notorious of all Mexican era land grants. For the past decade, the Sangre de Cristo Land Grant has been the site of an intense and heated environmental conflict over commercial logging and land use regulations.

If one could choose the most decisive turning point in the history of this land grant, it would undeniably be in 1960 when Jack Taylor bought the last remaining unfenced portion of the Culebra uplands. The alienation of the Culebra Range from the villages below was the equivalent of being orphaned. Soon after the purchase (for a reported $500,000, or about $6.40 per acre), Taylor built a series of barricades at various entry points to the property, fenced the land, and brutalized the locals to keep them from exercising their use rights. He then set out to remove the heirs' use rights from the title to the uplands through a Torrens Title Action.[53] This blatant erasure was only successful because Taylor used a controversial legal process and a calculated methodology that undermined the heirs' constitutional rights. Quite simply, the successful outcome in this case pivoted upon a conservative Colorado judiciary that was overtly hostile to land grant claims; why else would the court allow any quiet title action to omit naming 80 percent of the interested parties?[54] This enclosure led to a veritable "range war," as the land grant heirs repeatedly challenged Taylor's claims to "La Sierra," or Mountain Tract.

Taylor's use of violence and his continued racial rhetoric fell short of suppressing the more than one hundred years of local access to the upland commons for the exercise of use rights. The local challenge to the quiet title action originated with the Asociacíon de Derechos Cívicos and was later continued by the Land Rights Council, a local grassroots organization composed of land grant heirs that was established in 1979. In 1981, the LRC filed a lawsuit challenging two 1960 decisions in a case that has come to be known as *Rael v. Taylor*. At the heart of *Rael v. Taylor* is the coming to terms with the asymmetries of a contemporary legal procedure that denied locals their right to an impartial hearing. Among the underpinnings of discontent with the earlier proceedings is the perception that Taylor's extralegal activities helped him to perfect the quiet title action through bribery and chicanery. One may never know the true story. However, one salient fact remains: the people have not had their day in court.[55]

The effects of the enclosure on the land grant villages of the Culebra basin were immediate and dramatic.[56] A cursory examination of agricultural census records for Costilla County clearly demonstrates a downward spiral in the community's economic mainstay, the independent sheep trade. In 1959 there were 21,650 head of sheep; by 1964, this number declined to 14,190; and in 1969, the sheep population dropped to 13,410 head. By 1974, the pastoral economy was down to 5,513 head. The effects of the enclosure were also evident in the decline of the amount of irrigated acreage among Chicano farmers with water rights on bottom lands in the Culebra watershed.[57]

To further complicate the situation, Taylor initiated large-scale commercial logging operations on La Sierra between 1964 and 1981. As most of the commercial quality timber is located at elevations ranging from 9,500 to 11,500 feet, a dense network of roads was constructed to facilitate high-altitude cutting. During these decades, Taylor Ranch management used a particularly harmful method called "road-terraced clearcutting." While terracing was thought to control erosion, logging on public lands had demonstrated that this technique damaged watersheds. In 1972, the USFS banned road terracing on public lands. Whereas the Taylor Ranch is private property, such evidence of damage was ignored during partial and clearcut timbering of 20 million board feet (mbf) on more than 2,500 acres in southern tributaries.[58] Today, aerial documentation indicates that forest regeneration on the Taylor Ranch is impaired or nonexistent at elevations where soils are thinner, the terrain is very rocky and steep, and persistently low temperatures stunt the growth of foliage.[59] Furthermore, during the spring snow melt, road cuts and exposed ground promoted erosion and finally sedimentation all along the watercourses. As road cuts lacked any type of water-barring structures (e.g., culverts, rip-rap barriers, or check dams), irrigation ditches were slowly choked with sediment. Ultimately, the *acequia* system was impaired since it cannot absorb or store excessive spring runoff. The consequences of three decades of logging have been evaluated by a renowned hydrologist, who found that *acequias* diverting water from the southern tributaries have experienced sedimentation and "flashy" spring snow melt. Technically, such problems produce "channel aggradation" and "disrupted stream hydrograph slope," phenomena that drastically reduced the length of the irrigation season by as much as 40–60 days.[60]

Throughout the 1980s and early 1990s, the community continued to pursue various avenues to resist Taylor's hegemonic control over La Sierra. PAES promoted solar energy as a means to avoid trespassing to gather wood and paying high fuel costs. The LRC continued its legal efforts to restore the historic use rights.[61] A key development in this contemporary cycle of struggle was the emergence, in 1988–89, of Battle Mountain Gold (BMG). The campaign against BMG mobilized several community groups to collectively develop the capacity to critique the use of science by the corporate lawyers, understand the technical aspects of the mining permit and operations, and acquire legal resources to monitor the mine during and after operations.[62]

In the interim, Jack Taylor died in 1987 as an embittered absentee owner scorned by the land grant heirs as a dictatorial robber baron.

Zachary (the youngest of Jack's two sons) emerged as the executor of the estate and family businesses that are based in the southeastern seaboard town of New Bern, North Carolina. Zachary inherited more than his father's view of locals as trespassers and an infamous legal team that would protect the client's interests at all costs; his patrimony included a small publishing company, real estate firm, lumber company, and La Sierra tract.[63] The need for a steady cash flow to deal with ongoing corporate indebtedness and perpetual tax problems have made the younger Zachary Taylor into a facsimile of his father.

By 1993, a new group, La Sierra Foundation (LSF), organized to purchase the Taylor Ranch and establish a self-managed community land trust.[64] Later that year, Governor Roy Romer signed an executive order establishing the Sangre de Cristo Land Grant Commission to discuss resolution of the Taylor Ranch conflict. After four months of public meetings, the commission proposed an unprecedented solution: the State of Colorado and the local community would cooperate to acquire the Taylor Ranch. Specifically, the commission recommended that: (1) the historic "use rights of the land grant heirs should be guaranteed in perpetuity"; (2) "the mountain should be managed in an environmentally and culturally sound manner"; (3) "management decisions . . . will be made in partnership between the state and the local people"; and (4) "each party will own an interest in the land in perpetuity with corresponding rights and obligations."[65]

In April 1994, after an independent appraisal, Ken Salazar, the Executive Director of the Colorado Department of Natural Resources (DNR) and chair of the commission, offered Zachary Taylor $15 million for the purchase of the land. Taylor firmly rejected the offer, stating that the property and its resources were worth at least $30 million. As a consequence, the commission went into a temporary hiatus, and some of the local members resigned in protest over undemocratic decision making and what they perceived as duplicitous behavior by Zack Taylor.

During the Land Grant Commission hearings, Taylor looked into a joint mining venture with BMG and actively sought to sell the timber rights to several West Coast lumber brokers. Meanwhile, LSF launched a national campaign to raise funds to acquire the land, while the LRC continued to pursue a legal, rather than a financial, resolution to the problem. In April 1995, LSF representatives met with Zachary Taylor to discuss the possibility of a local option to purchase the land. Two days after this meeting, Taylor announced that the family partnership had sold timber rights to Stone Container, Inc., a Chicago-based multinational corporation with a dismal environmental record.[66] Stone's logging operations began in the early fall of 1995 in the southern tributaries of La Sierra.

Initially, Taylor announced that he would "limit" logging to 12 percent of the ranch and that cutting would follow a modified version of Montana's "Best Management Practices" (BMPs) to ensure "watershed maintenance."[67] However, serious questions surfaced regarding Taylor's honesty when the Taylor Family Partnership announced three additional timber contracts for a minimum cut of 70 mbf on 34,000 acres.[68] The dubious distinction of the Taylor Ranch timber sales is that they represent the largest public or private cuts for a contiguous area in Colorado history.[69] According to Taylor's consulting foresters, 44 percent of the land contains approximately 210 mbf of merchantable timber, so logging could conceivably be expanded.[70] Whatever the eventual scope of cuts, to undertake this magnitude of logging, the contractors had to import equipment from the West Coast, including feller-forwarder and debranching machines.

Scientific critics observe that Stone's "forest management guidelines" are destroying large quantities of small trees and ground cover, exposing the remaining trees to damaging windthrows, increasing erosion from roads built on steep slopes, and damaging soil by prodigious ground-based skidding of logs.[71] The loggers have also failed to remove large slash piles from streamside zones. Hauling contractors have failed to mediate road vibrations that have destabilized several adobe homes or compensate the county for damages to local roads and bridges. The most serious complaints by villagers are that logging trucks ignore speed limits and fail to obey posted stop signs. Such reckless driving has caused near collisions with a school bus and several cars.[72] As Colorado does not regulate timbering operations and logging contractors often act irresponsibly, Taylor and Stone are adversely impacting the upland ecosystem and public property. The inequity of this situation is leading to more animosity and tension.

Other groups had foreseen the inevitability of logging and other types of extractive development within the Culebra Range. Any hope for local regulation of high impact development requires a land use master plan and enforceable code. For Costilla County, this was a major undertaking, as it is one of the last Colorado counties to regulate development. Early in the 1970s, the county was encouraged to create a land use code, but local officials were reluctant to sign off on any proposal. In 1991, an attorney, assigned to PAES by the Land and Water Fund of the Rockies, customized and updated the state Senate Bill 1041's county-level land use regulations.[73] In 1993, after two years of public hearings, regulations were adopted. In June 1995, the Costilla County Conservancy Distirct (CCCD) began working with the County Commissioners to establish a Watershed Protection District and a planning commission. Always resistant to any attempt to address land use in Costilla County, Taylor continues to protest that his private property and civil rights are being violated. Not only has Taylor written a series of letters threatening to sue the county if regulations are enacted, he currently has all public meetings videotaped to intimidate the planning commission and has demanded the resignation of one member.

As Taylor Ranch logging operations expanded, regional environmentalists began to take an active interest in La Sierra. In June 1996, the "Salva tu Sierra" (Save Your Mountain) Coalition joined local activist groups, clergy, some county officials, ranchers, and farmers with regional environmental groups. This later became the Culebra Coalition in October 1996. The birth of this coalition may mark the start of a trend by radical environmentalists to interface their "direct action campaigns" with Rio Arriba activists who are fighting to protect their village ecosystems. The direct action campaign launched by the coalition has so far included a half-dozen road blockades, lockdowns, public rallies, protest marches, and nonviolence workshops.[74] For the first time, a ranching community made the pages of the *Earth First! Journal* in a series of sympathetic articles portraying the direct action campaign as a struggle to protect a sustainable local culture and its endangered watershed ecosystem.

The political situation surrounding the Taylor Ranch is complex. Currently, the *Rael v. Taylor* case is slated for district court late in 1998. While there is always optimism for a return of use rights, activists understand that this lawsuit will not resolve the county's environmental problems. In this respect, the implementation of land use regulations is of paramount importance. The CCCD is exploring the possibility of legal action (through the Denver office of the Sierra Club Legal Defense Fund) to mediate logging damage to the *acequias*.

Meanwhile, in January 1997, the Sangre de Cristo Land Grant Commission reconvened to review a state-prepared "letter of intent" that was subsequently delivered to Taylor. The letter established a purchase price not to exceed $20 million, called for inspection of timber deeds and property by state foresters and scientists designated by the local community, and curtailed any future logging or mining contracts.[75] The state also added several "contingencies," including a requirement that locals and environmentalists withdraw from active opposition to the logging; the state officials insist that the acquisition and timber sales are separate issues.[76] A point of contention is the fact that the state is willing to buy the land with all current timber contracts in place, the last of which expires in 2003. The buy-out will have to overcome several other difficult hurdles including multiphased appropriations from a state legislature that is dominated by conservative Republicans, a difficult feat considering that the politically powerful hunting and outdoor recreation lobby has charged that the Land Grant Commission's 1993 recommendations were "promoting special rights for Hispanics."[77]

Needless to say, a key issue for the heirs is the status of *Rael v. Taylor*. The state is insisting that the LRC drop the lawsuit. Theoretically, dropping *Rael v. Taylor* will return a use rights easement to the community. However, LSF must raise $5 million to $7 million to purchase the easement despite the fact that the use rights litigation will reduce the cost of the land. This is not merely an economically difficult proposition, it is an inequitable predicament. Rather, the LRC would like to see the use rights transferred by Taylor to the heirs *as* property rights *before* the state obtains fee simple title to avoid "special rights" opposition. But the state not only wants the lawsuit dropped before it will agree to integrate the use rights into a management plan for La Sierra, it contends that the LRC must "quantify" the use rights. The fear here is that the level of use rights will be diminished under political pressure or will fall short of meeting the needs of the contemporary community. As the desire for a locally managed commons is strong, activists are concerned that Colorado's conservative legislature and various lobbyists will eventually take full control of the management of La Sierra.

There are legitimate reasons for local mistrust of the state. Under pressure from Forbes and Taylor, in 1989 the Colorado Division of Wildlife (DOW) coordinated a "poaching" sting operation to control locals who were allegedly trespassing to kill elk and other wildlife. A federal undercover agent encouraged locals to hunt illegally. Over 270 federal and state agents conducted a predawn commando-style raid on the villages to serve 53 warrants. Not only was there a helicopter, the agent commandos controlled the perimeter around the villages and in military fashion sought out individuals to serve a simple warrant while terrorizing innocent residents. Most disturbing was the stereotyping produced by the national media coverage which continues to be repeated out of context by Tom Wolf.[78] It is no wonder that many heirs mistrust the state and doubt whether it will deal fairly with their claims to the Taylor Ranch.

Taylor's duplicitous behavior continued to resurface during the course of the 1993–97 negotiations, usually in the form of contractual conditions deemed unacceptable to the state and locals alike. While Taylor resists any appraisal that includes the value of the use rights, he usually haggles over survey costs and tax loopholes. However, the full appraisal process (which includes reductions for logging damage and the value of the heirs' use rights) remains higly contested. Taylor has not only resisted a scientific site inspection of the watershed, he has threatened to withhold parts of the timber contracts, and steadfastly refused to allow an appraisal that includes calculating the value of the use rights. For the people of the Culebra villages, the objective is not just restoring the use rights but reclaiming the watershed commonwealth tradition of local management of an upland habitat that has long sustained the agroecological traditions of an important bioregional community.

CONCLUSION

The struggles of Chicanos in the Tierra Amarilla and Sangre de Cristo Land Grants hold valuable lessons for the enviromental movement.

The case of Ganados demonstrates that newly arriving, predominantly white, middle-class environmentalists are often disinterested in the environmental and economic history of the bioregion. This insensitive stance disregards the actions by locals to protect the environment against corporate exploitation; it also ignores the land tenure struggles of village cultures.[79] In the decade preceding the 1891 enclosure of the common lands and the expansion of extractive industries, Tierra Amarilla was transformed in ways that the community has yet to accept. Just after the turn of the century, the federal government expropriated several land grants in Taos and Rio Arriba counties to create the Carson National Forest. While some may applaud President Theodore Roosevelt for this conservation effort, in the context of Taos and Rio Arriba counties, this was an illegal and unjust taking of Spanish–Mexican land grants. Not only did the National Forest system sever the agropastoralists from the upland commons, it also imposed a biased policy that incrementally reduced *ranchero* husbandry in favor of extractive commercial industries.

Western resources have long been controlled by a corporate collective that has effectively implemented state-subsidized multiple-use policies in the management of public lands. As extractive resource colonialism became dominant, the former self-reliant villagers were transformed into a rural proletarian work force. Over three generations (between 1891 and the 1960s), *mexicanos* worked for the railroad or logging, mining, livestock, and agricultural industries. The great environmental paradox for the Chicanos of the Rio Arriba is that enclosure brought ecological ruin to watersheds that had been sustainably managed for numerous generations. And yet, as waged workers, rural Hispanos may have inadvertently participated in the processes that degraded their own upland commons.

Tourism emerged as a mainstay of the regional economy of New Mexico at the turn of the century with the establishment of the Taos Art Colony.[80] After decades of tourism development, by the 1970s, the cultural landscapes of northern New Mexico had become part of the exploitable "spectacle" that led to the economic and cultural transformation of the bioregion. And it was once again Chicanos who were the preferred labor pool for the tourist amenity and outdoor recreation service sectors. Since these are largely part-time, seasonal, and minimum-wage jobs, Chicanos remained at the bottom of the labor market and experienced diminishing earning power in the face of the higher cost of living associated with a tourism economy. As wealthy newcomers (regardless of their environmental ethics) settle in the Rio Arriba, they are gentrifying the village cores and in the process are driving up the price of real estate and subsequently increasing property taxes. This process is accelerating the rate at which the traditional villagers are displaced from their homes and the remaining land base.[81]

This is the historical context that environmentalists overlooked in the ill-advised conflict with Ganados. Moreover, the struggle between the descendants of displaced agropastoralists and state land managers (influenced by the powerful conservation and outdoor sports and recreation interest groups and their lobbies) has created a lingering and deep resentment because the plight of Ganados in its search for grazing range was ignored. As a major regional employer for the villages, it seems inconceivable that the state land managers and progressives in the environmental movement would not want to do everything they could to encourage Ganados in its quest for self-sufficiency. The political blunders by The Nature Conservancy, the underhanded dealings of the Sierra Club Foundation, and the biased writings of environmentalists like Tom Wolf demonstrate that some sectors of the nature conservation movement are insensitive and elitist. Their myopic views, which exclude humans from their place in the ecosystem, have damaged the prospects for a meaningful coalition of villagers and environmentalists in New Mexico. It will take years to heal this rift.

The case of the Culebra villages demonstrates that land-based communities are sources of independent environmental organizing. Because environmentalists often only pay lip service to communities of color that are threatened by extractive industries, the villagers have to self-organize the environmental battles (often without the adequate technical and financial support available to middle-class white ecoactivists). But Costilla County is a good example of how a core group of multigenerational residents cultivated alliances and formed effective coalitions with various sectors of the environmental movement. We posit that this was, in part, a direct result of the highly polarized "Taylor versus the Culebra villagers" political millieu.[82] The resistance of the community was an act of self-preservation, and it created an exception to the pattern of conflict between land grant activists and environmentalists. This explanation also underscores the volatile issue of private property rights versus community welfare. In this context, the *acequias* have become a powerful symbol of the struggle to protect an endangered watershed ecosystem and the equally threatened local culture it sustains from damage by private exploitation of natural resources. The community (and ecological) value of the *acequias* is counterposed against the extractive imperatives of private property.[83] The struggle for La Sierra shows that environmentalists can recognize and respect local claims to the upland commons, work with historic *acequia* communities on ecological protection and restoration, and restore the place of humans in nature and the place of nature in sustaining right livelihoods. Here, environmental and social justice ethics intersect with amazing clarity.

In the case of La Sierra, the common land was never converted to the public domain. Effective enclosure by a private landowner did not occur until well past the mid-twentieth century, unusually late for the intermountain West. Because the villagers had access to the upland commons for so long, they were able to sustain an economy of independent agropastoralists and a corresponding network of artisans and tradespeople. As a result, heirs who remained in the villages never became, en masse, waged workers. Bound together by an agropastoral economy and a legacy of fighting to defend water rights and the land base, it is little wonder that the Culebra villagers have forged their own strong "sense of place" and a "deep interest in the watershed."[84] While sympathetic observers view this as "very rare," it is in fact a common denominator of village social fabric in the Rio Arriba. It is the tradition of the *acequia* commonwealth form of self-governance that encouraged the heirs of the Sangre de Cristo Land Grant to find common ground with environmentalists. This fact can become the basis for unity between Chicanos and environmentalists, if we can educate the green movement about the unique cultural and ecological history of the Rio Arriba and the important role traditional local cultures play as stewards of the upland commons.

On the periphery of this struggle is none other than Tom Wolf, whose research supports the brokering of a plan for the management of the Sangre de Cristo Mountain Range (including La Sierra) to "honor and reward private owners who achieve ecological excellence," as if a 100 mbf timber cut on a mere 34,000 acres is a sound example of conservation biology. Two of these owners, singled out by Wolf as "unlikely environmental heroes," are none other than the "cantakerous Jack Taylor and his feisty neighbor, Malcolm Forbes." Wolf proposes that the "new ecology," coupled with the "market economy, individual liberty, and private property rights," will solve everyone's problems. The presence of such self-styled "environmental interlopers" is another example in which the politically correct, better-educated ecoactivist shines the guiding light of the enlightened market upon the darkened niches of a bioregion despoiled by ecologically backward local cultures.[85] In the context of state and local negotiations for the acquisition of La Sierra, other community members continue to question the openness of the entire process. Whatever the outcome, a continuing problem with this entire political milieu is that discussion and opinion framing are often dominated by nonheirs and even the apologists of corporate domination; there is only limited opportunity for input by the heirs or the LRC board.

Both case studies pose challenges for environmentalists and land grant activists hoping to integrate local communities into land use management decisions. There are precedents for this. For example, the 48,000-acre Blue Lake Wilderness, in the Carson National Forest above Taos Pueblo, is a prototype for the transfer of the responsibility for the management of public lands to native peoples with long-standing historical and cultural ties to the land. More recently, there are plans by the New Mexico Wilderness Coalition to integrate Native American and Hispano communities in the management of any new areas proposed for wilderness designation. Getting to local self-management of watershed ecosystems will require more than an expression of solidarity from environmentalists and wilderness advocates. Land grant activists have called for special congressional hearings to reexamine the status of Spanish and Mexican land grants in New Mexico and Colorado.[86]

When native peoples and local cultures are excluded from maintaining sustainable traditions and management of the upland commons, the results are overbearingly destructive: local economies are shattered, poverty and unemployment increase, artisan and ethnoscientific knowledge traditions are undermined, and multigenerational farming families move out of rural areas and into cities.

The political failure here derives from the discouragement of genuine alternatives for sustainable development with cultural integrity under the rules of a common property regime. If we as Americans do not defend culturally protected land use spaces, we will commit several strategic errors. First, we will negate the value of traditional land use practices that can contribute to the ecological health of wild lands and watersheds. Second, we displace autochthonous land stewards and erode local knowledge.[87] Third, we ignore the legal standing of common property rights (like those embodied in the *acequias*), which are an overlooked yet powerful tool in the political struggle over "takings" that has come to dominate so much of the land use discourse in the 1990s. We must not neglect the role of land-based communities in strengthening the defense of the rules of the natural resource commons against the environmentally damaging effects of privatization and commodification. For example, one legal theorist has argued that some "consumptive uses" (like logging, mining, and subdiviving) can be banned as a "wasteful" use of private property, especially when a strong local standard of "good husbandry" of resources can be invoked.[88]

Finally, by overemphasizing the protection of artificially managed wilderness remnants, we continue to promote a false sense of security about our environmental objectives. Landscapes under National Forest, Wilderness, or Wildlife Refuge systems have been and continue to be exploited by multinational corporate capital. Witness, for example, the 830 million tourists who visited National Forests in 1996, not to mention the federal logging and mining subsidies that promoted overextraction. In other words, the designation of public and private lands into the national wilderness and wildlife management systems does not resolve the more holistic problem of the political economic contradictions within rural and urban development patterns in the intermountain West. Nor does it address the impact of such policies on rural ethnic communities. The strategy might "save" a few remnants that can never truly fulfill the unrequited dreams of wilderness enthusiasts for an experience involving some "original unity" with nature, but it cannot be the guiding premise of a more holistic model for sustainable development, cultural survival, and protection of wild lands. It also removes from rural areas those who are the strongest "natural" allies of environmentalists, native land-based peoples. The mainstream, nature conservation position can be commended for seeking to protect wild lands and biodiversity. However, biodiversity conservation strategies must address the way in which narrowly construed approaches (that exclude humans) can penalize rural land-based communities by denying them access to common sovereign space they need to sustain their cultural traditions and historic usufructuary rights.[89]

The implications for the future of the environmental movement in the U.S. are clear: either conserve culture and nature together and establish a long-term blueprint for cultural and ecological renewal, or enjoy the temporary delusion of a victory for the preservation of a (nonexistent) pristine nature, and ignore the broader threat of advanced capitalism on our watersheds. If the conservation movement opts to be proactive, it will refocus priorities on empowering effective local land use management and promoting employment of rural communities in well-funded efforts to restore damaged ecosystems. However, if the movement prefers to pursue a more narrowly defined agenda and ignores injustice and inequality, it will merely reduce nature and culture to the simulacra of "gardens in the midst of blight"—a model of ecopolitics that insults and trivializes the land ethics and naturalist sensibilities of the Chicanos of the Tierra Amarilla and Sangre de Cristo Land Grants.

The unfolding dilemma in the Upper Chama and Culebra watersheds should behoove environmentalists to recognize the intersections of class and racial differences in the articulation of movement values, goals, and strategies. In the cases examined, two endangered local cultures show the wisdom of a struggle to maintain a sense of place and regain common property resources. This strategy emphasizes ecologically sustainable land use traditions as a communal alternative to capitalist industrialization and proletarianization. The environmental movement should take a closer look at such case studies to gain a better understanding of ways it can help rural communities in struggles to protect their watersheds. For example, they could funnel resources to local community groups in the Rio Arriba and elsewhere to promote ethnoscientific and ecologically sound planning.

The cases we outlined in this chapter offer hope for "reconciliation" by linking the struggle for environmental protection with the movement for social justice, workplace democracy, and rural self-determination—all contributions on the road to cultural and ecological democracy in the Southwest.

ACKNOWLEDGMENT

An earlier version of this chapter appeared in *Capitalism, Nature, Socialism, 3*(1), 1992, pp. 79–103.

NOTES

1. See Robert D. Bullard, ed., *Confronting Environmental Racism: Voices from the Grassroots* (Boston: South End Press, 1993); Robert Gottlieb, *Forcing the Spring: The Transformation of the American Environmental Movement* (Washington DC: Island Press, 1993); Bullard, ed., *Unequal Protection: Environmental*

Justice and Communities of Color (San Francisco: Sierra Club Books, 1994); B. Bryant, ed., *Environmental Justice: Issues, Policies, and Solutions* (Washington DC: Island Press, 1995); and Laura Pulido, *Environmentalism and Economic Justice: Two Chicano Struggles in the Southwest* (Tucson: University of Arizona Press, 1996).

2. The Group of Ten includes the Sierra Club, The Nature Conservancy, The National Wildlife Federation, The Wilderness Society, The World Wildlife Fund, The Natural Resources Defense Council, The Environmental Defense Fund, The Izaak Walton League, The National Audubon Society, and The National Parks and Conservation Association. See Gottlieb, *Forcing the Spring*, especially Chapter 4; and Pulido, *Environmentalism*, pp. 20–24.

3. The Environmental Careers Organization, *Beyond the Environment: Redefining and Diversifying the Environmental Movement* (Boston: Author, 1992), pp. 157–177.

4. Ibid., p. 159; see also Bullard, ed., *Confronting Environmental Racism.*

5. Alternative Policy Institute of the Center for Third World Organizing, *Toxics and Minority Communities*, Issue PAC #2 (Oakland, CA: Center for Third World Organizing, July 1986). See also General Accounting Office, United States Congress, *Siting of Hazardous Waste Landfills and Their Correlation with the Racial and Economic Status of Surrounding Communities* (Washington, DC: U.S. Government Printing Office, 1983); Commission for Racial Justice, *Toxic Wastes and Race in the United States* (New York: United Church of Christ, 1987); Bullard, ed., *Confronting Environmental Racism*; Bryant, ed., *Environmental Justice.*

6. A. Hurley, *Environmental Inequalities: Class, Race, and Industrial Pollution in Gary, Indiana, 1945–1980* (Chapel Hill: University of North Carolina Press, 1995), p. 172.

7. To quote a colleague at the First National Conference on Cultural Conservation, American Folklife Center, Library of Congress, Washington, DC, May 1990. Also see W. Cronon, ed., *Uncommon Ground: Towards Reinventing Nature* (New York: Norton, 1995).

8. While some rationalize ecotourism as an alternative method of preserving endangered habitats, data suggest that trekkers promote deforestation and trash removal and sewage problems. For an overview see Haunani Kay Trask, *From a Native Daughter: Colonialism and Sovereignty in Hawai'i* (Monroe, ME: Common Courage Press, 1993); see also *Cultural Survival Quarterly* (Spring 1990 issue on tourism).

9. There is a growing body of evidence showing that indigenous communities throughout the world are being displaced not merely by capitalist maldevelopment but by the "nature conservation and appreciation" industries. See, for example, Ramachandra Guha, "Radical American Environmentalism and Wilderness Preservation: A Third World Critique," *Environmental Ethics, 11*(1), 1989. See also Devon Peña, ed., *Chicano Culture, Ecology, Politics: Subversive Kin* (Tucson: University of Arizona Press, 1998), especially Chapters 1 and 2.

10. H. Rothman, "Cultural and Environmental Change on the Pajarito Plateau," *New Mexico Historical Review, 64*(2), 1989, pp. 211–212.

11. W. deBuys, *Enchantment and Exploitation: The Life and Hard Times of a New Mexico Mountain Range* (Albuquerque: University of New Mexico Press, 1985).

12. For detailed commentary related to this issue, see S. Rodriguez, "Land, Water, and Ethnic Identity in Taos," in C. Briggs and J. R. Van Ness, eds., *Land,*

Water, and Culture: New Perspectives on New Mexico Land Grants (Albuquerque: University of New Mexico Press, 1987), pp. 354–357.

13. T. Wolf, *Colorado's Sangre de Cristo Mountains* (Niwot: University Press of Colorado, 1996), p. 265; emphasis added.

14. Wolf, ibid., pp. xx, 265, and 280. For a detailed critique of the environmental history of the Sangre de Cristo Mountains of Colorado and New Mexico, see Devon Peña, *Gaia in Aztlan: The Politics of Place in the Rio Arriba* (Tucson: University of Arizona Press, in press). Interestingly, in a study for the USFS, Dortignac presents data indicating that the grasslands of the Sangre de Cristo Mountain Range were not historically overgrazed and that before the turn of the century (1891, when the USFS was established), 80 percent of the forest stands in the Sangres were old-growth or late successional communities; see E. J. Dortignac, *Watershed Resources and Problems of the Upper Rio Grande Basin* (Albuquerque, NM: Field Station Unit, Rocky Mountain Forest and Range Experiment Station, Department of Agriculture, USFS, 1956).

15. Wolf, *Colorado's Sangre de Cristo Mountains*, p. xx.

16. M. Ebright, *Land Grants and Lawsuits in Northern New Mexico* (Albuquerque: University of New Mexico Press, 1994), p. 90.

17. Ibid., p. 90.

18. For an incisive critique of the "tragedy of the commons," see Michael Goldman, "Tragedy of the Commons or the Commoners' Tragedy?" *Capitalism, Nature, Socialism, 4*(4), 1993.

19. See Arne Naess, "The Shallow and the Deep, Long Range Ecology Movement: A Summary," *Inquiry, 16*, 1973; W. DeVall and G. Sessions, *Deep Ecology* (Salt Lake City: Peregrine Smith, 1985); and W. DeVall, *Simple in Means, Rich in Ends* (Salt Lake City, UT: Peregrine Smith, 1988).

20. For early discussions of the concept of bioregion, see DeVall, *Simple in Means*, pp. 57–69; and Kirkpatrick Sale, *Dwellers in the Land: The Bioregional Vision* (San Francisco: Sierra Club Books, 1985).

21. J. Cheney, "Postmodern Environmental Ethics: Ethics as Bioregional Narrative," *Environmental Ethics, 11*(2), 1989. For discussions of place-centered environmental ethics in the context of the Upper Rio Grande watershed, see J. B. Jackson, *A Sense of Place, a Sense of Time* (New Haven: Yale University Press, 1994), especially Chapters 2, 3, and 4; C. H. Wilkinson, *The Eagle Bird: Mapping a New West* (New York: Pantheon, 1992), especially Chapter 10; and Peña, *Gaia in Aztlan.*

22. W. Berry, "The Work of Local Culture," in *What Are People For?* (San Francisco: North Point Press, 1990), p. 166.

23. On the intersections of reductionist science and technology with capitalist development, see Vandana Shiva, *Staying Alive: Women, Ecology and Development* (London: Zed Books, 1988).

24. On the *acequia* and riparian long-lot cultural landscapes of the Upper Rio Grande, see D. Peña, "Cultural Landscapes and Biodiversity: The Ethnoecology of a Watershed Commons in the Upper Rio Grande," in V. Narvaez-Sandoval, ed., *Ethnoecology: Located Knowledge, Situated Lives* (Tucson: University of Arizona Press, in press). For further discussion of the role of cultural landscapes in the formation of landscape linkages and biological corridors, see W. E. Hudson, ed., *Landscape Linkages and Biodiversity* (Washington, DC: Island Press and Defenders of Wildlife, 1991), especially Chapters 9–10.

25. See, for example, K. Oberg, "Cultural Factors and Land-Use Planning in Cuba Valley, New Mexico," *Rural Sociology*, *5*(2), 1940; A. G. Harper, A. Cordóva, and K. Oberg, *Man and Natural Resources in the Middle Rio Grande Valley* (Albuquerque: University of New Mexico Press, 1943); P. Van Dresser, "The Bio-Economic Community: Reflections on a Development Philosophy for a Semi-Arid Environment," in C. Knowlton, ed., *Indian and Spanish-American Adjustments to Arid and Semi-Arid Environments* (Lubbock: Texas Technological College Press, 1964); Knowlton, "Land Grant Problems among the State's Spanish-Americans," *New Mexico Business, 20*(6), 1967; Knowlton, "Changing Spanish-American Villages of Northern New Mexico," *Sociology and Social Research, 53*(2), 1969; O. Leonard, *The Role of the Land Grant in the Social Organization and Social Processes of a Spanish-American Village in New Mexico* (Albuquerque, NM: Calvin Horn, 1970); M. Rock, "The Change in Tenure New Mexico Supreme Court Decisions Have Effected upon the Common Lands of Community Land Grants in New Mexico," *The Social Science Journal, 13*(13), 1976; F. L. Brown and H. Ingram, *Water and Poverty in the Southwest* (Tucson: University of Arizona Press, 1987); Briggs and Van Ness, eds., *Land, Water, and Culture*; R. Martínez, "Chicano Lands: Acquisition and Loss," *Wisconsin Sociologist, 42*(2–3), 1987; Ebright et al., *Land Grants*; Peña, "Cultural Landscapes and Biodiversity."

26. Knowlton, "Land Grant Problems," pp. 3–4.

27. S. Deutsch, *No Separate Refuge: Culture, Class, and Gender on an Anglo-Hispanic Frontier in the American Southwest, 1880–1940* (New York: Oxford University Press, 1987), p. 21. The grasslands of the Upper and Middle Rio Grande watersheds were left largely intact during the Spanish and Mexican periods. Gelbach cites an 1846 eyewitness account (by William Emory and John Bartlett) just prior to annexation of Mexico by the United States to conclude that "rapid and extensive demise of grassland is the general outcome of Anglo settlement . . . in concert with the postwar and railroad era of Anglo ranching" (F. R. Gelbach, *Mountain Islands and Desert Seas: A Natural History of the U.S.–Mexican Borderlands*, 2nd ed. [College Station: Texas A&M University Press, 1993], pp. 109–111).

28. See U.S. Forest Service, *The Western Range* (74th Congress, 2nd Session, Document No. 199; Washington, DC: U.S. Government Printing Office, 1936); P. H. Roberts, *Hoof Prints on Forest Ranges: The Early Years of National Forest Range Administration* (San Antonio, TX: Naylor Publishing, 1963); J. R. Van Ness, "Hispanic Land Grants: Ecology and Subsistence in the Uplands of Northern New Mexico and Southern Colorado," in Briggs and Van Ness, eds., *Land, Water, and Culture*; Deutsch, *No Separate Refuge;* S. Forrest, *The Preservation of the Village: New Mexico's Hispanics and the New Deal* (Albuquerque: University of New Mexico Press, 1989); Dortignac, *Watershed Resources*; Harper, Cordova, and Oberg, *Man and Natural Resources*; Leonard, *Land Grant*; and Gelbach, *Mountain Islands and Desert Seas*.

29. V. Westphall, *The Public Domain in New Mexico, 1854–1891* (Albuquerque: University of New Mexico Press, 1965).

30. See Deutsch, *No Separate Refuge*; Forrest, *Preservation.*

31. For further discussion, see M. Stoller, "Grants of Desperation, Lands of Speculation: Mexican Period Land Grants in Colorado," in J. Van Ness and C. M. Van Ness, eds., *Spanish and Mexican Land Grants in New Mexico and Colorado* (Manhattan, KS: Sunflower University Press, 1980); V. Westphall, *Mercedes Reales* (Albuquerque: University of New Mexico Press, 1983); M. Ebright et al., "Spanish

and Mexican Land Grants and the Law," *Journal of the West* 27(3), 1988; Ebright, *Land Grants*; R. D. García and T. Howland, "Determining the Legitimacy of Spanish Land Grants in Colorado: Conflicting Values, Legal Pluralism, and Demystification of the Sangre de Cristo/Rael Case," *Chicano–Latino Law Review, 16*, Winter 1995, pp. 39–68; M. Mondragon-Valdéz, *El Mundo Aparte*, paper presented at the 1997 Annual Conference of the Western Social Science Association, Albuquerque, NM, April 1997.

32. For details on these federal programs, see P. van Dresser, *A Landscape for Humans: A Case Study of the Potentials for Ecologically Guided Development in an Uplands Region* (Santa Fe, NM: Lighting Tree Press, 1976); on federal intervention during the 1930s, see S. Forrest, *The Preservation of the Village: New Mexico's Hispanics and the New Deal* (Albuquerque: University of New Mexico Press, 1989).

33. Translated by D. Peña from a traditional New Mexican folktale, "El Espíritu del Bosque" (The Forest Spirit), as recorded in the Rubén Cobos New Mexican Folklore Collection (Reel 40.B.2d), Tutt Library, Special Collections, Colorado College. This folktale is particularly popular in the Rio Arriba (Rio Chama basin) and the regions around Santa Fe and Albuquerque.

34. On the "Santa Fe Ring," see M. Ebright, *The Tierra Amarilla Land Grant: A History of Chicanery* (Santa Fe, NM: The Center for Land Grant Studies, 1980); and V. Westphall, *Thomas Benton Catron and His Era* (Tucson: University of Arizona Press, 1973). For more on the history of the Tierra Amarilla Land Grant, with special reference to the region's cultural landscapes, see C. Wilson and D. Kammer, *Community and Continuity: The History, Architecture and Cultural Landscape of La Tierra Amarilla* (Santa Fe: New Mexico Historic Preservation Division, 1989).

35. Valentina Valdéz and Maria Varela, *25 Years of La Clínica's History* (Tierra Amarilla, NM: La Clínica del Pueblo, 1989), p. 3.

36. On the Tierra Amarilla Courthouse Raid, see R. Gardner, *Grito!: Reies Tijerina and the New Mexico Land Grant War of 1967* (New York: Harper Colophon, 1971); and R. J. Rosenbaum, *Mexicano Resistance in the Southwest: The Sacred Right of Self-Preservation* (Austin: University of Texas Press, 1981). After years of frustration (in the courts and through complaints to the state) regarding enclosure of common land, reduction in grazing permits for small share holders, and increased grazing by large land owners, the community organized La Alianza Federal de Mercedes (Federal Alliance of Land Grants). On June 5, 1967, La Alianza leaders were arrested for an alleged "unlawful assembly." Ironically, this supposed assembly was yet to occur. Tijerina and his followers attempted to make a citizens' arrest of the District Attorney and other local officials over the unconstitutional action. In the process, violence broke out at the courthouse and a "shoot-to-kill" order was issued. This "manhunt" involved 2,000 members of the National Guard, state police, game wardens, and sheriffs from a five-county area using helicopters, five armored tanks, and mounted patrols.

37. Valdéz and Varela, *25 Years*, p. 8.

38. Ganados received assistance from Utah State University in this regeneration program. *Churro* were introduced to New Mexico from Spain in the 1600s. The *churro* breed was prized for its hardiness in an arid environment and for its long, relatively greaseless, wool. After the Civil War, the *churro* was crossbred with merino sheep. By 1970, the *churro* was nearly extinct.

39. M. Varela, "Economic Growth versus Development of Rural Communities: It Means the Difference between Local and Outside Control," *The Workbook*,

19(2), 1994, pp. 58–59. Ganados has created several spin-off operations, including a general store, organic meat processor, and wool-washing operation. Ganados also sponsors three "programs" (Churro Breeding and Livestock Program, Children's Summer Arts, and the Milagro Fund, which provides seed money to promote other regional cooperatives). See Pulido, *Environmentalism*; J. Kutz, *Grassroots New Mexico: A History of Citizen Activism* (Albuquerque, NM: Inter-Hemispheric Education Resource Center, 1989); F. O. Sargent, P. Lusk, J. A. Rivera, and M. Varela, *Rural Environmental Planning for Sustainable Communities* (Washington, DC: Island Press, 1991).

40. See W. Dusenberry, *The Mexican Mesta: The Administration of Ranching in Colonial Mexico* (Urbana: University of Illinois, 1963).

41. Pulido, *Environmentalism*, p. 139.

42. Antonio Manzanares in personal communication to Devon Peña (June 1990, Los Ojos, NM).

43. Pulido, *Environmentalism*, p. 132.

44. See ibid., p. 159.

45. Tom Wolf, personal communication with Devon Peña (March 1990, Colorado Springs, CO).

46. Ganados del Valle, *A Proposal to Improve Wildlife Habitat in the Rio Chama, Humphries, and Sargent Wildlife Areas*, unpublished proposal, 1989. Also see D. F. Thomas, "The Use of Sheep to Control Competing Vegetation in Conifer Plantations on the Downeville Ranger District, Tahoe National Forest," in U.S. Forest Service, U.S. Department of Agriculture, *Proceedings of the Fifth Annual Forest Vegetation Conference* (Sacramento, CA: Author, 1985).

47. Tom Wolf, personal communication with Devon Peña (March 1990, Colorado Springs).

48. See D. D. Jackson, "Around Los Ojos, Sheep and Land Are Fighting Words," *Smithsonian, 22*(1), April 1991, pp. 36–47.

49. Maria Varela, "On the Road to a Healthy Environment," *Noticias Nuevas de Ganados del Valle: Here's What's New at Ganados* (Spring and Summer 1996), p. 13.

50. R. Ring and M. Frie, "In the Heart of the New West, the Sheep Win One," *High Country News, 27*(9), October 16, 1995, p. 14. Also see M. Varela, "Ganados Wins Sierra Club Foundation Settlement," *Noticias Nuevas de Ganados del Valle: Here Is What's New at Ganados* (Spring and Summer 1995).

51. Malcolm Forbes bought the northernmost half (or Trinchera Estate) of the Sangre de Cristo Land Grant in 1969 for a reported $3 million. Forbes subdivided 86,500 out of 267,368 acres of his tract into smaller plots of 5, 10, and 20 acres. He constructed some 400 miles of roads to provide access to his "Sangre de Cristo Ranches," "Forbes' Park," and "Wagon Creek" subdivisions; the rest of the land is managed as a "wildlife ranch" and timber plantation.

52. A. A. Valdéz and M. A. Valdéz, *Historic and Architectural Context for Costilla County, 1851–1941: Costilla County Adobe Survey* (San Luis, CO: Valdéz and Associates, 1991); A. A. Valdéz, *Historic Vernacular Architecture and Settlement Patterns of the Culebra River Villages of Southern Colorado (1850–1959),* master's thesis, University of New Mexico, Albuquerque, 1992; and Peña, "Cultural Landscapes and Biodiversity."

53. The Torrens Act is a legal procedure to "streamline" quiet title action lawsuits that many legal theorists argue violates the constitutional right to due

process. For more on the use of the Torrence Act and local usufructuary rights to the Sangre de Cristo Land Grant, see García and Howland, note 33, above.

54. Hatfield Chilson, who was involved in hearing the original quiet title action, stated in chambers that one of the reasons he ruled in favor of Taylor was that "he wanted to bring the Mexicans into the twentieth century." As quoted in Land Rights Council, *La Merced de Sangre de Cristo: El Valle de San Luis, Colorado* (Chama, CO, 1980), p. 16.

55. A controversial 1994 decision by the Colorado Supreme Court remanded *Rael v. Taylor* back to the district court to be heard by a specially appointed judge. While the case seems to change daily, the trial, tentatively scheduled for the fall of 1998, will address two central issues: first, due process violations affecting 1960 property owners; and second, a class action filing that will address the rights of all locals to use La Sierra.

56. See R. D. García and T. Howland, "Determining Legitimacy," pp. 47–48; also, see M. Stoller, *Preliminary Manuscript on the History of the Sangre de Cristo Land Grant and the Claims of the People of the Culebra River Villages on the Lands*, unpublished manuscript submitted as an affidavit to the district court of Costilla County, Colorado.

57. It was also after the enclosure by Taylor that the Chicano population of Costilla County experienced a dramatic increase in the rate of out-migration. In addition, the value of agricultural production declined by more than half in the aftermath of the enclosure; for a detailed discussion, see Peña, *Gaia in Aztlan*, Stoller, *Preliminary Manuscript*; and Valdéz and Valdéz, *Historic Context*.

58. M. L. Devere and S. Edmunds, *Taylor Ranch Forestry Management Position Statement*, unpublished consulting forester's report (revised April 1996), p. 1.

59. D. Peña, "Latest Aerial Inspection Confirms Timber Devastation," *La Sierra: Costilla County Edition*, 2(2), April 1996, pp. 1 and 4–5.

60. R. Curry, "The State of the Culebra Watershed: The Southern Tributaries," *La Sierra: National Quarterly Edition*, 2(1), Fall 1995.

61. The efforts of both these groups were highlighted in documentaries that aired nationally on public educational broadcasts under the titles of "La Tierra" and "Land of Cool Sun" (the text for the latter was written by John Nichols and narrated by Steve Allen).

62. On the struggle to confront the "cult of expertise" in the BMG campaign, see D. Peña and J. Gallegos, "Local Knowledge and Collaborative Environmental Action Research," in P. Nyden, A. Figert, M. Shibley, and D. Burrows, eds., *Building Community: Social Science in Action* (Thousand Oaks, CA: Pine Forge Press, 1997). D. Peña, R. Martínez, and L. McFarland ("Rural Chicana/o Communities and the Environment: An Attitudinal Survey of Residents of Costilla County, Colorado," *Perspectives in Mexican American Studies*, 4, August, 1993), conducted a survey during the summer of 1990 and found that close to 80 percent of the respondents were opposed to the mine. The underlying opposition was related to concern over damage to water rights and farm production, wildlife habitat, and public drinking water supplies. Also see D. Peña and J. Gallegos, "Nature and Chicanos in Southern Colorado," in Bullard, ed., *Confronting Environmental Racism*. PAES worked with the CES to critique BMG's plan. Later, PAES became a joint objector before Colorado's Mined Land Reclamation Board (with legal assistance from the Law and Water Fund of the Rockies of Boulder, Colorado) after high levels of cyanide killed migratory birds in BMG's tailings pond. A PAES and CCCD legal team forced BMG

to install a $2-million detoxification process. See M. Mondragon-Valdéz, "BMG Cyanide Still above Permitted Levels," *The Valley Voice,* Spring/Summer 1992.

63. According to the American Business Information Directory, the Taylor Family Partnership owns: (1) Taylor Lumber and Land Company; (2) Taylor Publications, Inc., which publishes *Carolina Business*, a monthly magazine for real estate developers; and (3) Oriental Plantations, a residential real estate development firm. Dun and Bradstreet citations indicate that the Taylor family's wealth is locked into ownership of La Sierra, which has an estimated $20 million to $30 million worth of merchantable timber.

64. See D. Peña and R. K. Green, *Revenue Potential and Ethical Issues in the Management of the Culebra Mountain Tract as a Common Property Resource* (San Luis, CO: La Sierra Foundation, 1993).

65. Sangre de Cristo Land Grant Commission, *Sangre de Cristo Land Grant Commission Report* (Denver, CO: Governor's Office, December 1993), p. 17.

66. According to Boulder Earth First!, Stone Container has hundreds of citations by the EPA for pollution. Between 1986 and 1996, Stone was fined for willful violation of federal health and safety standards. This habitual disregard for worker health and safety included three work-related deaths. Currently, the corporation is facing a $2-million fine for violation of the Clean Water Act.

67. See *Montana Best Management Practices: Forest Stewardship Guidelines for Water Quality* (Bozeman: Montana State University, 1991). For a critique, see R. Curry, M. Soulé, D. Peña, and M. McGowan, *Critical Analysis of Montana's Best Management Practices and Sustainable Alternatives*, unpublished report submitted to the Costilla County Land Use Planning Commission (La Sierra Foundation and Costilla County Conservancy District, San Luis, CO, 1996).

68. J. Lloyd, "150 Year-Old Land Dispute Intensifies in Colorado: Community Claims Logging on Private Property Hurts Water Supply for Entire Alpine Valley," *Christian Science Monitor*, March 3, 1997, p. 4.

69. J. Brooke, "In Colorado Valley, Hispanic Farmers Try to Stop a Timber Baron," *New York Times,* March 24, 1997, p. A-8.

70. Devere and Edmunds, *Taylor Ranch Forestry* . . . The actual extent of the logging cuts remains proprietary information. However, aerial inspections suggest a cut of more on the order of 100–110 mbf.

71. For example, see the analysis by Rocky Smith, forestry expert for Ancient Forest Rescue, at the Boulder Earth First! web site: http://bcn.boulder.co.us/environment/earthfirst/LaSierra.

72. Early in March 1997, an improperly licensed and underaged truck driver was killed when his "unchained" load trapped him in the cab after his speeding truck failed to negotiate a sharp and muddy curve and slid off a steep embankment. Other safety violations include balding tires, unlicensed rigs, and overloaded trucks.

73. This refers to legislation passed by the Colorado state legislature (S.B. 1041) enabling local county commissioners to regulate major development activities on private land that may impact wildlife habitat, geological hazards, archaeological or historical resources, and other areas of "state interest." However, mapping (a key aspect for enacting 1041 regulations) was left pending because of a lack of local funding and inaction by the state to assist the county.

74. The Salva tu Sierra Coalition included La Sierra Foundation, Ancient Forest Rescue, Greenpeace-Boulder, New Mexico Direct Action, Earth First!, Forest Guardians, The Wildlands Project, Forest Conservation Council, Sangre de Cristo

Parish, Costilla County Economic Development Council, San Luis Town Council, Vega Board, Land Rights Council, Costilla County Conservancy District, Committee for Environmental Soundness, and Acequia Advisory Board. This was the first time in the history of the environmental movement that local government officials, ranchers, and farmers joined with radical environmental groups like Earth First! and Ancient Forest Rescue (AFR). The coalition's first protest march (on June 10, 1996) was led by Sister Teresa Jaramillo, a San Luis native and medical missionary. For coverage of this historic protest, see " 'Bizarre' Coalition Fights Logging on the Taylor Ranch," *High Country News, 28*(16), September 2, 1996, p. 4. Part of this resistance involves regular reconnaissance flights by LightHawk, the "Environmental Air Force," which have helped to document the extent of old and new logging activities. Local farmers and ranchers act as observers during the lockdowns and road blockades. Salva Tu Sierra Coalition operates a web site at: http://bcn.boulder.co.us/environment/earthfirst/LaSierra.

75. While the purchase price would be based on an actual appraised value (presumably less than $20 million), to sweeten the deal, Taylor is offered a second inflated appraisal of $30 to $35 million to reduce the capital gains taxes by somewhere between $4.5 and $9.5 million.

76. The Colorado State Forest Service assisted Taylor Ranch to mark trees for logging and has justified inadequate timber practices "as a matter of style" while acknowledging that "some of the logging had been too heavy." See Bob Irvine, "Memorandum to the Division of Parks and Outdoor Recreation," Colorado State Forest (February 4, 1997). Many local people feel that the state of Colorado is not opposed to the logging since it opens new "habitat" (grazing and browsing range) for elk, which is a valuable "commodity" species worth millions of dollars in revenues for the Division of Wildlife.

77. Charlie Meyers, "Taylor Ranch: Lousy Value," *The Denver Post* (February 18, 1997), p. C–3.

78. Wolf, *Colorado's Sangre de Christo Mountains*, 266–267.

79. This situation has not gone unnoticed. Speaking before the Santa Fe Sierra Club in May 1996, John Nichols addressed the current animosity between Chicanos and environmentalists. He warned that the Sierra Club must show concern for the economic and social plight of Chicanos; without understanding and empathy "there will be no environmental movement here." In Nichols's view, coming together would require that environmentalists seek "support and agreement and approbation" from the villagers, or they will "be just like Americans trying to win Vietnam." J. Nichols, "Notes for a Sierra Club Dialogue," Unitarian Church, Santa Fe, New Mexico, (May 21, 1996), pp. 1–4.

80. See S. Rodriguez, "Art, Tourism, and Race Relations in Taos: Toward a Sociology of the Art Colony," *Journal of Anthropological Research, 45*(1), 1990, pp. 77–99.

81. See S. Rodriguez, "The Tourist Gaze, Gentrification, and the Commodification of Subjectivity in Taos," in R. Francaviglia, ed., *Essays on Changing Images of the Southwest* (College Station: Texas A&M University Press, 1994), pp. 105–126.

82. S. Horton, "Rich Man, Poor Town: A Classic Tale of the West," *The Reporter,* April 9–15, 1997, pp. 15–19.

83. For interesting commentary on the legal theory informing such a strategy (of natural use as against private property rights), see W. H. Rodgers, "Bringing

People Back: Toward a Comprehensive Theory of Taking in Natural Resources Law," *Ecology Law Quarterly, 10*(2), 1982, pp. 205–252.

84. See R. Curry, "Help Needed in the Beanfields," document on the web site of the Watershed Management Council; consult: http://www.watershed.org.

85. Wolf, *Colorado's Sangre de Cristo Mountains*, pp. xviii, 264. Wolf has a history of promoting "corporate" rights to exploit and "scientifically manage" the San Luis Valley and Sangre de Cristo Mountains. For example, his views on the mining of the valley's aquifer water, for export to urban areas, was recently criticized by Anglo and Chicano farmers. See C. Cannaly, "The Tangled Web We Weave: Undercurrents," *The Valley Voice, 6*(1), Spring 1997, p. 2.

86. This appeal was recently made to Congressman Bill Richardson of New Mexico by land grant activists at the "Wilderness and Culture in New Mexico" day of dialogue held on October 5, 1996 in Santa Fe. Unfortunately, Richardson recently left Congress to represent the U.S. in the United Nations and no one else has emerged to take leadership on this effort.

87. See Peña, *Gaia in Aztlan*, for more discussion.

88. Rodgers, "Bringing People Back," p. 249.

89. Likewise, we need to promote conservation easements to accommodate owners of large estates. Through federal and state tax incentives, private land owners can be encouraged to promote historic usufructuary rights in an environmentally sound fashion.

Index

Contributors

Paul Almeida is a PhD candidate in Sociology at the University of California at Riverside. He has been an active participant in both the ecology and Central American solidarity movements. His research interests center on political sociology and social movements. Currently, he is writing a dissertation on waves of popular insurgency in twentieth-century El Salvador.

Judi Bari was a prominent Earth First! organizer who sought to incorporate the concerns of timber workers into the struggle to protect the ancient redwoods of Northern California. On May 24, 1990, she received worldwide attention, when, because of her organizing efforts, a bomb which had been planted beneath the seat of her car exploded. Despite suffering permanently debilitating injuries from the blast, Judi continued her work. She organized the September 15, 1996, demonstration at Headwaters Forest, the largest mass-action of civil disobedience over a forest issue in the history of the United States. She died of breast cancer in March of 1997. Judi's voice lives on in her authored work *Timber Wars* (Common Courage Press, 1995), and the audio CD *Who Bombed Judi Bari?: Spoken Word with Music Too* (Alternative Tentacles, San Francisco, 1997).

Douglas Bevington is a forest activist in the Pacific Northwest.

Giovanna Di Chiro is Assistant Professor of Environmental Studies at Allegheny College, Meadville, Pennsylvania. Her work focuses on environmental and social justice issues in cross-cultural and international perspectives. Dr. Di Chiro is currently at work on a book examining the emergence of alternative forms of environmental expertise in transnational environmental justice struggles.

Michael Dreiling is Assistant Professor of Sociology at the University of Oregon in Eugene and is active in both the labor and environmental movements. His publications include a number of articles on the mobilization of labor and environmentalists against the North American Free Trade Agreement (NAFTA). Dr. Dreiling's current research continues the focus on the politics of trade, including the formulation of a network analysis of corporate political unity around NAFTA and its 1998 global counterpart, the Multilateral Agreement on Investment (MAI).

Daniel Faber is Assistant Professor of Sociology at Northeastern University in Boston, and is a longtime social and environmental activist. In 1984, he cofounded the Environmental Project on Central America at the Earth Island Institute, and served as the Project's Research Director until 1990. A cofounding editor and coordinator of the Boston Editorial Group of the journal *Capitalism, Nature, Socialism*, Dr. Faber is also a participating editor with the journals *Latin American Perspectives* and *Organization and Environment*. His publications include *Environment Under Fire: Imperialism and the Ecological Crisis in Central America* (Monthly Review Press, 1993).

Rodger C. Field is an environmental attorney in Chicago. He is a graduate of Indiana University School of Law where he was Editor-in-Chief of the *Indiana Law Journal*. His previous publications include "Children, Community and Pollution Control: Toward a Community-Oriented Environmentalism," *Childhood*, 2(1/2), 1994; "Post-Fordist Environmentalism," *EcoSocialist Review*, 7(2), Summer, 1993; and "Bringing Class Back In: Race, Class and Environmental Justice," *EcoSocialist Review*, 8(1), Spring, 1994.

John Bellamy Foster is Associate Professor of Sociology at the University of Oregon in Eugene. He is Coeditor of the journal *Organization & Environment*, a member of the editorial board of *Capitalism, Nature, Socialism*, and a member of the board of the Monthly Review Foundation. Dr. Foster is the author of *The Vulnerable Planet* (Monthly Review Press, 1994), as well as numerous books on the political economy of the United States and international capitalism.

Al Gedicks is Professor of Sociology at the University of Wisconsin, La Crosse, and a longtime environmental/Native rights activist in the upper Midwest. He has served as the Director of the Center for Alternative Mining Development Policy and as the Executive Secretary of the Wisconsin Resources Protection Council. Dr. Gedicks is also the author of *The New Resource Wars: Native and Environmental Struggles against Multinational Corporations* (South End Press, 1993). His latest video, *Keepers of the Water*, documents the Indian–environmental resistance to Exxon/Rio Algom's proposed mine next to the Mole Lake Chippewa reservation in Wisconsin.

Charles Levenstein is Professor of Work Environment Policy at the University of Massachusetts at Lowell and Editor of *New Solutions: A Journal of Environmental and Occupational Health Policy*. A leading authority on the political economy of occupational health issues in the United States and in Eastern and Central Europe, Dr. Levenstein also holds academic appointments at the Center for Occupational and Environmental Health Policy at De Montfort University in Leicester, England; the Central European University in Budapest, Hungary; the Harvard School of Public Health; and the Department of Community Health and Family Medicine at Tufts Medical School. As part of the "Democracy and Ecology" book series with The Guilford Press, his latest works (with John Wooding) are the edited collection *Work, Health, and Environment: Old Problems, New Solutions* (1997) and, forthcoming, tentatively titled *The Point of Production: The Political Economy of Work Environment*.

María Mondragon-Valdéz is Codirector of Peoples Alternative Energy Services, and is on the Board of Directors of the Land Rights Council in the Southwestern United States. She is also a sixth-generation heir to the Sangre de Cristo Land Grant and a doctoral student in American Studies at the University of New Mexico in Albuquerque.

Richard Moore is a key national leader of the environmental justice movement. For over 25 years he has worked as a community activist and organizer around such issues as welfare rights, police brutality, street-gang activities, drug abuse, low cost healthcare, child nutrition and the fight against racism. Of Puerto Rican descent, Richard is a founding member of the multiracial Southwest Organizing Project (SWOP) in Albuquerque; as well as a a founding member and Coordinator of the Southwest Network for Environmental and Economic Justice (SNEEJ), a coalition of grassroots organizations working in communities of color in the southwestern United States and Mexico. A planning committee member for the First National People of Color Environmental Leadership Summit in October of 1991, Richard now serves on the Board of Directors of the Environmental Support Center and is Chair of the National Environmental Justice Advisory Committee. He is also a longtime member of the Eco-Justice Working Group of the National Council of Churches.

Patrick Novotny is Assistant Professor of Political Science at Georgia Southern University in Statesboro. His teaching and research interests include the politics of the environmental justice movement, political culture and the politics of film, environmental politics, and social movements. His work has appeared in the journals *Capitalism, Nature, Socialism*; *New Political Science*; *Peace and Change*; and *Social Science Computer Review*. His forthcoming book is entitled *Where We Live, Work and Play: The Politics of the Environmental Justice Movement*.

Devon Peña is Director of the Rio Grande Bioregions Project and teaches Sociology at Colorado College. His most recent books include *The Terror of the Machine: Technology, Work, Gender, and Ecology on the U.S.–Mexico Border* (University of Texas Press, 1997) and *Chicano Culture, Ecology, Politics: Subversive Kin* (University of Arizona Press, 1998).

Laura Pulido is Assistant Professor of Geography at the University of Southern California in San Diego. She is the author of *Environmentalism and Economic Justice: Two Chicano Struggles in the Southwest* (University of Arizona Press, 1996), and has written numerous articles on issues of race, labor, and environmental justice in Los Angeles. Active in social and environmental justice struggles throughout southern California, Dr. Pulido is also a longtime member of the Labor/Community Strategy Center.

John Wooding is Chair of the Department of Regional Economic and Social Development and Adjunct Professor of Work Environment Policy at the University of Massachusetts at Lowell. Dr. Wooding is also on the Editorial Board of *New Solutions: A Journal of Environmental and Occupational Health Policy*, and a member of the Boston Editorial Group of the journal *Capitalism, Nature, Socialism*. A prominent scholar on occupational health and safety policy in the United States and Great Britain, his latest works (with Charles Levenstein) as part of the "Democracy and Ecology" book series with The Guilford Press include the edited collection *Work, Health, and Environment: Old Problems, New Solutions* (1997), and, forthcoming, tentatively titled *The Point of Production: The Political Economy of Work Environment*.